AF581621

Advances in Smart Materials and Structures

Advances in Smart Materials and Structures

Editors

Bing Wang
Tung Lik Lee
Yang Qin

Basel • Beijing • Wuhan • Barcelona • Belgrade • Novi Sad • Cluj • Manchester

Editors
Bing Wang
School of Mechancial Engineering and Automation
Fuzhou University
Fuzhou, China

Tung Lik Lee
ISIS Neutron and Muon Source
STFC Rutherford Appleton Laboratory
Didcot, United Kingdom

Yang Qin
School of Civil Aviation
Northwestern Polytechnical University
Xi'an, China

Editorial Office
MDPI
St. Alban-Anlage 66
4052 Basel, Switzerland

This is a reprint of articles from the Special Issue published online in the open access journal *Materials* (ISSN 1996-1944) (available at: https://www.mdpi.com/journal/materials/special_issues/ASMS).

For citation purposes, cite each article independently as indicated on the article page online and as indicated below:

Lastname, A.A.; Lastname, B.B. Article Title. *Journal Name* **Year**, *Volume Number*, Page Range.

ISBN 978-3-0365-9560-3 (Hbk)
ISBN 978-3-0365-9561-0 (PDF)
doi.org/10.3390/books978-3-0365-9561-0

Contents

About the Editors

Bing Wang

Bing Wang is a professor of Mechanical Engineering at the School of Mechanical Engineering and Automation at Fuzhou University. He is a Deputy Director of the Fujian Provincial Key Laboratory of Terahertz Functional Devices and Intelligent Sensing, China. His PhD is from the University of Hull (2013-2016), and his postdoc is from the University of Cambridge (2017-2020). Dr Wang is a Life Member of Clare Hall Cambridge, a Premium Member for both the Chinese Mechanical Engineering Society and Chinese Society for Composite Materials, a Topic Editor and Guest Editor for *Materials*, a Guest Editor for *Advances in Mechanical Engineering*, a Youth Editor for *Journal of Fuzhou University*, and a recognized peer reviewer for more than twenty international journals, including *Nature*, *Advanced Materials*, *NDT & E International*, etc. Bing has been awarded as an outstanding reviewer for *Archives of Civil and Mechanical Engineering*. He served as Chair, co-Chair, and TPC Chair for more than 10 international conferences. His main research focuses on smart materials and structures, composite mechanics, non-destructive testing and evaluation, as well as aerospace structural design and optimization. He has published about 50 papers, with 2 ESI Highly Cited Papers and 1 Featured Article, more than 10 patents, and 3 monographs/book chapters. He has developed a novel flexible mechanical hinge and applied it for a Tier 1 aerospace supplier and has been awarded by EPSRC. Bing has been invited about 20 times to deliver Keynote Speeches or Invited Seminars at international conferences or institutes.

Tung Lik Lee

Dr Tung Lik Lee is an Instrument Scientist responsible for the ENGIN-X neutron diffractometer at the ISIS Neutron and Muon Source, Science & Technology Facilities Council (STFC), UK, specializing in non-destructive neutron-based stress/strain mapping of engineering materials. Tung Lik is a topic editor of the journal *Materials* and also a Professional Member of the Institute of Materials, Minerals & Mining. Tung Lik completed his PhD at the University of Hull (2011-2016) and served as a junior academic visitor at Oxford University (Begbroke Science Park) in 2013.

Yang Qin

Yang Qin got his PhD from the University of Hull and is now an Associate Professor at the School of Civil Aviation, Northwestern Polytechnical University. His main research focuses on composite manufacturing, nature/environmentally friendly materials, and smart materials and structures, as well as damage/failure mechanisms of aircraft materials. Dr Qin is leading a youth project funded by the National Natural Science Foundation of China, and he has published seven research papers.

 *materials*

Editorial

Advances in Smart Materials and Structures

Bing Wang [1,*], Tung Lik Lee [2] and Yang Qin [3]

1 Fujian Provincial Key Laboratory of Terahertz Functional Devices and Intelligent Sensing, School of Mechanical Engineering and Automation, Fuzhou University, Fuzhou 350108, China
2 ISIS Neutron and Muon Source, STFC Rutherford Appleton Laboratory, Harwell Campus, Didcot OX11 0QX, UK; tung-lik.lee@stfc.ac.uk
3 School of Civil Aviation, Northwestern Polytechnical University, Xi'an 710072, China; yang_qin@nwpu.edu.cn
* Correspondence: b.wang@fzu.edu.cn

Smart materials and structures are capable of active or passive changes in terms of shapes (geometries), properties, and mechanical or electromagnetic responses, in reaction to an external stimulus, such as light, temperature, stress, moisture, and electric or magnetic fields. They have attracted increasing interest for their enhanced performance and efficiency over a wide range of industrial applications, especially in the field of aerospace. These applications require novel engineering approaches and design philosophy in order to integrate the actions of sensors, actuators, and control circuit elements into a single system that can respond adaptively to the surrounding changes.

In this Special Issue, we have collected the most recent advances in smart materials and structures, including seven original research papers and three review articles, co-authored by 65 scientists and engineers from 18 institutions and 3 industries. The research topics mainly cover advanced materials, applications of smart materials and structures, and recent development in sensing techniques.

Advanced materials and structures have attracted growing interest for their design and manufacturing flexibilities, high specific strength, and stiffness, which are superior in terms of reducing structural weight and functional complexities. Smart morphing composite technologies are developed to design and manufacture structures that can sense and respond to ambient environmental changes. Wang et al. [1] focus on reviewing recent progress in biomimetic Venus flytrap structures based on smart composites. The biomechanics of real Venus flytraps was first introduced to reveal the underlying mechanisms. Smart composite technology was then discussed by covering mainly the principles and driving mechanics of various bistable composite structures, followed by research progress on the smart composite-based biomimetic flytrap structures, concentrating mainly on the bionic strategies in terms of sensing, responding, and actuation, as well as the rapid snap-trapping process, aiming to enrich the diversities and reveal the fundamentals in order to further advance the multidisciplinary science and technological development into composite bionics.

Guo et al. [2] investigated the microstructural and mechanical performance of the $Co_{32}Cr_{28}Ni_{32.94}Al_{4.06}Ti_3$ high-entropy alloy. It was found that the alloy had a single-phase, disordered, face-centered, cubic solid-solution structure and was strengthened through solid solution. Furthermore, cholesteric liquid crystals (CLCs) are molecules that can self-assemble into helicoidal superstructures exhibiting circularly polarized reflection. This is promising for an array of industrial applications, including reflective displays, tunable mirror-less lasers, optical storage, tunable color filters, and smart windows. Lee et al. [3] present a review on the electro-optic response of polymer-stabilized CLCs, i.e., PSCLCs. They exhibited dynamic optical responses that can be induced by external stimuli, including electric fields, heat, and light. Their multiple electro-optic responses and potential mechanisms were discussed in great detail.

Citation: Wang, B.; Lee, T.L.; Qin, Y. Advances in Smart Materials and Structures. *Materials* **2023**, *16*, 7206. https://doi.org/10.3390/ma16227206

Received: 4 November 2023
Accepted: 16 November 2023
Published: 17 November 2023

For their unique responsive performance, smart materials and structures have been applied to reduce impact damages and improve aeroelastic stability. The vibration and impact of a humanoid bipedal robot during movements such as walking, running, and jumping may cause potential damage to the robot's mechanical joints and electrical systems. Huang et al. [4] developed a composite bidirectional vibration isolator based on magnetorheological elastomer (MRE) for the cushioning and damping of a humanoid bipedal robot under foot contact forces. The vibration isolation performance of the vibration isolator was tested experimentally, and a vibration isolator dynamics model was developed. The MRE vibration isolator hardware-in-the-loop-simulation experiment platform based on dSPACE was built to verify the vibration reduction control effect of the fuzzy PID algorithm. It was found that the vibration amplitude was significantly attenuated, which verifies the effectiveness of the fuzzy PID damping control algorithm.

Liu et al. [5] applied topologically optimized piezoelectric smart material to improve the aeroelastic stability of bladed disks. Piezoelectric materials were embedded or bonded to each blade and use different shunt capacitance on each blade as the source of mistuning. When the shunt capacitance varies from zero (open-circuit, OC) to infinity (short-circuit, SC), the stiffness of each blade changes within a relatively small interval. In this way, the required small difference of stiffness among blades can be altered into a relatively larger difference of the shunt capacitance.

Structural health monitoring plays a vital role in determining the structural integrity of advanced materials and structures. Perfetto et al. [6] proposed a guided wave-based artificial neural network (ANN) to determine the positions of damages. The ANN was developed using FE model and trained on an aluminum plate, which was subsequently verified in a composite plate, as well as under different damage configurations. The ANN allowed detection and localization of damages with high accuracy.

Smart materials and structures are often applied in sensing techniques. Recent advances in terms of textile-based sensors, stretchable sensors, and multifunctional sensors are leading the technological development in many industries. Textile-based sensors have drawn great interest since they are superior in terms of flexibility, comfort, low cost, and wearability. They are often tied to certain parts of the human body to collect mechanical, physical, and chemical stimuli to identify and record human health and exercise. Zhou et al. [7] reviewed the recent advances in the textile-based mechanical sensors (TMSs), where sensing mechanisms, textile structure, and novel fabrication strategies for implementing TMSs were summarized. The critical performance criteria such as sensitivity, response range, response time, and stability were also investigated. Finally, the challenges and prospects were proposed in order to provide meaningful guidelines and directions for flexible sensing techniques.

As an indispensable part of wearable devices and mechanical arms, stretchable conductors have received extensive attention in recent years. The design of a high-dynamic-stability, stretchable conductor is the key technology to ensure the normal transmission of electrical signals and electrical energy of wearable devices under large mechanical deformation. Liu et al. [8] designed a stretchable conductor with a linear bunch structure and prepared by combining numerical modeling and simulation with 3D printing technology. The stretchable conductor consists of a 3D-printed bunch-structured equiwall elastic insulating resin tube and internally filled free-deformable liquid metal, showing good mechanical and electrical properties with great application potential.

Martowicz et al. [9] developed a novel measurement approach to determine the thermomechanical properties of a gas foil bearing using a specialized sensing foil made of Inconel alloy. The strain and temperature distributions were identified based on the readings from the strain gauges and integrated thermocouples mounted on the top foil. The measurements' results were obtained for experiments that represent the arbitrarily selected operational conditions of the tested bearing. Specifically, the considered measurement scenario relates to the operation at a nominal rotational speed, i.e., during the stable process, as well as the run-up and run-out stages.

The measurement of pH has received great attention in diverse fields, such as clinical diagnostics, environmental protection, and food safety. Optical fiber sensors are widely used for pH sensing because of their great advantages. Long et al. [10] fabricated and studied a rapid-response and wide-range pH sensor enabled by a self-assembled functional polyaniline/polyacrylic acid (PAni/PAA) layer on no-core fiber. The developed sensor is high in sensitivity and quick in both response and recovery time, which is promising for detecting pH in the liquid phase with temperature variation.

We hope that this Special Issue will contribute to disseminating the latest progress in smart materials and structures, as well as stimulating the interest of its audiences to work in this important and vibrant area to promote and benefit the multidisciplinary scientific communities. Owing to the word limit on this Editorial, audiences are recommended to refer to the original papers for further information on their specific interests.

Funding: This work was supported by the National Natural Science Foundation of China (52005108, 12202361), and the Start-up Funding from Education Department of Fujian Province and Fuzhou University (GXRC-20066). We also thank the technical staff and aegis of the Fuzhou University International Joint Laboratory of Precision Instruments and Intelligent Measurement & Control.

Conflicts of Interest: The authors declare that they have no known competing financial interest or personal relationship that could have appeared to influence the work reported in this paper.

References

1. Wang, B.; Hou, Y.; Zhong, S.; Zhu, J.; Guan, C. Biomimetic Venus Flytrap Structures Using Smart Composites: A Review. *Materials* **2023**, *16*, 6702. [CrossRef] [PubMed]
2. Guo, J.; Tang, C.; Lai, H.S. Microstructure and Mechanical Properties of $Co_{32}Cr_{28}Ni_{32.94}Al_{4.06}Ti_3$ High-Entropy Alloy. *Materials* **2022**, *15*, 1444. [CrossRef] [PubMed]
3. Lee, K.M.; Marsh, Z.M.; Crenshaw, E.P.; Tohgha, U.N.; Ambulo, C.P.; Wolf, S.M.; Carothers, K.J.; Limburg, H.N.; McConney, M.E.; Godman, N.P. Recent Advances in Electro-Optic Response of Polymer-Stabilized Cholesteric Liquid Crystals. *Materials* **2023**, *16*, 2248. [CrossRef]
4. Huang, X.; Zhai, Y.; He, G. Research on Vibration Control Technology of Robot Motion Based on Magnetorheological Elastomer. *Materials* **2022**, *15*, 6479. [CrossRef] [PubMed]
5. Liu, X.; Fan, Y.; Li, L.; Yu, X. Improving Aeroelastic Stability of Bladed Disks with Topologically Optimized Piezoelectric Materials and Intentionally Mistuned Shunt Capacitance. *Materials* **2022**, *15*, 1309. [CrossRef]
6. Perfetto, D.; De Luca, A.; Perfetto, M.; Lamanna, G.; Caputo, F. Damage Detection in Flat Panels by Guided Waves Based Artificial Neural Network Trained through Finite Element Method. *Materials* **2021**, *14*, 7602. [CrossRef] [PubMed]
7. Zhou, Z.; Chen, N.; Zhong, H.; Zhang, W.; Zhang, Y.; Yin, X.; He, B. Textile-Based Mechanical Sensors: A Review. *Materials* **2021**, *14*, 6073. [CrossRef] [PubMed]
8. Liu, C.; Wang, Y.; Wang, S.; Xia, X.; Xiao, H.; Liu, J.; Hu, S.; Yi, X.; Liu, Y.; Wu, Y.; et al. Design and 3D Printing of Stretchable Conductor with High Dynamic Stability. *Materials* **2023**, *16*, 3098. [CrossRef] [PubMed]
9. Martowicz, A.; Roemer, J.; Zdziebko, P.; Żywica, G.; Bagiński, P.; Andrearczyk, A. A Novel Measurement Approach to Experimentally Determine the Thermomechanical Properties of a Gas Foil Bearing Using a Specialized Sensing Foil Made of Inconel Alloy. *Materials* **2023**, *16*, 145. [CrossRef] [PubMed]
10. Long, G.; Wan, L.; Xia, B.; Zhao, C.; Niu, K.; Hou, J.; Lyu, D.; Li, L.; Zhu, F.; Wang, N. Rapid-Response and Wide-Range PH Sensors Enabled by Self-Assembled Functional PAni/PAA Layer on No-Core Fiber. *Materials* **2022**, *15*, 7449. [CrossRef] [PubMed]

Review

Biomimetic Venus Flytrap Structures Using Smart Composites: A Review

Bing Wang [1,2], Yi Hou [1,2], Shuncong Zhong [2,*], Juncheng Zhu [1,2] and Chenglong Guan [2]

1 School of Advanced Manufacturing, Fuzhou University, Fuzhou 362251, China; b.wang@fzu.edu.cn (B.W.)
2 Fujian Provincial Key Laboratory of Terahertz Functional Devices and Intelligent Sensing, School of Mechanical Engineering and Automation, Fuzhou University, Fuzhou 350108, China; guancl@fzu.edu.cn
* Correspondence: sczhong@fzu.edu.cn

Abstract: Biomimetic structures are inspired by elegant and complex architectures of natural creatures, drawing inspiration from biological structures to achieve specific functions or improve specific strength and modulus to reduce weight. In particular, the rapid closure of a Venus flytrap leaf is one of the fastest motions in plants, its biomechanics does not rely on muscle tissues to produce rapid shape-changing, which is significant for engineering applications. Composites are ubiquitous in nature and are used for biomimetic design due to their superior overall performance and programmability. Here, we focus on reviewing the most recent progress on biomimetic Venus flytrap structures based on smart composite technology. An overview of the biomechanics of Venus flytrap is first introduced, in order to reveal the underlying mechanisms. The smart composite technology was then discussed by covering mainly the principles and driving mechanics of various types of bistable composite structures, followed by research progress on the smart composite-based biomimetic flytrap structures, with a focus on the bionic strategies in terms of sensing, responding and actuation, as well as the rapid snap-trapping, aiming to enrich the diversities and reveal the fundamentals in order to further advance the multidisciplinary science and technological development into composite bionics.

Keywords: biomimetic; Venus flytrap; smart; composite; mechanics

Citation: Wang, B.; Hou, Y.; Zhong, S.; Zhu, J.; Guan, C. Biomimetic Venus Flytrap Structures Using Smart Composites: A Review. *Materials* **2023**, *16*, 6702. https://doi.org/10.3390/ma16206702

Academic Editor: Maja Dutour Sikirić

Received: 18 September 2023
Revised: 5 October 2023
Accepted: 7 October 2023
Published: 16 October 2023

1. Introduction

Biomimetic structures are inspired by elegant and complex architectures of natural creatures, which can be lightweight and offer combinations of mechanical properties that often surpass those of their components by orders of magnitude [1]. Biomimetic structures are often conceived to achieve specific functions, drawing inspiration from biological structures [2]. A typical example is the flexible solar array, the core part of a spacecraft power generation system, which is limited by space during transportation; whilst a deployable solar array similar to the convolvulaceae increases the stowed-to-deployed ratio [3]. For the biomimetic convolvulaceae deployable structure, petals are spirally folded around the center to form a floral bud, and then slowly open along the radial direction of the flower until full bloom. Another attraction is the fast nastic motion in nature. The ballistic projection of a chameleon's tongue is a typical example of rapid motion due to elastic forces [4]. It involves a complex interplay between the internal organization, collagen fibrous, stress release and geometry [5]. Hummingbirds can prey on insects because their beaks have the ability to impede quickly, the high-speed capture motion cannot be explained only by direct muscle action, possibly powered by the sudden release of stored elastic energy [6]. The rapid closure of a Venus flytrap leaf is one of the fastest motions in plants, once a principal natural curvature of the leaf changes, there will be a fast snap-through phenomenon [7]. These unique biological structural mechanics have gone through a lengthy evolution, providing optimized solutions for desired functional systems. Scientists and engineers have explored and imitated for decades and continue to investigate in order to reveal the underlying mechanisms, aiming to benefit and facilitate practical engineering designs and applications for industries [8].

Composite materials have attracted growing interest for their design and manufacturing flexibilities, high specific strength and stiffness, which are superior to reducing structural weight and functional complexities [9]. Smart morphing composite technologies are developed to design and manufacture structures that can sense and respond to ambient environmental changes [10]. In particular, a bistable composite is a thin shell laminate structure, which is able to undergo repeated reversible changes between two stable shapes under external stimuli [11], analogous to rapid morphing mechanisms in nature.

The biomimetic Venus flytrap structures are found mostly by employing the morphing mechanics of smart composites. These contribute to their main characteristics and advantages in terms of (i) structural design freedom, since organisms are variable and complex in nature; (ii) simplicity and lightweight, the conventional deformable bioinspired structure connected by multiple rigid modules through motion pairs is complex and heavy, whilst the shape-changing process of a thin-shelled smart composite is similar to rapid snap-trapping of a Venus flytrap leaf, simplifying the design of biomimetic structures; (iii) energy conservation and flexibility, the smart composite structure stores shapes through binary states, the shape-changing process is reversible without additional input energy, which offers efficient and flexible energy-saving rapid morphing mechanics driven by stored elastic energy.

There have been reviews concerning the bionics of the Venus flytrap. The research progress on plant-inspired adaptive structures and materials for morphing and actuation has been reported in [2], where the Venus flytrap was introduced as one of the imitated creatures for impulsive and reversible movements. The fast active motion principle inspired by the Venus flytrap was summarized in [12] with a focus on the application to soft machine systems. Here, we focus on reviewing the most recent development in the smart composite-based biomimetic Venus flytrap structures. Firstly, we gave an overview of the biomechanics of the Venus flytrap in order to reveal the underlying mechanisms. The smart composite technology was then introduced by covering mainly the principles and driving mechanics of various types of smart composite structures, followed by the research progress on the smart composite-based biomimetic flytrap structures, with a focus on the bionic strategies in terms of sensing, responding and actuation, as well as the rapid snap-trapping, aiming to enrich the diversities and reveal the fundamentals in order to further advance the multidisciplinary science and technological development into composite bionics.

2. Biomechanics of Venus Flytrap

Venus flytrap is a magic insect-trapping plant that catches agile insects by the rapid closure of a pair of symmetrical shell-like leaves [13]. When the Venus flytrap is stimulated, the geometry of flytrap leaves changes from convex to concave shapes within 100 ms, see Figure 1a, similar to the snapping of a bistable composite structure [7]. The biomechanics of the Venus flytrap does not rely on muscle tissues to produce rapid shape-changing, which is significant for engineering applications.

The trigger hairs on the inner surface of the flytrap leaves are highly sensitive, see Figure 1b, they bend when sensing moving insects [14]. The sensitive cells produce initial action potential, and their structural characteristics contribute to the rapid propagation of action potential [15]. The action potential is the form of electrical signal transmission, which propagates between plant cells through the vasculature system [16]. Figure 1c shows a side view of the vascular structure within a Venus flytrap leaf. The propagation of action potential depends on the transmembrane flow of ions, with a speed of up to 10 m/s [17]. A hypothesis on the rapid trapping movement is then proposed: when the action potential is diffused in the motor tissue, a fast permeability change leads to massive leakage of ions; the rapid decrease in internal osmotic pressure causes the Venus flytrap leaves to close rapidly [18]. Thus, cell dimensional changes are amplified in the movement of attached organs.

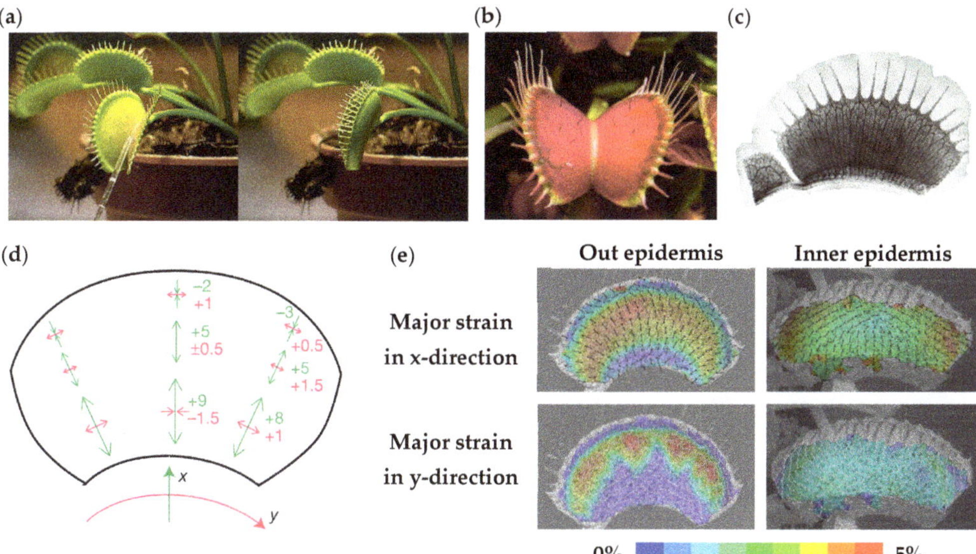

Figure 1. Biomechanics of Venus flytrap, showing: (**a**) open and closed status of a flytrap leaf [7]; (**b**) structural characteristics of the highly sensitive trigger hairs on the inner surface; (**c**) side view of the vascular structure; (**d**) internal stress field of a leaf following the single tissue layer model [17]; (**e**) the strain distribution of a leaf following the bi-layer model [19].

Osmotic driven (active water transport) is crucial for active motion but cannot explain the rapid trapping process of the Venus flytrap. An important reason is that the snap-trap is not caused by the bending of the entire leaves, but a snap-buckling instability through active control [7]. The geometry of the doubly-curved leaf is bistable, which provides mechanisms through elastic energy storage and release. It is the passive motion of the trap, and manifested by the release of prestress. The internal stress field of flytrap leaves is considered to follow the geometry of the leaves but is locally orthotropic. Figure 1d shows the internal stress field and distribution [17]. The variation of curvature in the x-direction caused by the biochemical reaction is much larger than that in the y-direction, which is considered to be the actuating factor of the snapping. To dig further into the triggering mechanics of the passive motion at a macroscopic scale, a more detailed bi-layer model has been developed to simulate the internal stress/strain field, see Figure 1e. It is confirmed that the trapping process is the interaction between the shrinking/swelling processes of the various tissue layers, as well as the release of stored elastic strain energy [19].

Therefore, the biomechanics of the Venus flytrap consists of two mechanisms: (i) the biochemical reaction through changes in osmotic pressure as explicit sensing and trigger actuation; (ii) the bistable shape transition as the implicit fast trapping motion. Thus, the progress and development of smart composite-based biomimetic flytrap structures are also concentrated on mimicking the smart sensing and actuation strategies, as well as the rapid snap-trapping phenomenon.

3. Smart Composite Technology

Smart materials and composite technology are derived from the 1980s. It refers to a material or structural system that is capable of sensing external changes, and responding to stimuli by changing their specific properties, geometric configurations, etc. [20]. They have been widely applied to medical devices, flexible electronics, as well as aerospace engineering [21]. A smart material can transform from a temporary configuration to its original configuration in response to external stimuli: this is well-known as the shape memory effect. Especially, shape-memory materials have attracted great interest due to their unique shape-memory effects, these mainly include shape-memory alloys (SMAs) [22], as well as shape-memory polymers (SMPs) [23]. SMAs are usually low in recovery rate, whilst SMPs are high in recovery rate, low in density, and possess large shape morphing ability that can be beneficial for intelligent morphing designs [24]. Commonly used thermoplastic SMPs include polylactic acid (PLA), polycaprolactone (PCL), polyester (PE), polyurethane (PU), etc. [25]; shape memory effects have also been identified in thermoset polymers such as epoxy and photo-reactive resins [26,27], as well as in hydrogels [28].

Smart composites are developed based on smart polymers with improved mechanical performance. Depending on the specific material and structure employed to produce a smart composite, it can be driven by heat [29], light [30], magnetism [31], electricity [32], or water/moisture [33], etc. They are superior in terms of diversity in smart driving strategy, flexibility in structural programming, highly recoverable ability, large deformation capability, as well as good biocompatibility [34]. Therefore, they are ideal for applications to sensors and actuators within a biomimetic flytrap to sense and initiate shape-changing actuation.

A prelude of smart composite structures also includes those that are bistable, enabling elastic strain energy-based rapid morphing of the biomimetic flytrap. In producing a composite, it is known that thermal residual stress would be generated during cooling, which may lead to deflection, local buckling, and even premature failure of a composite structure in service [35,36]. Although they are usually detrimental, they can also be beneficial to produce bistable or multistable composite structures. Hyer [37] first discovered the bistability from producing unsymmetric thin laminates, where orthogonal deflections result in a bistable structure, and the Classical Laminate Theory (CLT) was found not valid to predict the saddle shapes. It was further expanded by considering the geometric nonlinearity of the stress-strain relationship and analyzing the size effects on bistability [38,39].

There are mainly two types of bistable composite structures considering the curvature directions of the two stable shapes, namely (i) opposite-sense bistable (OSB) structure, and (ii) equal-sense bistable (ESB) structure, see Figure 2. Although composites have also been used to explore bistable twisted structures [40,41], these are out of the scope of the current focus. It is reported that for (i), OSB structures can be produced by adjusting in-plane stress levels, these are mainly achieved by employing thermal residual stress [37,42], piezoelectric actuation [43,44], continuous elastic fiber prestressing [45,46], as well as continuous viscoelastic fiber prestressing [11,47]; whilst for (ii), ESB structures are mainly produced by exploring the geometric curvature effects [48], also commonly known as composite tape-springs [49,50].

The thermal residual stress-induced OSB structure is the most established, which is produced by employing unsymmetric composite layups. A typical example is shown in Figure 2a. Owning to a mismatch in thermal expansion between fiber and matrix materials, thermal residual stresses are developed during cooling, leading to unsymmetric in-plane stresses, resulting in out-of-plane deflections in order to generate bistability. However, these are found sensitive to temperature and humidity conditions, and difficult to control precisely [51,52]. For the piezoelectric actuation-based OSB structure, in-plane stress is altered by using piezoelectric strips which are responsive to temperature changes and controlled through voltage [43], see Figure 2b. The continuous fiber prestressing techniques employ elastic (see Figure 2c) or viscoelastic (see Figure 2d) recovery within a composite structure, which would introduce compressive stresses and interact with the intrinsic thermal residual stress in order to induce out-of-plane deflections [53,54]. As for ESB structure, a composite tape-spring structure explores the positive Gaussian curvature effects, see Figure 2e, the governing factors of its bistability depend on the material constitutive behavior, initial geometric proportions, as well as the geometrically non-linear structural behavior [55].

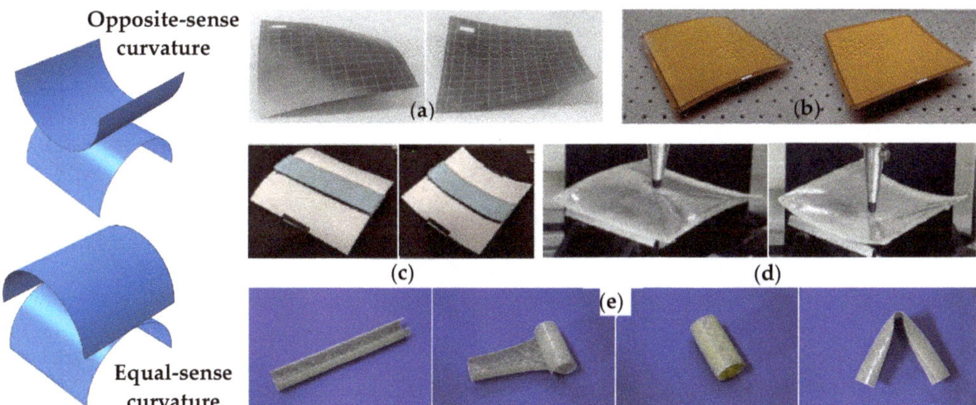

Figure 2. Bistable composite structures developed by employing (**a**) thermal residual stress [42]; (**b**) piezoelectric actuation [43]; (**c**) continuous elastic fiber prestressing [45]; (**d**) continuous viscoelastic fiber prestressing [47]; (**e**) geometrical curvature effects [49].

4. Biomimetic Venus Flytrap Structures

Plants that harvest energy from nature through photosynthesis accomplish perception and response to changing conditions without brain control. They provide possibilities for constructing advanced biomimetic structures that rely on adaptive material systems without control centers. The sense and response of a material system to changing environmental conditions are related to the corresponding biological reaction mechanisms. Therefore, plant-inspired biomimetic structures have focused on the achievement of plant functional principles, rather than simply imitating the behavior of plant movement in past decades [12]. These biomimetic structures, which appeared closely to the biological model, are to achieve self-growth and self-repair and respond correctly when sensing changes in ambient conditions. Since these strategies are still in their infancy stage, the existing Venus flytrap biomimetic structures are still not capable of self-growing and self-repairing. Therefore, the main biomimetic characteristics in terms of sensing, and actuation, as well as the rapid snap-trapping phenomenon of the biomimetic flytrap structures are mainly summarized here, with a focus on exploring the smart composite technology.

4.1. Electric-Driven Sensing and Actuation

The function of sensing for biomimetic flytrap structure is usually achieved by integrating smart material-based sensors. To date, the sensing characteristics of ionic polymer metal composites (IPMCs) are the most popular to mimic the trigger hairs of flytrap leaves. IPMC is an intelligent material superior in terms of bidirectional motion, fast response, and low driving voltage. When it is bent by an external force, the solvent is replaced, and the resulting charge polarisation generates a voltage on both sides of the material. Based on this acting principle, the IPMC has been applied to mimetic the trigger hairs, i.e., bending sensor. Under the influence of an external electric field, the redistribution of water molecules in IPMC leads to its bending deformation, involving a series of energy conversions, including electrical energy, chemical energy and mechanical energy.

The sensing characteristics and mechanical bending of an IPMC system are similar to the sensing and actuation of the flytrap. Figure 3a shows a small biomimetic flytrap-based robot employing the IPMC-based sensor and actuator [56]. The IPMC-based bristles initiate signals via an amplifying circuit when subjected to bending, and trigger the actuation circuit of the IPMC lobes, which are then bent rapidly and close within nearly 0.3 s. Although the sensing characteristics of IPMC related to internal ion transfers are similar to those of a real flytrap, the output of IPMC brush as a sensor within a biomimetic structure is rather weak and unstable, its bending is also limited. The infrared proximity sensors were then developed to replace the IPMC bristles, Figure 3b,c show typical examples [57]: in order to further improve the capability in sensing, number of proximity sensors can be increased; to obtain maximal shape-changing, three IPMC lobes were used to reduce gaps.

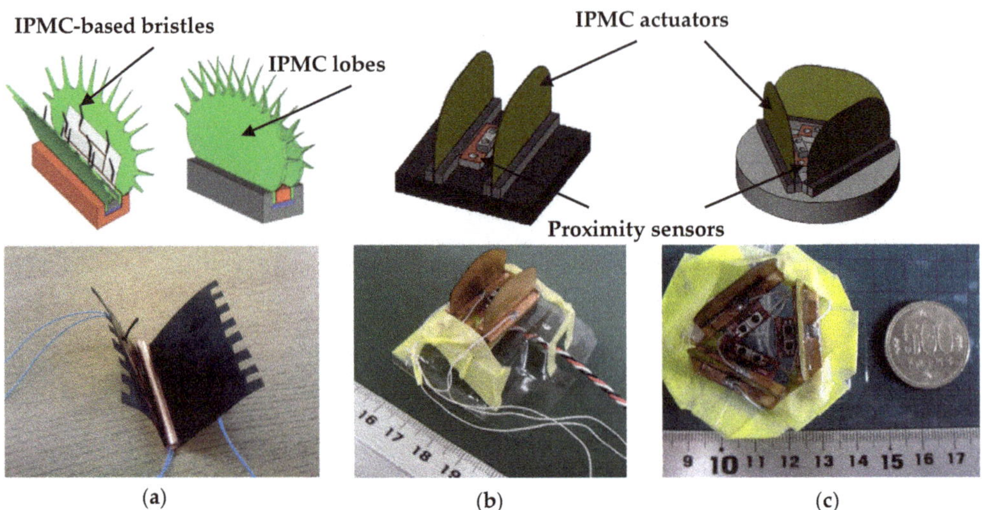

Figure 3. Electric-driven ionic polymer metal composite-based biomimetic flytrap structure, showing (**a**) a small biomimetic flytrap-based robot [58]; (**b**,**c**) proximity sensors are applied to imitate the trigger hairs to sense adjacent approaching objects [57].

4.2. Water-Driven Sensing and Actuation

The sensing function of biomimetic structures is often achieved by using stimulus-responsive materials, which are sensitive to changes in environmental conditions and deformed accordingly. A notable feature of hydrogels is that their volume changes during a wide range of environmental conditions. However, the leaves of a flytrap can quickly close within a tenth of a second to catch insects, significantly different from the usual slow shape transition of hydrogels. Therefore, the ingenious combination of sensing external stimuli and improving responsive speed needs to be implemented in hydrogel-based biomimetic designs. Figure 4a shows a hydrogel-based doubly bending structure incorporating elastic instability [59]. The hydrogel leaf protrudes outward, and three microfluidic channels are embedded on its inner surface for solvent transportation. The expansion of hydrogel is controlled in the same way as in the deformation of a flytrap leaf, only the curvature in one direction is actively manipulated by changes in solvent through the microfluidic channels, and the curvature in the other direction remains passive. During the swelling of the hydrogel, bending–stretching coupling of the doubly-curved geometry stores elastic strain energy along the axis. With further expansion, the stored elastic potential energy is released instantaneously after passing through the energy barrier, leading to the snap-buckling of the hydrogel leaf. During the drying or de-swelling process, reverse movements occur and quickly return to the original shape of the leaf. The whole shape-changing cycle can be controlled within 5 s.

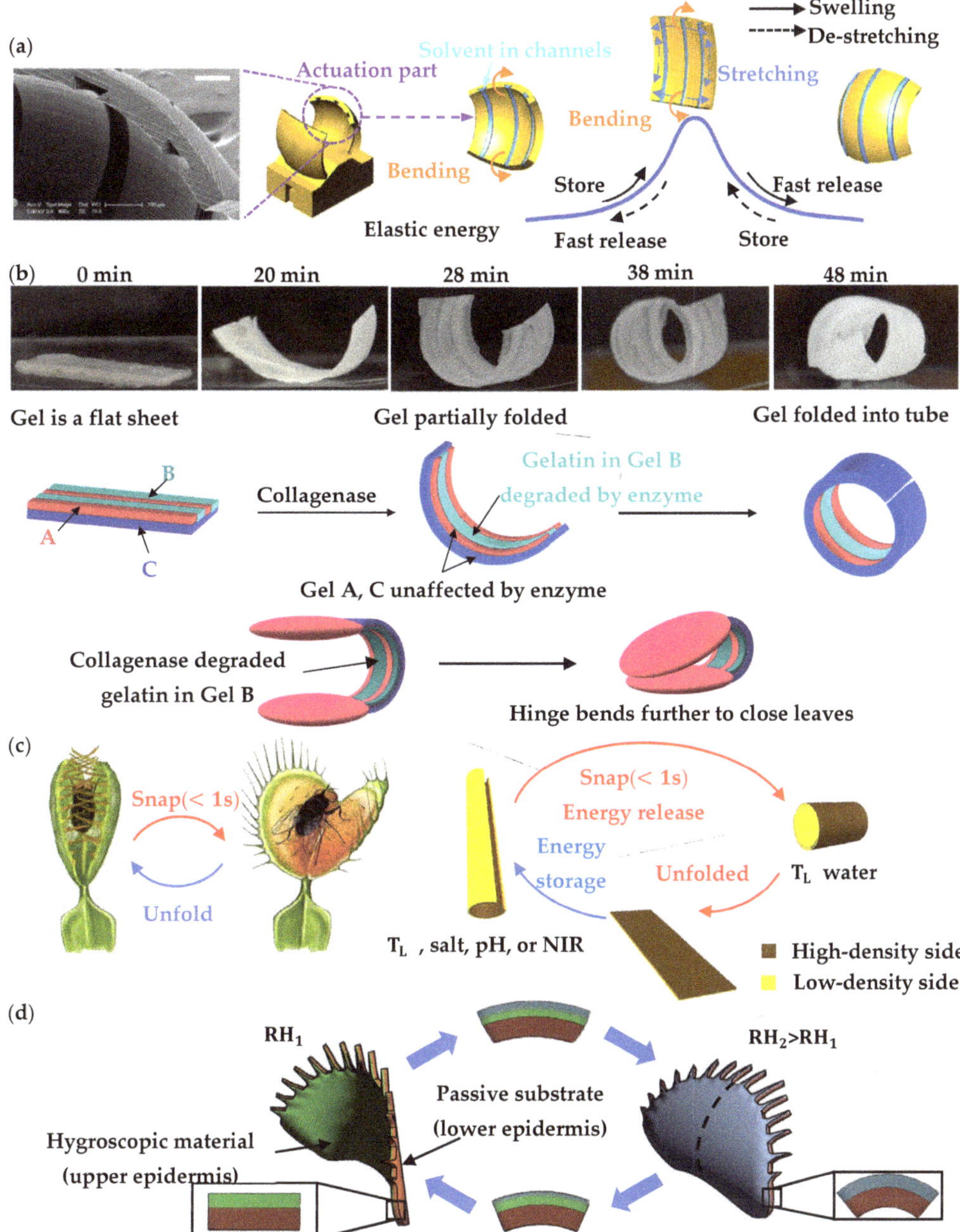

Figure 4. Water-driven hydrogel-based biomimetic flytrap structure, showing (**a**) a hydrogel-based doubly bending structure incorporating elastic instability [59]; (**b**) a bi-layered biomimetic flytrap based on a gel consisting of a mixture of three different components that is triggered by using enzyme [60]; (**c**) temperature response process of a rGO/PDMAEMA composite hydrogel sheet-based snap-trap structure [61]; (**d**) a pure humidity-responsive artificial flytrap leaf based on hygroscopic bistability [62].

It is worth noting that soft materials in nature are inhomogeneous. They have multiple functional regions with different chemical and mechanical compositions. Therefore, the composition and distribution of different materials could also be learned for biomimetic response structural systems. Figure 4b shows a bi-layered biomimetic flytrap, which is designed based on a mixture of three different gels that is triggered by an enzyme in the solution in order to change shapes [60]. The gels are constructed into a bilayer structure, which consists of two layers: a gel A/B layer is sandwiched above a layer of gel C. Two elliptical leaves made of gel A are connected by a hinge made of a mixed gel. Initially, when the biomimetic flytrap is placed in water, the gel C layer swells more than the gel A/B layer. Despite the mismatch in swelling speed, the hinge remains flat since the gel A/B layer is produced to be stiffer. In addition to the collagenase enzyme in the water, it cleaves the gelatin chains in gel B, which then reduces the stiffness of the A/B layer. The swollen gel C layer is now able to fold over the A/B layer, leading to the hinge being transformed into a specific shape. The reaction time for the leaf to be fully folded is about 50 min for 50 U/mL of enzyme. It should be noted that it is necessary to include energy storage and release mechanisms in order to improve the response period.

Another concept of snap-trap based on the rGO/PDMAEMA composite hydrogel sheet is shown in Figure 4c, associated with the temperature response process [61]. The composite hydrogel sheet has a double gradient along the thickness direction, i.e., chain density and cross-linking density gradient, which is able to accumulate elastic energy and release the stored energy rapidly through ultrafast snapping deformation. The composite gel is flat when immersed in 20 °C water. When immersed in 60 °C water, the composite sheet is bent along the longitudinal axis to the higher gradient side and transferred into a tubular structure, with stored energy. When replaced with the 20 °C water, the hydrogel sheet does not follow exactly the opposite path of the original shape transition, but a third state appears. During flattening, the tubular hydrogel snaps rapidly and flattens after curling. Trigger conditions for gel plates are complex and time-consuming, and the snapping velocity, angle, and location of the sheet can be tuned by modulating the magnitude and location of stored energy within the hydrogel.

A pure humidity-responsive artificial flytrap leaf is shown in Figure 4d. It is obtained by bonding pre-stretched poly dimethyl siloxane (PDMS) layers prior to depositing electrospun polyethylene oxide nanofibers to induce hygroscopic bistability [62]. The moisture absorption capacity of the electrospun material is combined with the mechanical advantages of the preloaded structure to increase the actuating speed. When polyethylene oxide expands with increasing ambient humidity, its coupling with the passive layer causes the curvature of the artificial leaf to decrease until it snaps within 0.5 s. When the humidity is reduced, the initial state is restored.

4.3. Light-Driven Sensing and Actuation

Compared with the above hydrogel-based driving systems that can only perceive and react in a liquid environment, light-driven smart material has lower environmental requirements. Figure 5a shows a light-driven biomimetic flytrap structure using a thin layer of light-responsive liquid crystal elastomer (LCE) as an actuator [63]. The open-aligned LCE actuator is integrated with the fiber tip and leaves a window in the center for light injection. When the structure is subjected to light, the molecular alignment arrangement changes in LCE, sufficient optical feedback (reflected or scattered light) then generates strains, leading to expansion and shrinkage on different surfaces; the flytrap structure is then closed by the induced strain difference on the surfaces. Figure 5b shows a liquid crystal network-based biomimetic flytrap structure, which is controlled and modified by light and humidity [64]. It can be closed in low humidity levels and high light levels, and open in no light and high humidity levels. The sensing strategies of the light-driven structures are flexible and diverse, they are expected to be used in future adaptive and intelligent biomimetic structures.

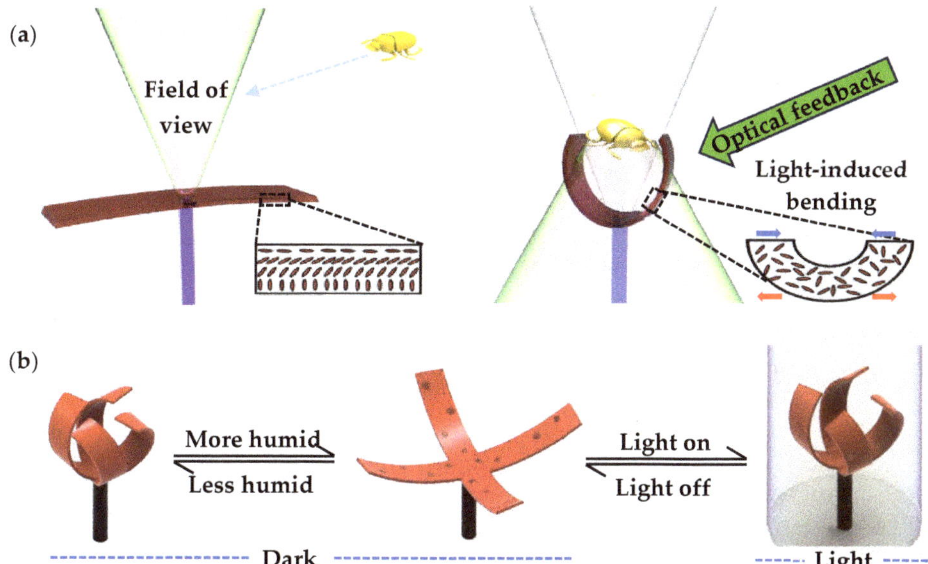

Figure 5. Light-driven smart material-based biomimetic flytrap structure, showing (**a**) a light-driven biomimetic flytrap leaf using a thin layer of light-responsive liquid crystal elastomer (LCE) as actuator [63]; (**b**) a liquid crystal network-based biomimetic flytrap structure, which can be controlled and modified by light and humidity [64].

4.4. Elastic Energy-Driven Rapid Snap-Trapping

The biomimetic strategies stated above mostly rely on the sensing capability to initiate snap-trapping. They lack elastic deformation-induced actuation, and the smart responses are usually in slow motion. Integration of the sensing capability into a bistable system may be more suitable for the biomechanics of the flytrap structure. The flytrap leaf is regarded as an elastic irregular curved shape. The macroscopic response of the flytrap after sensing the prey is that the leaves deform and pass through the energy barrier, and the stored elastic energy is instantly released and converted into kinetic energy, thereby forming a rapid snap-trapping action [65]. To date, the biomimetic structure close to this reaction is basically bistable. There are two ways to achieve the bistable characteristics of the biomimetic flytrap structure. These are not able to sense and are usually triggered by manual input. One strategy is to bind the bistable actuator between the artificial leaves, shape transitions between the two steady states of the actuator and drive the artificial leaves to morph. Although these structures have energy storage and release processes, they lack curvature changes as in a real leaf, and their reaction is similar to grasping rather than snap-trapping. The other strategy is to explore the curvature changes during the morphing of a bistable composite structure.

Therefore, the morphing process of an OSB- or ESB-based bistable structure can be designed to introduce elastic energy-driven mechanisms into the flytrap structure. Figure 6a,b shows an OSB-based biomimetic flytrap structure, which is produced using asymmetric composite laminate, i.e., thermal residual stress-induced bistability, and actuated by SMA [66,67]. The bistable structure is used to imitate the artificial leaves, providing similar shapes to a real flytrap leaf. The two stable configurations correspond to the open and curled closed shapes of the real leaf with different structural curvatures. The SMA coil spring is used to induce snap-through of artificial leaves after being electrically heated, similar to the active motion of a flytrap, which is embedded on both surfaces of the artificial leaves to be repeatable of the rapid actions. Since the SMA-based actuator requires relaxation time to cool down, the morphing frequency of the structure is limited. It is verified by experiment that this type of bistable biomimetic flytrap structure is able to close within 100 ms, and the macroscopic rapid snapping deformation mechanism is similar to that of real flytrap leaves. The performance of the bistable biomimetic structure is then improved by designing the bistable characteristics of artificial leaves and changing the geometry and locations of the embedded SMA [68].

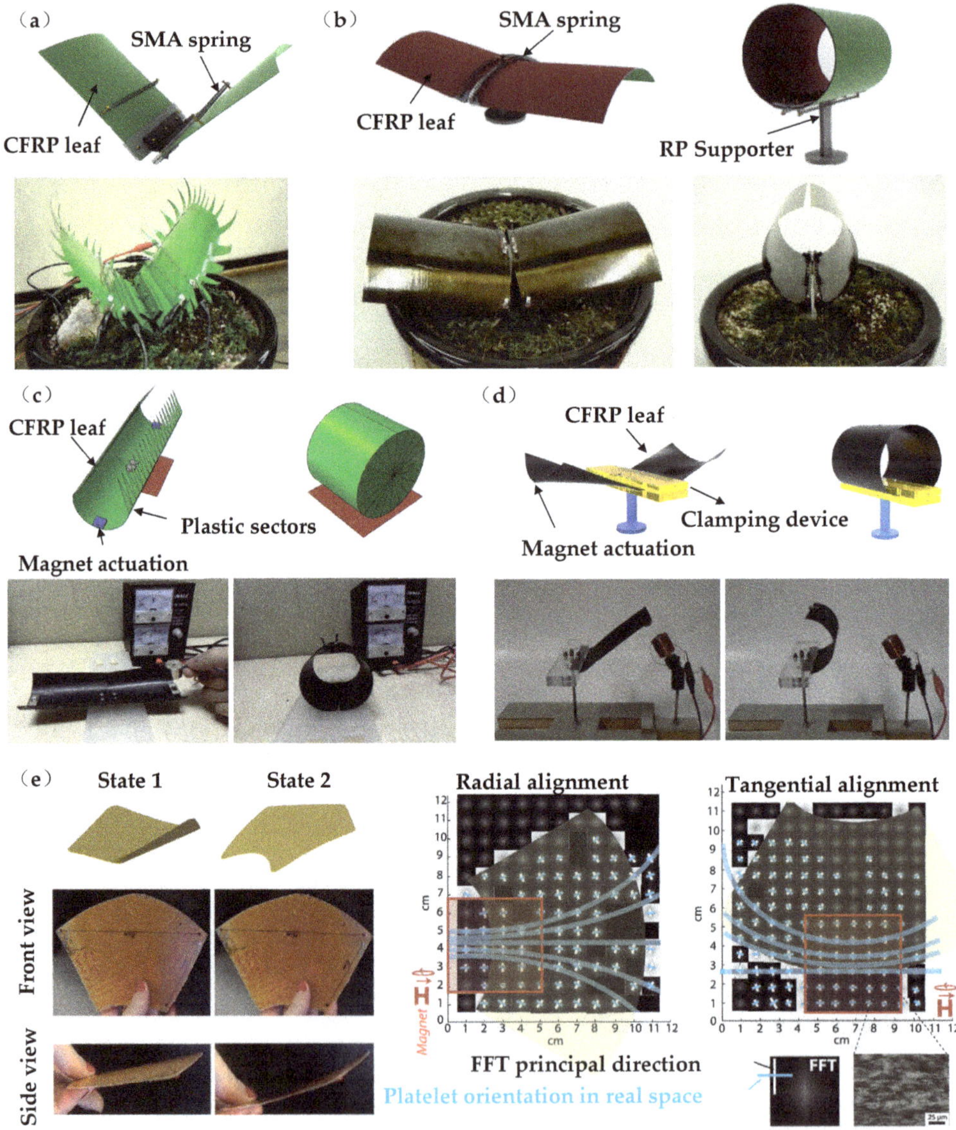

Figure 6. Elastic energy-driven bistable composite-based biomimetic flytrap structure, showing (**a**,**b**) OSB composite-based biomimetic flytrap structures actuacted by using shape memory alloys (SMAs), [66,67]; (**c**) an ESB-based biomimetic flytrap structure actuated by using magnets [69]; (**d**) an ESB-based flytrap structure with clamped boundaries actuated by using a clamping device [70]; (**e**) a magnet oriented particle-reinforced epoxy resin composite-based flytrap leaf structure [71].

In addition, attempts have also been explored by using ESB-based structures, which also depict the real curvature changes of the flytrap leaves. Figure 6c shows an ESB-based biomimetic flytrap structure, actuated by using magnets. The bistability is derived by using the antisymmetric composite layup-induced geometry curvature effects [69]. On the upper side of an artificial leaf, the iron sheet is attached to the middle part of the outer curved edge to be attracted by magnets. The electromagnet is placed just above the iron patch in order to generate a suitable trigger force for the shape-changing activation. When the electromagnet is activated, the curve edge is subjected to magnetic force and produces a bending moment on the artificial leaves, which makes the biomimetic structure shift to the second stable state. This non-contact actuation simplifies the biomimetic structure and actuation design, with adjustable magnetic force through current control. A further improvement was then carried out to reduce the actuation force required to trigger the morphing action. Figure 4d shows an ESB-based flytrap structure with clamped boundaries [70]. The inner curved edges are clamped, and the snap-through of artificial leaves is constrained by a clamping device. Controlling the width of the clamping edge can effectively reduce the actuation force required to trigger the morphing actuation.

Although the shape-changing characteristics of the bistable artificial leaves are similar to those of the real flytrap leaves, the internal stress field of the real flytrap leaf is locally orthotropic and follows the geometric shape of the leaf as a whole, see details as discussed in Section 2. A further trial is shown in Figure 6e, where a magnet-oriented particle-reinforced epoxy resin composite was applied to mimic the internal stress field of a real flytrap leaf [71]. The microstructural changes are controlled by a magnetic field using locally oriented rigid anisotropic magnetic particles, in order to adjust the local prestrain and stiffness anisotropy of the composite. Compared with carbon fiber reinforced composites, the local residual strain of this biomimetic composite structure is more controllable, with a certain performance and shape gradient, and can achieve rapid shape-changing actuation, approaching the true shape transition mechanisms of the real flytrap.

A further strategy is considering multiple driving methods to mimic the biomechanics of the flytrap. Figure 7 shows a design scheme using multi-stimulus and multi-temporal responsive composites, enabling both fast morphing through structural bistability and slow morphing through diffusion processes using hydrogel [72]. The multi-responsive composites consist of a hydrogel layer and an architected particle-reinforced epoxy bilayer. The spatial distribution orientation of the magnetic responsive plates in each epoxy layer is achieved by using a magnetic field to induce in-plane mechanical properties and shrinkage. The epoxy double layer is used to adjust the prestress in the material, while the hydrogel layer controls the time response according to the hydration level. This smart composite-based biomimetic structure exhibits rapid and slow deformation in response to mechanical, magnetic, thermal and hydration stimuli. The integration of bistable and stimuli-responsive materials or smart materials can sense a variety of specific environmental conditions and react quickly.

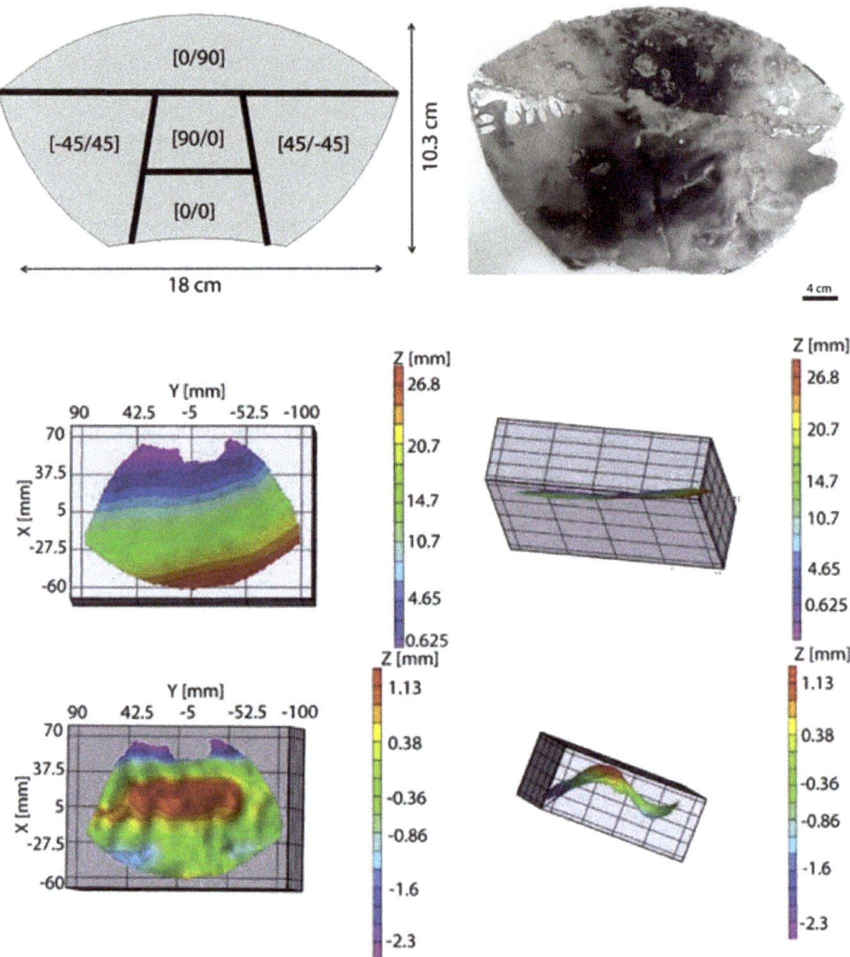

Figure 7. A multi-stimulus and multi-temporal responsive composites-based biomimetic flytrap structure, enabling fast snap-trapping through structural bistability and slow motion through diffusion processes using hydrogels [72].

5. Conclusions and Future Outlook

The biggest challenge for future smart composite bionics is to create a biomimetic flytrap structure that can pass the Turing test, which is designed to assess whether a model can accurately mimic the properties of a biological system. This means that a biomimetic structure is able to grow, and recover, and also has similar sense and response capabilities to that of a real Venus flytrap. Although some investigations have proposed self-growing robot concepts, the current biomimetic structures for flytraps still remain at the initial quickly growing stage. Therefore, this review focuses on the bionic strategies in terms of sensing, responding and actuation, as well as the rapid snap-trapping of the biomimetic Venus flytrap structures by employing the smart composite technology, aiming to enrich the diversities and reveal the fundamentals in order to further advance the multidisciplinary science and technological development into composite bionics.

The in-depth analysis of the biomechanics of a real Venus flytrap is helpful and essential in order to further promote and advance biomimetic structures. Flytrap leaves are not fully folded when stones, raindrops, etc., come into contact with trigger hairs, but only when prey is getting in contact. This energy-saving mechanism is still magical and worth further investigation. By using a multi-stable structure or driving control strategy, a multifunctional fly-catching biomimetic structure may be achieved. Compared with the enclosed trapping of the real flytrap, it is more realistic to mimic the bending or clamping motion for smart composite-based flytrap structures, and their driving strategies are diverse. The precision control of the express snapping and the load output range are still limited and depend on further development into smart composite technology. The sensing characteristics of IPMC are most similar to those of flytraps; whilst it lacks the elastic energy-derived rapid snapping. IPMC is promising to be applied as an actuator in the future owing to its sensing and fast response capabilities, in order to trigger the bistable artificial leaves. An exciting solution is a programmable microstructural bistable composite with multi-stimuli and multi-temporal responsive mechanics. The combination of bistability and bionics provides a new concept for developing the biomimetic structure with rapid shape morphing capability. In particular, the bistable composite structures offer similar shape-changing characteristics to a real flytrap leaf, and the elastic energy-driven actuation enables rapid capture motion. In the meantime, imitating the real curvature changes and internal stress field of a flytrap leaf improves the capture effectiveness of the biomimetic flytrap structure.

Author Contributions: B.W.: Investigation, Supervision, Formal analysis, Writing—original draft, Writing—review and editing, Funding acquisition; Y.H.: Investigation, Data curation, Software, Formal analysis, Writing—original draft; S.Z.: Supervision, Writing—review and editing, Funding acquisition; J.Z.: Validation, Writing—review and editing; C.G.: Supervision, Writing—review and editing. All authors have read and agreed to the published version of the manuscript.

Funding: The authors are grateful for financial support from the National Natural Science Foundation of China (52005108, 52275096), Fuzhou-Xiamen-Quanzhou National Independent Innovation Demonstration Zone High-end Equipment Vibration and Noise Detection and Fault Diagnosis Collaborative Innovation Platform Project, Fujian Provincial Major Research Project (2022HZ024005), as well as the Start-up Funding from Education Department of Fujian Province and Fuzhou University (GXRC-20066). We also thank the technical staff and aegis of the Fuzhou University International Joint Laboratory of Precision Instruments and Intelligent Measurement & Control.

Institutional Review Board Statement: Not applicable.

Informed Consent Statement: Not applicable.

Data Availability Statement: Data are available on request to the corresponding author.

Conflicts of Interest: The authors declare that there is no conflict of interest.

References

1. Wegst, U.G.K.; Bai, H.; Saiz, E.; Tomsia, A.P.; Ritchie, R.O. Bioinspired Structural Materials. *Nat. Mater.* **2015**, *14*, 23–36. [CrossRef] [PubMed]
2. Li, S.; Wang, K.W. Plant-Inspired Adaptive Structures and Materials for Morphing and Actuation: A Review. *Bioinspir. Biomim.* **2017**, *12*, 011001. [CrossRef] [PubMed]
3. Zirbel, S.A.; Lang, R.J.; Thomson, M.W.; Sigel, D.A.; Walkemeyer, P.E.; Trease, B.P.; Magleby, S.P.; Howell, L.L. Accommodating Thickness in Origami-Based Deployable Arrays1. *J. Mech. Des. Trans. ASME* **2013**, *135*, 111005. [CrossRef]
4. Roberts, T.J.; Azizi, E. Flexible Mechanisms: The Diverse Roles of Biological Springs in Vertebrate Movement. *J. Exp. Biol.* **2011**, *214*, 353–361. [CrossRef]
5. Moulton, D.E.; Lessinnes, T.; O'Keeffe, S.; Dorfmann, L.; Goriely, A. The Elastic Secrets of the Chameleon Tongue. *Proc. R. Soc. A Math. Phys. Eng. Sci.* **2016**, *472*, 20160030. [CrossRef]
6. Smith, M.L.; Yanega, G.M.; Ruina, A. Elastic Instability Model of Rapid Beak Closure in Hummingbirds. *J. Theor. Biol.* **2011**, *282*, 41–51. [CrossRef] [PubMed]
7. Forterre, Y.; Skotheim, J.M.; Dumais, J.; Mahadevan, L. How the Venus Flytrap Snaps. *Nature* **2005**, *433*, 421–425. [CrossRef] [PubMed]
8. Guo, Q.; Dai, E.; Han, X.; Xie, S.; Chao, E.; Chen, Z. Fast Nastic Motion of Plants and Bioinspired Structures. *J. R. Soc. Interface* **2015**, *12*, 20150598. [CrossRef]

9. Cao, Y.; Derakhshani, M.; Fang, Y.; Huang, G.; Cao, C. Bistable Structures for Advanced Functional Systems. *Adv. Funct. Mater.* **2021**, *31*, 2106231. [CrossRef]
10. Chi, Y.; Li, Y.; Zhao, Y.; Hong, Y.; Tang, Y.; Yin, J. Bistable and Multistable Actuators for Soft Robots: Structures, Materials, and Functionalities. *Adv. Mater.* **2022**, *34*, 2110384. [CrossRef]
11. Wang, B.; Ge, C.; Fancey, K.S. Snap-through Behaviour of a Bistable Structure Based on Viscoelastically Generated Prestress. *Compos. Part B Eng.* **2017**, *114*, 23–33. [CrossRef]
12. Esser, F.J.; Auth, P.; Speck, T. Artificial Venus Flytraps: A Research Review and Outlook on Their Importance for Novel Bioinspired Materials Systems. *Front. Robot. AI* **2020**, *7*, 75. [CrossRef] [PubMed]
13. Poppinga, S.; Zollfrank, C.; Prucker, O.; Rühe, J.; Menges, A.; Cheng, T.; Speck, T. Toward a New Generation of Smart Biomimetic Actuators for Architecture. *Adv. Mater.* **2018**, *30*, 1703653. [CrossRef]
14. Scherzer, S.; Federle, W.; Al-Rasheid, K.A.S.; Hedrich, R. Venus Flytrap Trigger Hairs Are Micronewton Mechano-Sensors That Can Detect Small Insect Prey. *Nat. Plants* **2019**, *5*, 670–675. [CrossRef] [PubMed]
15. Williams, M.E.; Mozingo, H.N. The Fine Structure of the Trigger Hair in Venus's Flytrap. *Am. J. Bot.* **1971**, *58*, 532–539. [CrossRef]
16. Fabricant, A.; Iwata, G.Z.; Scherzer, S.; Bougas, L.; Rolfs, K.; Jodko-Władzińska, A.; Voigt, J.; Hedrich, R.; Budker, D. Action Potentials Induce Biomagnetic Fields in Carnivorous Venus Flytrap Plants. *Sci. Rep.* **2021**, *11*, 1438. [CrossRef] [PubMed]
17. Volkov, A.G. Signaling in Electrical Networks of the Venus Flytrap (*Dionaea muscipula* Ellis). *Bioelectrochemistry* **2019**, *125*, 25–32. [CrossRef]
18. Hill, B.S.; Findlay, G.P. The Power of Movement in Plants: The Role of Osmotic Machines. *Q. Rev. Biophys.* **1981**, *14*, 173–222. [CrossRef]
19. Sachse, R.; Westermeier, A.; Mylo, M.; Nadasdi, J.; Bischoff, M.; Speck, T.; Poppinga, S. Snapping Mechanics of the Venus Flytrap (*Dionaea muscipula*). *Proc. Natl. Acad. Sci. USA* **2020**, *117*, 16035–16042. [CrossRef]
20. Oliveira, J.; Correia, V.; Castro, H.; Martins, P.; Lanceros-Mendez, S. Polymer-Based Smart Materials by Printing Technologies: Improving Application and Integration. *Addit. Manuf.* **2018**, *21*, 269–283. [CrossRef]
21. Mahmood, A.; Akram, T.; Shenggui, C.; Chen, H. Revolutionizing Manufacturing: A Review of 4D Printing Materials, Stimuli, and Cutting-Edge Applications. *Compos. Part B Eng.* **2023**, *266*, 110952. [CrossRef]
22. Mohd Jani, J.; Leary, M.; Subic, A.; Gibson, M.A. A Review of Shape Memory Alloy Research, Applications and Opportunities. *Mater. Des.* **2014**, *56*, 1078–1113. [CrossRef]
23. Mather, P.T.; Luo, X.; Rousseau, I.A. Shape Memory Polymer Research. *Annu. Rev. Mater. Res.* **2009**, *39*, 445–471. [CrossRef]
24. Lin, X.; Wang, B.; Zhong, S.; Chen, H.; Liu, D. Smart Driving of a Bilayered Composite Tape-Spring Structure. *J. Phys. Conf. Ser.* **2022**, *2403*, 012042. [CrossRef]
25. Zhao, Q.; Qi, H.J.; Xie, T. Recent Progress in Shape Memory Polymer: New Behavior, Enabling Materials, and Mechanistic Understanding. *Prog. Polym. Sci.* **2015**, *49–50*, 79–120. [CrossRef]
26. Li, Y.; Xiao, Y.; Yu, L.; Ji, K.; Li, D. A Review on the Tooling Technologies for Composites Manufacturing of Aerospace Structures: Materials, Structures and Processes. *Compos. Part A Appl. Sci. Manuf.* **2022**, *154*, 106762. [CrossRef]
27. Yao, Y.; Zhou, T.; Wang, J.; Li, Z.; Lu, H.; Liu, Y.; Leng, J. 'Two Way' Shape Memory Composites Based on Electroactive Polymer and Thermoplastic Membrane. *Compos. Part A Appl. Sci. Manuf.* **2016**, *90*, 502–509. [CrossRef]
28. Yang, H.; Lu, H.; Miao, Y.; Cong, Y.; Ke, Y.; Wang, J.; Yang, H.; Fu, J. Non-Swelling, Super-Tough, Self-Healing, and Multi-Responsive Hydrogels Based on Micellar Crosslinking for Smart Switch and Shape Memory. *Chem. Eng. J.* **2022**, *450*, 138346. [CrossRef]
29. Deng, H.; Zhang, C.; Sattari, K.; Ling, Y.; Su, J.W.; Yan, Z.; Lin, J. 4D Printing Elastic Composites for Strain-Tailored Multistable Shape Morphing. *ACS Appl. Mater. Interfaces* **2021**, *13*, 12719–12725. [CrossRef]
30. Fang, L.; Chen, S.; Fang, T.; Fang, J.; Lu, C.; Xu, Z. Shape-Memory Polymer Composites Selectively Triggered by near-Infrared Light of Two Certain Wavelengths and Their Applications at Macro-/Microscale. *Compos. Sci. Technol.* **2017**, *138*, 106–116. [CrossRef]
31. Ze, Q.; Kuang, X.; Wu, S.; Wong, J.; Montgomery, S.M.; Zhang, R.; Kovitz, J.M.; Yang, F.; Qi, H.J.; Zhao, R. Magnetic Shape Memory Polymers with Integrated Multifunctional Shape Manipulation. *Adv. Mater.* **2020**, *32*, 1906657. [CrossRef] [PubMed]
32. Zhang, F.; Xia, Y.; Wang, L.; Liu, L.; Liu, Y.; Leng, J. Conductive Shape Memory Microfiber Membranes with Core-Shell Structures and Electroactive Performance. *ACS Appl. Mater. Interfaces* **2018**, *10*, 35526–35532. [CrossRef] [PubMed]
33. Zhang, F.; Xiong, L.; Ai, Y.; Liang, Z.; Liang, Q. Stretchable Multiresponsive Hydrogel with Actuatable, Shape Memory, and Self-Healing Properties. *Adv. Sci.* **2018**, *5*, 1800450. [CrossRef] [PubMed]
34. Xia, Y.; He, Y.; Zhang, F.; Liu, Y.; Leng, J. A Review of Shape Memory Polymers and Composites: Mechanisms, Materials, and Applications. *Adv. Mater.* **2021**, *33*, 2000713. [CrossRef]
35. Wang, B.; Fancey, K.S. Viscoelastically Prestressed Polymeric Matrix Composites: An Investigation into Fibre Deformation and Prestress Mechanisms. *Compos. Part A Appl. Sci. Manuf.* **2018**, *111*, 106–114. [CrossRef]
36. Wang, B.; Zhong, S.; Lee, T.-L.; Fancey, K.S.; Mi, J. Non-Destructive Testing and Evaluation of Composite Materials/Structures: A State-of-the-Art Review. *Adv. Mech. Eng.* **2020**, *12*, 1687814020913761. [CrossRef]
37. Hyer, M.W. Some Observations on the Cured Shape of Thin Unsymmetric Laminates. *J. Compos. Mater.* **1981**, *15*, 175–194. [CrossRef]

38. Hyer, M.W. Calculations of the Room-Temperature Shapes of Unsymmetric Laminates. *J. Compos. Mater.* **1981**, *15*, 296–310. [CrossRef]
39. Hyer, M.W. The Room-Temperature Shapes of Four-Layer Unsymmetric Cross-Ply Laminates. *J. Compos. Mater.* **1982**, *16*, 318–340. [CrossRef]
40. Lachenal, X.; Daynes, S.; Weaver, P.M. A Non-Linear Stiffness Composite Twisting I-Beam. *J. Intell. Mater. Syst. Struct.* **2013**, *25*, 744–754. [CrossRef]
41. Xu, B.; Wang, B.; Fancey, K.S.; Zhong, S.; Zhao, C.; Chen, X. A Bistable Helical Structure Based on Composite Tape-Springs. *Compos. Commun.* **2023**, *43*, 101723. [CrossRef]
42. Dano, M.-L.; Hyer, M.W. Thermally-Induced Deformation Behavior of Unsymmetric Laminates. *Int. J. Solids Struct.* **1998**, *35*, 2101–2120. [CrossRef]
43. Lee, A.J.; Moosavian, A.; Inman, D.J. A Piezoelectrically Generated Bistable Laminate for Morphing. *Mater. Lett.* **2017**, *190*, 123–126. [CrossRef]
44. Lee, A.J.; Moosavian, A.; Inman, D.J. Control and Characterization of a Bistable Laminate Generated with Piezoelectricity. *Smart Mater. Struct.* **2017**, *26*, 085007. [CrossRef]
45. Chillara, V.S.C.; Dapino, M.J. Stability Considerations and Actuation Requirements in Bistable Laminated Composites. *Compos. Struct.* **2018**, *184*, 1062–1070. [CrossRef]
46. Daynes, S.; Potter, K.D.; Weaver, P.M. Bistable Prestressed Buckled Laminates. *Compos. Sci. Technol.* **2008**, *68*, 3431–3437. [CrossRef]
47. Wang, B.; Fancey, K.S. A Bistable Morphing Composite Using Viscoelastically Generated Prestress. *Mater. Lett.* **2015**, *158*, 108–110. [CrossRef]
48. Iqbal, K.; Pellegrino, S. Bi-Stable Composite Shells. In Proceedings of the 41st Structures, Structural Dynamics, and Materials Conference and Exhibit, Atlanta, GA, USA, 3–6 April 2000; American Institute of Aeronautics and Astronautics: Reston, VA, USA, 2000.
49. Wang, B.; Seffen, K.A.; Guest, S.D.; Lee, T.-L.; Huang, S.; Luo, S.; Mi, J. In-Situ Multiscale Shear Failure of a Bistable Composite Tape-Spring. *Compos. Sci. Technol.* **2020**, *200*, 108348. [CrossRef]
50. Yang, C.; Wang, B.; Zhong, S.; Zhao, C.; Liang, W. On Tailoring Deployable Mechanism of a Bistable Composite Tape-Spring Structure. *Compos. Commun.* **2022**, *32*, 101171. [CrossRef]
51. Wang, B.; Fancey, K.S. Towards Optimisation of Load-Time Conditions for Producing Viscoelastically Prestressed Polymeric Matrix Composites. *Compos. Part B Eng.* **2016**, *87*, 336–342. [CrossRef]
52. Daynes, S.; Weaver, P.M. Stiffness Tailoring Using Prestress in Adaptive Composite Structures. *Compos. Struct.* **2013**, *106*, 282–287. [CrossRef]
53. Fancey, K.S. Viscoelastically Prestressed Polymeric Matrix Composites: An Overview. *J. Reinf. Plast. Compos.* **2016**, *35*, 1290–1301. [CrossRef]
54. Chen, H.; Yu, F.; Wang, B.; Zhao, C.; Chen, X.; Nsengiyumva, W.; Zhong, S. Elastic Fibre Prestressing Mechanics within a Polymeric Matrix Composite. *Polymers* **2023**, *15*, 431. [CrossRef] [PubMed]
55. Seffen, K.A.; Wang, B.; Guest, S.D. Folded Orthotropic Tape-Springs. *J. Mech. Phys. Solids* **2019**, *123*, 138–148. [CrossRef]
56. Shahinpoor, M. Biomimetic Robotic Venus Flytrap (*Dionaea muscipula* Ellis) Made with Ionic Polymer Metal Composites. *Bioinspir. Biomim.* **2011**, *6*, 046004. [CrossRef]
57. Shi, L.; Guo, S. Development and Evaluation of a Venus Flytrap-inspired Microrobot. *Microsyst. Technol.* **2016**, *22*, 1949–1958. [CrossRef]
58. Colombani, M.; Forterre, Y. Biomechanics of Rapid Movements in Plants: Poroelastic Measurements at the Cell Scale. *Comput. Methods Biomech. Biomed. Eng.* **2011**, *14*, 115–117. [CrossRef]
59. Lee, H.; Xia, C.; Fang, N.X. First Jump of Microgel; Actuation Speed Enhancement by Elastic Instability. *Soft Matter* **2010**, *6*, 4342–4345. [CrossRef]
60. Athas, J.C.; Nguyen, C.P.; Zarket, B.C.; Gargava, A.; Nie, Z.; Raghavan, S.R. Enzyme-Triggered Folding of Hydrogels: Toward a Mimic of the Venus Flytrap. *ACS Appl. Mater. Interfaces* **2016**, *8*, 19066–19074. [CrossRef]
61. Fan, W.; Shan, C.; Guo, H.; Sang, J.; Wang, R.; Zheng, R.; Sui, K.; Nie, Z. Dual-Gradient Enabled Ultrafast Biomimetic Snapping of Hydrogel Materials. *Sci. Adv.* **2019**, *5*, eaav7174. [CrossRef]
62. Lunni, D.; Cianchetti, M.; Filippeschi, C.; Sinibaldi, E.; Mazzolai, B. Plant-Inspired Soft Bistable Structures Based on Hygroscopic Electrospun Nanofibers. *Adv. Mater. Interfaces* **2020**, *7*, 1901310. [CrossRef]
63. Wani, O.M.; Zeng, H.; Priimagi, A. A Light-Driven Artificial Flytrap. *Nat. Commun.* **2017**, *8*, 15546. [CrossRef] [PubMed]
64. Wani, O.M.; Verpaalen, R.; Zeng, H.; Priimagi, A.; Schenning, A.P.H.J. An Artificial Nocturnal Flower via Humidity-Gated Photoactuation in Liquid Crystal Networks. *Adv. Mater.* **2019**, *31*, e1805985. [CrossRef] [PubMed]
65. Wang, B.; Seffen, K.A.; Guest, S.D. Folded Strains of a Bistable Composite Tape-Spring. *Int. J. Solids Struct.* **2021**, *233*, 111221. [CrossRef]
66. Kim, S.W.; Koh, J.S.; Cho, M.; Cho, K.J. Towards a Bio-Mimetic Flytrap Robot Based on a Snap-through Mechanism. In Proceedings of the 2010 3rd IEEE RAS and EMBS International Conference on Biomedical Robotics and Biomechatronics, BioRob 2010, Tokyo, Japan, 26–29 September 2010; pp. 534–539.

67. Kim, S.W.; Koh, J.S.; Cho, M.; Cho, K.J. Design & Analysis a Flytrap Robot Using Bi-Stable Composite. In Proceedings of the 2011 IEEE International Conference on Robotics and Automation, Shanghai, China, 9–13 May 2011; pp. 215–220.
68. Kim, S.W.; Koh, J.S.; Lee, J.G.; Ryu, J.; Cho, M.; Cho, K.J. Flytrap-Inspired Robot Using Structurally Integrated Actuation Based on Bistability and a Developable Surface. *Bioinspir. Biomim.* **2014**, *9*, 036004. [CrossRef]
69. Zhang, Z.; Chen, D.; Wu, H.; Bao, Y.; Chai, G. Non-Contact Magnetic Driving Bioinspired Venus Flytrap Robot Based on Bistable Anti-Symmetric CFRP Structure. *Compos. Struct.* **2016**, *135*, 17–22. [CrossRef]
70. Zhang, Z.; Li, X.; Yu, X.; Chai, H.; Li, Y.; Wu, H.; Jiang, S. Magnetic Actuation Bionic Robotic Gripper with Bistable Morphing Structure. *Compos. Struct.* **2019**, *229*, 111422. [CrossRef]
71. Schmied, J.U.; Le Ferrand, H.; Ermanni, P.; Studart, A.R.; Arrieta, A.F. Programmable Snapping Composites with Bio-Inspired Architecture. *Bioinspir. Biomim.* **2017**, *12*, 026012. [CrossRef]
72. Le Ferrand, H.; Riley, K.S.; Arrieta, A.F. Plant-Inspired Multi-Stimuli and Multi-Temporal Morphing Composites. *Bioinspir. Biomim.* **2022**, *17*, 046002. [CrossRef]

Article

Microstructure and Mechanical Properties of $Co_{32}Cr_{28}Ni_{32.94}Al_{4.06}Ti_3$ High-Entropy Alloy

Jinquan Guo [1,2], Chaozhongzheng Tang [1] and Huan Sheng Lai [1,3,*]

1 School of Mechanical Engineering and Automation, Fuzhou University, Fuzhou 350108, China; megjq@fzu.edu.cn (J.G.); n190227094@fzu.edu.cn (C.T.)
2 Fujian Key Laboratory of Force Measurement, Fujian Metrology Institute, Fuzhou 350108, China
3 Sino-French Institute of Nuclear Engineering and Technology, Sun Yat-sen University, Guangzhou 510275, China
* Correspondence: laihsh@mail.sysu.edu.cn

Abstract: High-entropy alloys have good application prospects in nuclear power plants due to their excellent mechanical properties and radiation resistance. In this paper, the microstructure of the $Co_{32}Cr_{28}Ni_{32.94}Al_{4.06}Ti_3$ high-entropy alloy was researched using metallurgical microscopy, X-ray diffraction, and scanning electron microscopy. The mechanical properties were tested using a Vickers microhardness tester and a tensile testing machine, respectively. The results showed that $Co_{32}Cr_{28}Ni_{32.94}Al_{4.06}Ti_3$ had a single-phase, disordered, face-centered, cubic solid-solution structure and was strengthened by solid solution. The alloy lattice parameter and density were estimated as 0.304 nm and 7.89 g/cm^3, respectively. The test results indicated that the alloy had satisfactory mechanical properties with yield stress and tensile strength of about 530 MPa and 985 MPa, respectively.

Keywords: microstructure; high-entropy alloy; mechanical property

Citation: Guo, J.; Tang, C.; Lai, H.S. Microstructure and Mechanical Properties of $Co_{32}Cr_{28}Ni_{32.94}Al_{4.06}Ti_3$ High-Entropy Alloy. *Materials* **2022**, *15*, 1444. https://doi.org/10.3390/ma15041444

Academic Editor: Jan Frenzel

Received: 22 December 2021
Accepted: 2 February 2022
Published: 15 February 2022

Publisher's Note: MDPI stays neutral with regard to jurisdictional claims in published maps and institutional affiliations.

1. Introduction

High-entropy alloys (HEAs) are a type of alloys which contain five or more principle alloying elements to stabilize solid solution phases by maximizing configurational entropy. HEAs have been extensively studied since they were first reported in 2004 [1]. The novel design idea underlying HEAs has greatly expanded the composition range and research areas of metallic materials. Owing to the high entropy effect, the alloys tend to form simple microstructures, such as body-centered cubic (BCC), face-centered cubic (FCC), and densely packed hexagonal (HCP) [2,3]. It has been found that some HEAs not only have excellent mechanical properties [4,5], but also excellent thermal stability [5], wear resistance [4–7], and irradiation resistance. Therefore, HEAs are promising materials for application in nuclear power plants.

The phase composition and mechanical properties of HEAs are determined by the element types and their atomic ratios in the alloy [8]. For transition HEAs, $AlCoCrFeNiZr_x$ is known to possess a BCC structure when $x = 0$ [9]; with an increase in the Zr content, the mechanical properties are significantly improved, because the phase composition of the alloy changes from an ordered BCC solid-solution phase to an ordered Laves + BCC phase. The phase composition of the $AlCoCrFeNiB_x$ alloy changes from a BCC phase to BCC + FCC two-phase solid solution with an increase in the B content; meanwhile, the hardness and fracture strength increase first and then decrease [10]. $CoCrFeNiW_x$ is known to possess a single-phase structure for $x = 0.2$ [11]; the phase composition of the alloy changes from a single-phase to a hypoeutectic phase, and the mechanical properties are significantly improved when the W content is increased. The phase composition of $Al_xCoCrFeNiTi$ changes from α + Al grains to fine equiaxed crystals when the Al content is increased; in this way, the yield stress is increased, but the elongation is decreased [12].

AlFeCrCoNi only exhibits an FCC structure for a low ratio of Al/Ni. However, for a large ratio of Al/Ni, the alloy exhibits a BCC structure. The hardness increases with the increase in the ratio of Al/Ni [13].

The three-component alloy of CoCrNi has a single FCC structure. Its mechanical properties are better than other alloys in the alloy system, especially at liquid nitrogen temperature [14–16], but the strength is relatively low at chamber temperature [16]. The alloy used in this study is a new, unreported alloy with Al and Ti elements added. Therefore, in this paper, the microstructure and tensile properties of $Co_{32}Cr_{28}Ni_{32.94}Al_{4.06}Ti_3$ were researched in order to better understand the microstructure and mechanical properties of CoCrNi HEAs.

2. Experimental Materials and Methods

A $Co_{32}Cr_{28}Ni_{32.94}Al_{4.06}Ti_3$ plate with dimensions of 90 mm × 80 mm × 20 mm was prepared using vacuum arc melting equipment with pure metal Co, Cr, Ni, Al, and Ti with purities greater than 99.9 wt%. In order to ensure that the material was not oxidized in the melting process, pure metal Ti was melted firstly to absorb the residual gas of the electric arc furnace. The composition (atomic fraction and weight ratio, %) of the $Co_{32}Cr_{28}Ni_{32.94}Al_{4.06}Ti_3$ used in this study is shown in Table 1. The crystal structure was identified using a DY1602/Empyrean multifunctional X-ray polycrystalline diffractometer (PAnalytical, Alemlo, Holland) with Cu target radiation scanning in the range from 20° to 90° at a rate of 2°/min in a working voltage of 40 kV, a working current of 100 mA, and a characteristic wavelength of 1.5406 Å. The angle measurement accuracy was 0.02°. The HEA density was measured using a drainage method. In the drainage method, the mass of the HEA was measured by a scale; a measuring cylinder was used to measure the volume of water, and the volume of the HEA was the changed volume of the water in the cylinder before and after the HEA was put into the cylinder. The metallography of the HEA was characterized using a MV5000 metallographic microscope (Nanjing Lianchuang Analytical Instrument Manufacturing, Nanjing, China). The Vickers hardness of the HEA was measured using a THV-1MD microhardness tester (Teshi Detection Technology, Shanghai, China) under a load of 200 gf applied for 15 s. The tensile specimens with dimensions of 51.83 mm × 14 mm × 2.5 mm were used, as shown in Figure 1. Three tensile tests were carried out at air room temperature with a strain rate of 1×10^{-3} mm/s. The fracture morphology of the tensile specimen was analyzed using a Mira3 ultra-high-resolution field emission scanning electron microscope (FE-SEM) (TESCAN, Shanghai, China).

Table 1. Atomic and weight ratios of the principal elements of $Co_{32}Cr_{28}Ni_{32.94}Al_{4.06}Ti_3$.

Metal	Co	Cr	Ni	Al	Ti
Atomic ratio (%)	32	28	32.94	4.06	3
Weight ratio (%)	34.11	26.34	34.97	1.98	2.60

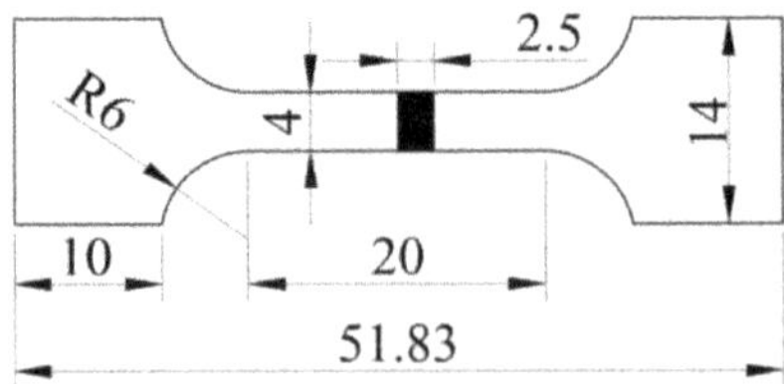

Figure 1. Dimensions of the tensile specimen (mm).

3. Results and Discussion

Figure 2 shows the X-ray diffraction (XRD) pattern of $Co_{32}Cr_{28}Ni_{32.94}Al_{4.06}Ti_3$, wherein three diffraction peaks corresponding to the (111), (200), and (220) peaks of the FCC structure were observed. There were no diffraction peaks corresponding to other structures. The result implies that the HEA was only composed of FCC solid solution.

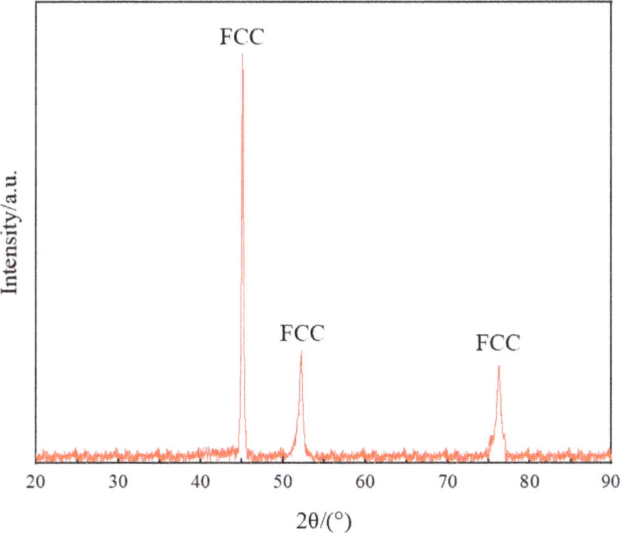

Figure 2. X-ray diffraction (XRD) results of $Co_{32}Cr_{28}Ni_{32.94}Al_{4.06}Ti_3$.

Generally, $\Omega > 1.1$ and $\delta < 6.5\%$ are the criteria for determining whether a solid solution phase can be formed [13]. Furthermore, when valence electron concentration (VEC) ≥ 8, the FCC solid solution is considered to be relatively stable [17]. The relevant parameters are calculated as follows:

$$\Delta H_{\text{mix}} = \sum_{i=1, i \neq j}^{n} \Delta H_{ij}^{\text{mix}} c_i c_j \tag{1}$$

$$\delta = \sqrt{\sum_{i=1}^{n} c_i \left(1 - r_i / \sum_{i=1}^{n} c_i r_i\right)^2} \tag{2}$$

$$\text{VEC} = \sum_{i=1}^{n} c_i (\text{VEC})_i \tag{3}$$

$$\Omega = \frac{T_m \Delta S_{\text{mix}}}{|\Delta H_{\text{mix}}|} \tag{4}$$

$$T_m = \sum_{i=1}^{n} c_i (T_m)_i \tag{5}$$

$$\Delta S_{\text{mix}} = -R \sum_{i=1}^{n} (c_i \ln c_i) \tag{6}$$

where ΔH_{mix} is the total enthalpy of mixing for the system, $\Delta H_{ij}^{\text{mix}}$ is the mixing enthalpy of the atomic pair of the *i*th and *j*th atoms, c_i and c_j are the atomic percentages of elements *i* and *j*, respectively, δ is the difference in atomic radius between the two atoms, r_i is the atomic radius of the *i*th element, VEC is the total VEC of the system, $(\text{VEC})_i$ is the VEC of the ith element, Ω is the disorder of the system, T_m is the mixed melting point of the system, ΔS_{mix} is the mixing entropy of the system, $(T_m)_i$ is the metal melting point of element *i*, and *R* is the gas constant.

Table 2 lists the characteristic parameters of the elements of $Co_{32}Cr_{28}Ni_{32.94}Al_{4.06}Ti_3$, and Table 3 lists the mixing enthalpy values of the atom pairs of the HEA elements [18].

According to the calculations, the difference in the atomic size (δ) of the HEA was 4.26%. The disorder of the system (Ω) of the HEA was 3.91, and the VEC of the HEA was 8.1. Therefore, the HEA considered here should be an FCC solid-solution structure in theory.

Table 2. Characteristic parameters of the principal elements of $Co_{32}Cr_{28}Ni_{32.94}Al_{4.06}Ti_3$.

Metal	Co	Cr	Ni	Al	Ti
Melting point (K)	1768.15	2132.15	1728.15	933.15	1941.15
Atomic radius [19,20] (nm)	0.125	0.128	0.123	0.143	0.147
VEC	9	6	10	3	4

Table 3. Mixed enthalpies among the principal elements of $Co_{32}Cr_{28}Ni_{32.94}Al_{4.06}Ti_3$ (kJ/mol).

Metal	Co	Cr	Ni	Al	Ti
Co	/	−4	0	−19	−28
Cr	−4	/	−7	−10	−7
Ni	0	−7	/	−22	−35
Al	−19	−10	−22	/	−30
Ti	−28	−7	−35	−30	/

The lattice constants of the five elements are listed in Table 4. The lattice constants of alloys can be calculated using the disorder principle [21]:

$$a_{\text{mix}} = \sum_{i=1}^{n} c_i a_i \tag{7}$$

where c_i is the atomic percentage of element i, and a_i is the lattice constants of element i. According to the calculations, the lattice constant of $Co_{32}Cr_{28}Ni_{32.94}Al_{4.06}Ti_3$ was 0.304 nm, which was consistent with the lattice constant obtained as per XRD analysis (Table 4).

Table 4. Densities and lattice constants of the component metals of $Co_{32}Cr_{28}Ni_{32.94}Al_{4.06}Ti_3$.

Metal	Co	Cr	Ni	Al	Ti	Alloy (Calculated)	Alloy (Measured)
Lattice constant (nm)	0.25	0.29	0.35	0.41	0.35	0.30	0.31
Density (g/cm^3)	8.90	7.19	8.90	2.70	4.54	7.85	7.84

The theoretical density of the alloy was calculated using the following formula [22]:

$$\rho_{\text{mix}} = \frac{\sum_{i=1}^{n} c_i A_i}{\sum_{i=1}^{n} c_i A_i / \rho_i} \tag{8}$$

where c_i is the atomic percentage of element i, A_i is the atomic weight of element i, and ρ_i is the density of element i. From the calculation results, we noted that the theoretical density (7.85 g/cm^3) of the HEA was almost identical to the measured value (7.84 g/cm^3). It was also found that the HEA results were consistent with the rule of mixtures upon comparing the lattice constant and theoretical density of the alloy, which was a disordered FCC solid-solution structure.

Figure 3 shows a light microscopy image of the metallographic structure of $Co_{32}Cr_{28}Ni_{32.94}Al_{4.06}Ti_3$. As shown in Figure 3, the alloy had a uniform, single-phase, equiaxial crystal structure, and the grain size was about 163 μm as measured using the transection method.

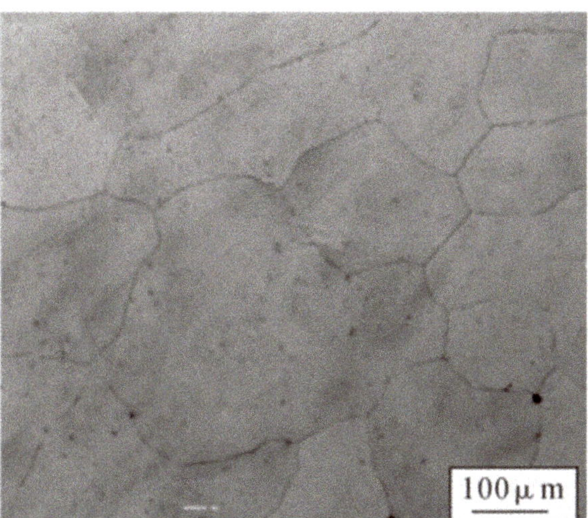

Figure 3. Microstructure of $Co_{32}Cr_{28}Ni_{32.94}Al_{4.06}Ti_3$.

Figure 4 shows the tensile stress–strain curve of $Co_{32}Cr_{28}Ni_{32.94}Al_{4.06}Ti_3$. As shown in Figure 4, the yield strength was 530 ± 6 MPa, the tensile strength was 985 ± 7 MPa, and the elongation was 37.16 ± 0.17%. Compared with the CoCrNi alloy, the yield strength and the tensile strength increased by 103% and 14%, while the elongation decreased by 7%. Obviously, the increase in Al and Ti elements improved the tensile properties of the CoCrNi baselime alloy. Moreover, from Table 5, which compares the uniaxial tensile test results of similar alloys, it can be seen that the alloy affords better strength and plasticity. Moreover, the microhardness of the alloy was 313 HV. Therefore, $Co_{32}Cr_{28}Ni_{32.94}Al_{4.06}Ti_3$ exhibits satisfactory tensile mechanical properties and hardness.

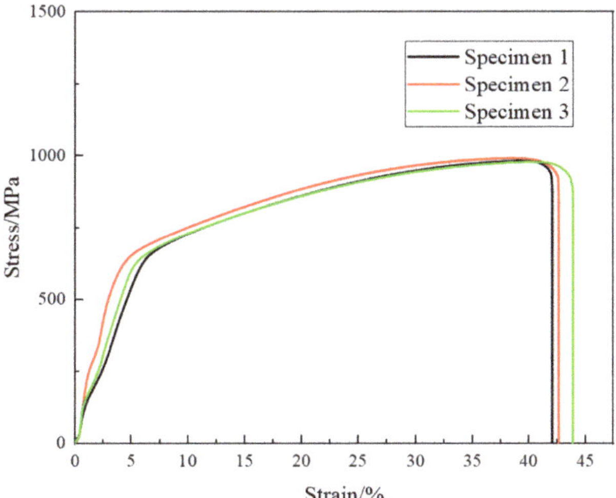

Figure 4. Stress–strain curve of $Co_{32}Cr_{28}Ni_{32.94}Al_{4.06}Ti_3$.

Table 5. Comparison of the tensile mechanical properties.

Alloy	Yield Strength (MPa)	Tensile Strength (MPa)	Elongation to Failure (%)
$Co_{32}Cr_{28}Ni_{32.94}Al_{4.06}Ti_3$	530 ± 6	985 ± 7	37.2 ± 0.17
CoCrNi [15]	260	870	40
$(Fe_{50}Mn_{30}Co_{10}Cr_{10})_{94}C_6$ [23]	450	700	18
$CoCrFeNiW_{0.4}$ [10]	525	970	11
$Al_3CoCrFeNiTi$ [11]	115	152	26
CoCrFeMnNi [24]	468	590	30
N18 Zircaloy [25]	390	420	38
N36 Zircaloy [25]	310	520	27

The tensile properties of the $Co_{32}Cr_{28}Ni_{32.94}Al_{4.06}Ti_3$ studied in this paper are obviously improved compared with N18 and N36 Zircaloy currently used in the nuclear industry, as shown in Table 5. Therefore, the alloy studied in this paper is could be used in the nuclear industry.

Generally, the main strengthening mechanism of a single disordered FCC solid-solution structure is solid-solution strengthening, which mainly originates from the interaction between solute atoms [26].

Figure 5 shows an SEM micrograph of the fractured tensile specimen of $Co_{32}Cr_{28}Ni_{32.94}Al_{4.06}Ti_3$. A large number of dimples and holes were observed in the fracture. The observed dimples were the traces left on the fracture after micropore nucleation and aggregation, which was the characteristic of micropore aggregation fractures.

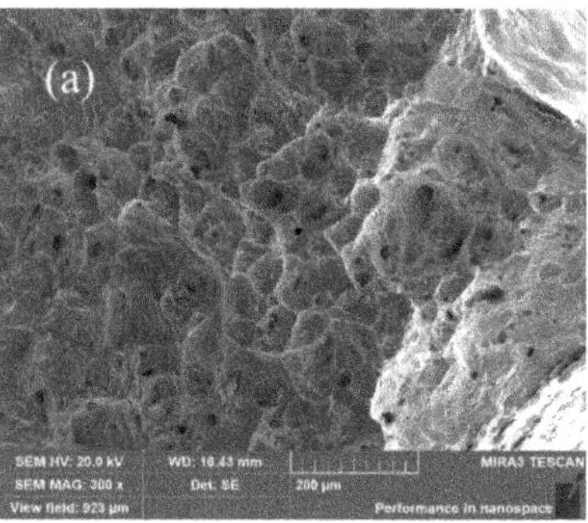

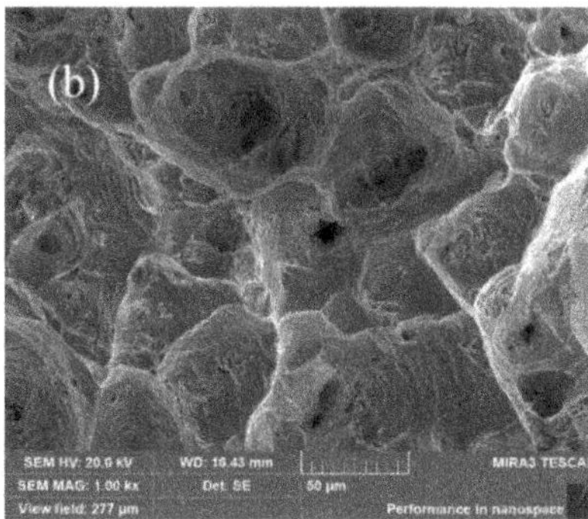

Figure 5. Scanning electron microscopy (SEM) images of the fracture surface of $Co_{32}Cr_{28}Ni_{32.94}Al_{4.06}Ti_3$. (**a**) magnification of 300 times; (**b**) magnification of 1000 times.

The results provide a reference for the study of microstructure and properties of CoCrNi HEAs. In order to fully explain the comprehensive mechanical properties of the HEA, further studies on CoCrNiAlTi HEAs are being carried out, such as the effects of different atomic ratios of major elements on the creep, fatigue properties, and irradiation resistance of CoCrNiAlTi HEAs.

4. Conclusions

In this paper, the microstructure and tensile-fracture characteristics of $Co_{32}Cr_{28}Ni_{32.94}Al_{4.06}Ti_3$ were researched. The following conclusions can be drawn:

(1) The $Co_{32}Cr_{28}Ni_{32.94}Al_{4.06}Ti_3$ exhibits a single disordered FCC solid-solution structure with a density of 7.89 g/cm^3.

(2) The microstructure of $Co_{32}Cr_{28}Ni_{32.94}Al_{4.06}Ti_3$ is equiaxed, with a grain size about 163 μm.

(3) The yield strength, tensile strength, and elongation of the $Co_{32}Cr_{28}Ni_{32.94}Al_{4.06}Ti_3$ are about 530 MPa, 985 Mpa, and 37.2%, respectively. The microhardness of the alloy is 313 HV.

Author Contributions: Conceptualization, J.G. and H.S.L.; methodology, J.G. and C.T.; experiment, C.T.; writing—original draft preparation, C.T.; writing—review and editing, J.G. and H.S.L. All authors listed have made a substantial, direct, and intellectual contribution to the work. All authors have read and agreed to the published version of the manuscript.

Funding: The authors are grateful to the National Natural Science Foundation of China (No. 11972005 and No. 51675103), the 2021 Independent Innovation Fund of Tianjin University-Fuzhou University (Grant Number TF2021-5), and the Open Fund of Fujian Key Laboratory of Force Measurement (Fujian Metrology Institute) (FJLZSYS202102) for the financial support to this study.

Institutional Review Board Statement: Not applicable.

Informed Consent Statement: Not applicable.

Data Availability Statement: Not applicable.

Conflicts of Interest: No potential conflict of interest was reported by the authors.

References

1. Yeh, J.W.; Chen, S.K.; Lin, S.J.; Gan, J.Y.; Chin, T.S.; Shun, T.T.; Tsau, C.H.; Chang, S.Y. Nanostructured high-entropy alloys with multiple principal elements: Novel alloy design concepts and outcomes. *Adv. Eng. Mater.* **2004**, *6*, 299–303. [CrossRef]
2. Schuh, B.; Mendez-Martin, F.; Völker, B.; George, E.P.; Clemens, H.; Pippan, R.; Hohenwarter, A. Mechanical properties, microstructure and thermal stability of a nanocrystalline CoCrFeMnNi high-entropy alloy after severe plastic deformation. *Acta Mater.* **2015**, *96*, 258–268. [CrossRef]
3. Lu, Y.P.; Gao, X.Z.; Jiang, L.; Zhen, Z.; Tang, T.; Jie, J.; Kang, H.; Zhang, Y.; Guo, S.; Ruan, H. Directly cast bulk eutectic and near-eutectic high entropy alloys with balanced strength and ductility in a wide temperature range. *Acta Mater.* **2017**, *124*, 143–150. [CrossRef]
4. Chou, Y.L.; Wang, Y.C.; Yeh, J.W.; Shih, H.C. Pitting corrosion of the high-entropy alloy $Co_{1.5}CrFeNi_{1.5}Ti_{0.5}Mo_{0.1}$ in chloride-containing sulphate solutions. *Corros. Sci.* **2010**, *52*, 3481–3491. [CrossRef]
5. Varalakshmi, S.; Kamaraj, M.; Murty, B.S. Processing and properties of nanocrystalline CuNiCoZnAlTi high entropy alloys by mechanical alloying. *Mater. Sci. Eng. A* **2010**, *527*, 1027–1030. [CrossRef]
6. Senkov, O.N.; Woodward, C.F. Microstructure and properties of a refractory $NbCrMo_{0.5}Ta_{0.5}TiZr$ alloy. *Mater. Sci. Eng. A* **2011**, *529*, 311–320. [CrossRef]
7. Varalakshmi, S.; Appa Rao, G.; Kamaraj, M.; Murty, B.S. Hot consolidation and mechanical properties of nanocrystalline equiatomic AlFeTiCrZnCu high entropy alloy after mechanical alloying. *J. Mater. Sci.* **2010**, *45*, 5158–5163. [CrossRef]
8. Guo, S.; Liu, C.T. Phase stability in high entropy alloys: Formation of solid-solution phase or amorphous phase. *Prog. Nat. Sci. Mater.* **2011**, *21*, 433–446. [CrossRef]
9. Chen, J.; Niu, P.Y.; Liu, Y.Z.; Lu, Y.K.; Wang, X.H.; Peng, Y.L.; Liu, J.N. Effect of Zr content on microstructure and mechanical properties of AlCoCrFeNi high entropy alloy. *Mater. Design* **2016**, *94*, 39–44. [CrossRef]
10. Chen, Q.S.; Lu, Y.P.; Dong, Y.; Wang, T.M.; Li, T.J. Effect of minor B addition on microstructure and properties of AlCoCrFeNi multi-compenent alloy. *Trans. Nonferr. Met. Soc.* **2015**, *25*, 2958–2964. [CrossRef]
11. Wang, L.; Wang, L.; Tang, Y.C.; Luo, L.; Luo, L.S.; Su, Y.Q.; Guo, J.J.; Fu, H.Z. Microstructure and mechanical properties of $CoCrFeNiW_x$ high entropy alloys reinforced by μ phase particles. *J. Alloys Compd.* **2020**, *845*, 155997. [CrossRef]
12. Li, Q.L.; Zhao, S.; Bao, X.P.; Zhang, Y.S.; Zhu, Y.Q.; Wang, C.Z.; Lan, Y.F.; Zhang, Y.X.; Xia, T.D. Effects of AlCoCrFeNiTi high-entropy alloy on microstructure and mechanical properties of pure aluminum. *J. Mater. Sci. Technol.* **2020**, *52*, 1–11. [CrossRef]
13. Harihar, S.; Joseph, W.N.; Frank, L.F. Microstructural characterization and mechanical properties of laser deposited high entropy alloys. *Mater. Sci. Forum.* **2014**, *3129*, 2370–2375.
14. Gludovatz, B.; Hohenwarter, A.; Thurston KV, S.; Bei, H.B.; Wu, Z.G.; George, E.P.; Ritchie, R.O. Exceptional damage-tolerance of a medium-entropy alloy CrCoNi at cryogenic temperatures. *Nat. Commun.* **2016**, *7*, 10602. [CrossRef]
15. Wu, Z.; Bei, H.; Pharr, G.M.; George, E.P. Temperature dependence of the mechanical properties of equiatomic solid solution alloys with face-centered cubic crystal structures. *Acta Mater.* **2014**, *81*, 428–441. [CrossRef]
16. Wu, Z.; Bei, H.; Otto, F.; Pharr, G.M.; George, E.P. Recovery, recrystallization, grain growth and phase stability of a family of FCC-structured multi-component equiatomic solid solution alloys. *Intermetallics* **2014**, *46*, 131–140. [CrossRef]
17. Guo, S.; Ng, C.; Lu, J.; Liu, C.T. Effect of valence electron concentration on stability of bcc or fcc phase in high entropy alloys. *J. Appl. Phys.* **2011**, *109*, 103505. [CrossRef]
18. Takeuchi, A.; Inoue, A. Classification of bulk metallic glasses by atomic size difference, heat of mixing and period of constituent elements and its application to characterization of the main alloying element. *Mater. Trans.* **2005**, *46*, 2817–2829. [CrossRef]
19. Pauling, L. Atomic Radii and Interatomic Distances in Metal. *J. Am. Chem. Soc.* **1947**, *69*, 542–553. [CrossRef]

20. Petrucci, R.H.; Harwood, W.S.; Geoffery, F.H.; Jeffry, D. Madura. In *General Chemistry*, 9th ed.; Pearsin Prentice Hall: Hoboken, NJ, USA, 2007.
21. Juan, C.C.; Tsai, M.H.; Tsai, C.W.; Lin, C.M.; Wang, W.R.; Yang, C.C.; Chen, S.K.; Liu, S.J.; Yeh, J.W. Enhanced mechanical properties of HfMoTaTiZr and HfMoNbTaTiZr refractory high-entropy alloys. *Intermetallics* **2015**, *62*, 76–83. [CrossRef]
22. Huang, H.; Wu, Y.; He, J.; Wang, H.; Liu, X.; An, K.; Wu, W.; Lu, Z. Phase-transformation ductilization of brittle high-entropy alloys via metastability engineering. *Adv. Mater.* **2017**, *29*, 1701678. [CrossRef] [PubMed]
23. Liu, X.L.; Zhao, X.R.; Chen, J.; Lv, Y.K.; Wang, X.H.; Liu, B.; Liu, Y. Effect of C addition on microstructure and mechanical properties of as-cast HEAs $(Fe_{50}Mn_{30}Co_{10}Cr_{10})_{100-x}C_x$. *Mater. Chem. Phys.* **2020**, *254*, 123501. [CrossRef]
24. Listyawan, T.A.; Lee, H.; Park, N.; Lee, U. Microstructure and mechanical properties of CoCrFeMnNi high entropy alloy with ultrasonic nanocrystal surface modification process. *J. Mater. Sci. Technol.* **2020**, *57*, 123–130. [CrossRef]
25. Nikulina, A.V. Zirconium alloy in nuclear power engineering. *Met. Sci. Heat Treat.* **2004**, *46*, 458–462. [CrossRef]
26. Tian, F.Y.; Varga, L.K.; Chen, N.X.; Shen, J.; Vitos, L. Empirical design of single phase high-entropy alloys with high hardness. *Intermetallics* **2015**, *58*, 1–6. [CrossRef]

Review

Recent Advances in Electro-Optic Response of Polymer-Stabilized Cholesteric Liquid Crystals

Kyung Min Lee [1,2,*], Zachary M. Marsh [1,2], Ecklin P. Crenshaw [1,2], Urice N. Tohgha [1,2], Cedric P. Ambulo [1,2], Steven M. Wolf [1,2], Kyle J. Carothers [1,2], Hannah N. Limburg [1,3], Michael E. McConney [1] and Nicholas P. Godman [1,*]

1 Air Force Research Laboratory, Materials and Manufacturing Directorate, Wright-Patterson Air Force Base, Dayton, OH 45433, USA
2 Azimuth Corporation, Beavercreek, OH 45431, USA
3 Department of Materials Science and Engineering, Texas A&M University, College Station, TX 77843, USA
* Correspondence: kyungmin.lee.3.ctr@us.af.mil (K.M.L.); nicholas.godman.2@us.af.mil (N.P.G.); Tel.: +1-937-656-4695 (K.M.L.); +1-312-785-9674 (N.P.G.)

Abstract: Cholesteric liquid crystals (CLC) are molecules that can self-assemble into helicoidal superstructures exhibiting circularly polarized reflection. The facile self-assembly and resulting optical properties makes CLCs a promising technology for an array of industrial applications, including reflective displays, tunable mirror-less lasers, optical storage, tunable color filters, and smart windows. The helicoidal structure of CLC can be stabilized via in situ photopolymerization of liquid crystal monomers in a CLC mixture, resulting in polymer-stabilized CLCs (PSCLCs). PSCLCs exhibit a dynamic optical response that can be induced by external stimuli, including electric fields, heat, and light. In this review, we discuss the electro-optic response and potential mechanism of PSCLCs reported over the past decade. Multiple electro-optic responses in PSCLCs with negative or positive dielectric anisotropy have been identified, including bandwidth broadening, red and blue tuning, and switching the reflection notch when an electric field is applied. The reconfigurable optical response of PSCLCs with positive dielectric anisotropy is also discussed. That is, red tuning (or broadening) by applying a DC field and switching by applying an AC field were both observed for the first time in a PSCLC sample. Finally, we discuss the potential mechanism for the dynamic response in PSCLCs.

Keywords: cholesteric liquid crystals; electro-optic response; polymer stabilization; ion-trapping mechanism

Citation: Lee, K.M.; Marsh, Z.M.; Crenshaw, E.P.; Tohgha, U.N.; Ambulo, C.P.; Wolf, S.M.; Carothers, K.J.; Limburg, H.N.; McConney, M.E.; Godman, N.P. Recent Advances in Electro-Optic Response of Polymer-Stabilized Cholesteric Liquid Crystals. *Materials* **2023**, *16*, 2248. https://doi.org/10.3390/ma16062248

Academic Editors: Bing Wang, Tung Lik Lee and Yang Qin

Received: 6 February 2023
Revised: 1 March 2023
Accepted: 6 March 2023
Published: 10 March 2023

1. Introduction

Calamitic liquid crystals (LCs) are rod-shaped organic molecules that exist between amorphous liquids and solid crystals. These anisotropic rods spontaneously self-assemble into a variety of periodic structures known as mesogenic phases. LCs are, arguably, one of the most well-studied systems for manipulating light, as their unique optical properties arise from a positional order that endows these materials with anisotropy. There are several types of LC phases and they can be distinguished by their textures and optical properties. Nematic LCs in particular display 1D long-range order, wherein the mesogens align along their long axis, but each individual compound is free flowing and randomly distributed throughout the bulk. By adding small amounts of chiral molecules into the nematic LCs, a chirality transfer to the mesogens is induced, creating a chiral nematic or cholesteric phase. The anisotropic nature and inherent dielectric anisotropy of LC materials enables the natural ordering to be manipulated using external stimuli. The electro-optic behavior of LCs has been utilized for display technology since the discovery of the first LC display in the 1960s [1]. Since then, LCs have been utilized in other technologies including visors, smart windows, medical devices, and thermometers. Cholesteric liquid crystals (CLC) are 1D photonic crystals that exhibit a circular rotation of the LC director along a helical axis.

The chiral nature of the self-assembled CLC is determined by the handedness of the chiral dopant, thus the helical rotation of a CLC is either right-handed or left-handed depending on the optical activity of the chiral dopant [2,3]. When unpolarized light is exposed to a CLC, a maximum of 50% circularly polarized light (CPL) of the same handedness as the CLC is reflected and 50% CPL of the opposite handedness is transmitted, which is called selective reflection (Scheme 1). The position and bandwidth of the selective reflection are related to the pitch (P_0), and are expressed as $\lambda_0 = \overline{n} \cdot P_0$ and $\Delta\lambda = \Delta\text{n} \cdot P_0$, where λ_0 is the center of the wavelength, $\Delta\lambda$ is bandwidth, P_0 is the natural cholesteric pitch length, $\overline{n} = (n_e + 2n_0)/3$ is the average refractive index, and $\Delta\text{n} = n_e - n_0$ is the birefringence of the LC. It should be noted that n_e and n_o are the extraordinary and ordinary refractive indexes of the LC, respectively. Since the CLC is formed by mixing a chiral dopant with a nematic LC, the position of the reflection band can be tuned by the chiral dopant concentration added to the LC mixture. The P_0 of the CLC is the distance along the helical axis that the director changes by 2π and is inversely proportional to the concentration of the chiral dopant. The reflection bandwidth of CLCs in the visible wavelength region is tens of nanometers (50–100 nm) and broadens at longer wavelengths; e.g., near infrared and beyond.

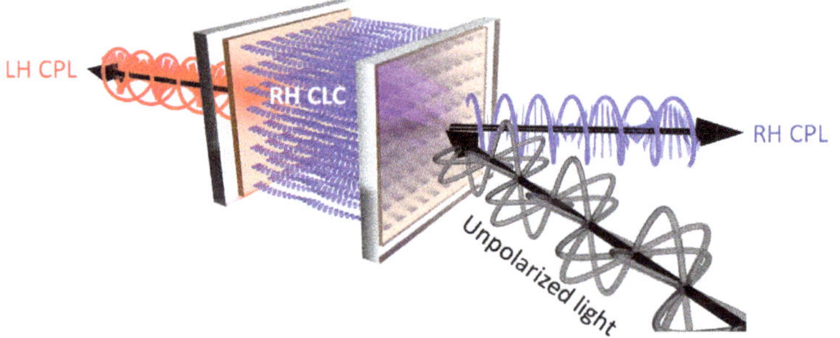

Scheme 1. Schematic description of selective reflection.

LCs are classified by their dielectric anisotropy ($\Delta\varepsilon$), where $\Delta\varepsilon = \varepsilon_{||} - \varepsilon_{\perp}$. This is representative of the difference between the permittivity parallel to the long axis ($\varepsilon_{||}$) and the permittivity of the short axis ($\varepsilon_{\perp}$) of the liquid crystal. LCs with a positive $\Delta\varepsilon$ will orient parallel to the direction of an electric field, while LCs with a negative $\Delta\varepsilon$ align perpendicular to the electric field direction [4]. The properties of a CLC system are highly dependent on the $\Delta\varepsilon$ of the chosen LCs. An example is a planar-aligned CLC system, which can be prepared to have an initial state of either reflective or scattering [5–21]. A CLC with positive $\Delta\varepsilon$ LCs can be switched to an optically transparent state by applying an electric field, which orients the liquid crystal molecules in the homeotropic orientation. After the field is removed, the CLCs in the homeotropic state relax back to the initial planar cholesteric state through the metastable focal conic state [5–7]. Since this relaxation process is very slow, polymer stabilization methods are used to improve the relaxation kinetics through strong anchoring between the polymer network and the CLC medium [8–10]. In contrast, CLCs with a negative $\Delta\varepsilon$ do not experience rotation in the planar state when applying an electric field.

Another advantage of LCs is the ability to stabilize the molecular orientation through polymer stabilization. Polymer-stabilized liquid crystals (PSLCs) consist of liquid crystals and a polymer network that is formed via in situ polymerization of reactive monomers dispersed throughout the LC system [22–24]. When the LCs are in the cholesteric phase, the LC polymer-composites are called polymer-stabilized cholesteric liquid crystals (PSCLCs). In both PSLCs and PSCLCs, small amounts of polymer are used to form the stabilizing network, typically below 10 wt.%. The polymerization is performed in an ordered phase of the mixture, preserving the initial alignment of the reactive mesogens in the resulting polymer network. Therefore, the polymer chains can act as alignment surfaces distributed throughout the thickness direction to stabilize the LC. This is evident when the temperature increases near the clearing temperature of the LC in PSLCs; at these temperatures, the order parameter sharply decreases [25,26], but complete isotropization is not achieved. Order is maintained due to the LC molecules strong anchoring to the surface of the polymer network. Therefore, these LC-polymer composites exhibit temperatures and electro-optic responses that differ from the bulk LC phase [22,24,27,28]. For example, in PSLCs and PSCLCs with positive dielectric anisotropy, the LC molecules can be reoriented by applying an electric field, the threshold voltage of which is higher than that of bulk CLC [3,5,25,27,29,30]. Due to the restoring force exerted by the polymer network, recovery to the initial state is much faster in the polymer-stabilized CLCs than in the bulk system.

Polymer networks can be used as three-dimensional scaffolds for aligning low molecular weight LC molecules. In this process, the LC molecules can be removed from the PSCLC composite using organic solvents, leaving structurally chiral insoluble polymer fractions. Then, the system can be refilled with other liquid crystal molecules [3,22,31]. These ordered polymer networks can strongly align non-mesogenic molecules or overcome the initial handedness of chiral LCs. When the polymer scaffold is refilled with CLC of the opposite handedness, the refilled PSCLC has the handedness of the polymer template, indicating that the refilled CLC fluid follows the handedness of the polymer template [32]. This washout/refill method can be used to prepare CLC-polymer composite systems with unique properties that are not observed in natural systems, such as hyper-reflective devices that reflect both right-handed and left-handed circularly polarized lights in a single cell [31,33–35].

The polymer network of PSCLCs can also lock the pitch of a CLC, and it affects various optical responses. Stabilizing low molar mass CLCs with positive $\Delta\varepsilon$ using polymer networks allows for fast and bistable switching between reflective and scattering textures [6–17,36–38]. These composite systems are described as PSCLCs, but, in the past, these systems were also known as polymer-stabilized cholesteric textures (PSCTs). PSCTs have two types of switching modes: normal mode and reverse mode. In normal mode, the switching from scattering to clear or scattering to reflective states occurs [5,18,19], and in reverse mode, the switching from reflective to scattering state is observed [5,20,21]. The normal or reverse operating mode is mainly determined by the sample preparation conditions. Applying an AC electric field during photopolymerization to prepare the normal mode samples aligns the LCs in the direction perpendicular to the cell. When the applied electric field is turned off after polymerization, the PSCLC forms the hazy focal conic texture. Therefore, the initial state of this sample, the off-state, has a scattering state. Upon applying an electric field, the positive $\Delta\varepsilon$ LCs reorient to the homeotropic state, also referred to as the clear state. To prepare the reverse mode sample, the mixture is polymerized in the off-state to obtain a sample with a stable reflective state in the off-state. However, when the reflection band is outside the visible range, the sample is optically transparent. Intermediate controls of scatter, or various gray scales, have also been reported [6,7,11,12].

In contrast, negative $\Delta\varepsilon$ LCs in a planar state do not respond to an applied electric field and have no switching response. Over the past decade, the dynamic response of PSCLCs with negative $\Delta\varepsilon$ has been extensively studied by the Air Force Research Laboratory (AFRL) liquid crystal team, including bandwidth broadening [39–43], red tuning [44–47], blue tuning [48], and switching [49,50]. The proposed mechanism for the electro-optic responses is based on the ion-mediated deformation of the polymer network that stabilized the CLC medium. The application of a DC field induces the migration of ions trapped in the polymer network, leading to the deformation of the polymer network toward the negative electrode. The LC host anchored on the deformed polymer network is also deformed, resulting in a cholesteric pitch variation across the cell [39,41,46]. Recently, a reconfigurable EO response of a single positive $\Delta\varepsilon$ PSCLC sample, such as switching and red-shifting tuning or switching and bandwidth broadening responses, by properly adjusting the AC and DC fields has been reported [51]. The redshift of the reflection band of the positive $\Delta\varepsilon$ PSCLC is due to strong anchoring of the LCs on the polymer network. This review summarizes the current understanding and potential mechanisms for the dynamic response of PSCLCs studied by the AFRL liquid crystal research team over the past decade.

2. Electro-Optic Response in PSCLCs

2.1. Bandwidth Broadening

Bandwidth broadening in PSCLCs has been an intensive research topic over the last decade from both an academic and practical application perspective. While a number of approaches exist to induce bandwidth broadening, electrically induced responses in PSCLCs are of great interest [42,52–60]. PSCLCs employed for such studies are typically comprised of CLCs with negative $\Delta\varepsilon$ and an anisotropic polymer network with a strong alignment effect on the LC [58,61]. The polymer network entraps ions during photopolymerization, and the movement of these ions via an applied electric field leads to deformation of the polymer network. The deformation leads to a change in pitch for the anchored LCs. The bandwidth broadening phenomenon is generally attributed to the non-uniform changes in these pitches.

In 2011, Bunning and his team first reported the electrically induced bandwidth broadening response of PSCLCs. Electro-optic studies have highlighted the effect of trapped charges on the polymer network of PSCLCs [39]. The distortion of the polymer network upon the application of the electric field led to the variation in pitch length and subsequent bandwidth broadening. Detailed studies by Yang and coworkers [58] showed that polymer networks and ion density were most influential on bandwidth broadening in PSCLCs. The authors also reported that the choice of photoinitiator had a large effect on the generated ion density, but not the polymer network, as evidenced by Scanning Electron Microscopy studies. The chiral dopant and monomer functionality were found to influence the polymer network as well. An optimal choice of the aforementioned parameters is thus needed for the desired bandwidth broadening in PSCLCs.

Duan et al. reported on a step-wise photopolymerization process utilizing two kinds of UV light, specifically 254 nm and 365 nm, as well as cationic and free radical photo-initiators [62]. The intensity of the light source and irradiation time were used to control bandwidth broadening in PSCLCs. In one case, broadening was observed from 916 nm to 1460 nm due to an increase in 254 nm UV light intensity from 0.5 mW cm^{-2} to 3.5 mW cm^{-2}.

A study by Nemati et al. found a correlation between the electrical resistivity and reflection bandwidth broadening of a PSCLC [61]. Higher ion densities led to a low resistivity in the cell, resulting in bandwidth broadening. The authors subsequently explored the impact that alignment layer thickness has on impedance in the cell and experimented with different additives to increase conductivity. They found that a thinner alignment layer led to bandwidth broadening, while increasing the thickness of the alignment layer was detrimental to bandwidth broadening. This was attributed to the polyimide acting as an insulating layer, which increased the impedance of the cell. The authors also reported that additives with a phenyl functional group were found to enhance broadening, while an additive containing two carboxylic acid functional groups suppressed broadening. The incorporation of esters into the reactive mesogen monomer may also improve the broadening performance, as it has previously been reported that the ester moieties of RM257 act as effective ion trapping groups [63].

Khandelwal et al. demonstrated electrically induced large-magnitude bandwidth broadening using a long and flexible ethylene glycol crosslinker twin molecule. This is due to the molecule enhancing the ability of the polymer-stabilizing networks to trap cationic impurities, and these ion-trapped PSCLCs show a nine-fold enhanced reflection bandwidth [56].

Lee et al. reported color-tunable mirrors based on reflection bandwidth broadening in PSCLCs [42]. The authors prepared a PSCLC device that switches from being selectively reflective to forming a broadband reflection mirror upon the application of an 80 V DC field. This is visually represented in Figure 1a, where, in the off-state (left), the word "Mirror" is visible behind the PSCLC device. When the device is turned on (right), "Mirror" becomes obscured and a reflection of the image is observed. It is important to note that the reversible behavior of the PSCLC takes several seconds to occur. The authors also carried out the effect of viscoelastic behavior of a polymer structure prepared from LC monomers with various alkyl chain lengths as a function of the electric field. Bandwidth broadening from 100 nm to nearly 600 nm, symmetric about the center of the reflection notch, was realized. The transmission spectra in Figure 1b show the broadening effect upon application of a DC field in the range of 0–140 V. The length of the flexible methylene spacers in the LC monomer unit (n = 3, 6, and 11) also influences the magnitude of the voltage required to achieve bandwidth broadening. Interestingly, the threshold voltage for bandwidth broadening was lower for LC monomers with longer spacers, and this was attributed to the viscoelastic properties of the polymer network. The effect of alkyl chain length on the magnitude of bandwidth broadening is summarized in Figure 1b(iv).

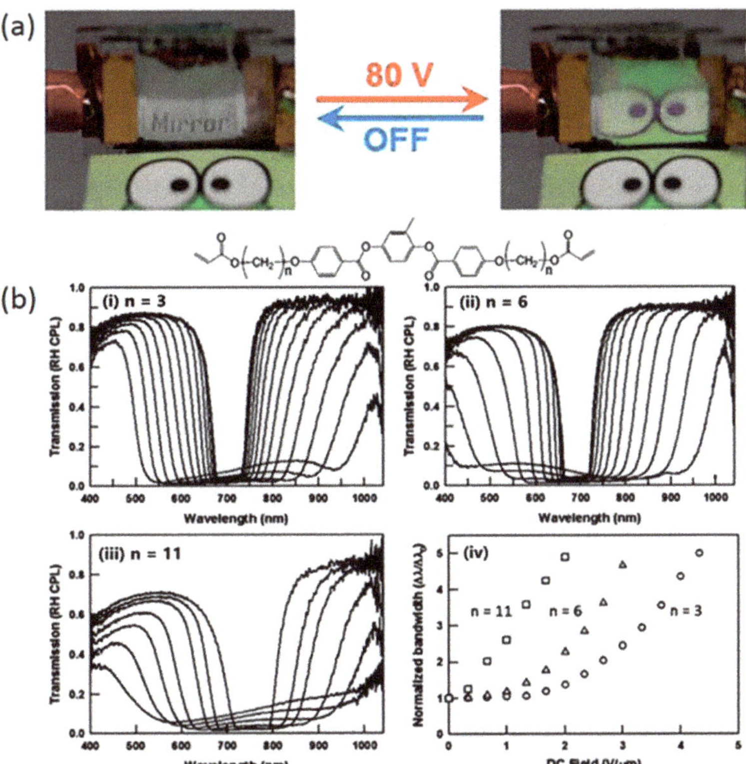

Figure 1. (**a**) Photographs illustrating the transmission and reflection of PSCLC with a thickness of 15 μm at 0 V (left) and 80 V (right). (**b**) Transmission spectra of PSCLC from (i) C3M (n = 3), (ii) C6M (n = 6), (iii) C11M (n = 11) before and during application of DC voltages, and (iv) a summary of the bandwidth change as a function of the DC field. Adapted from ref. [42].

The reports on reflection bandwidth broadening in PSCLCs support the hypothesis that the behavior is controlled by the polymer network. The polymer network both anchors the low molecular weight LCs and entraps ions. When a strong field is applied to a PSCLC, the movement of trapped ions leads to the deformation of the polymer network, which in turn expands and contracts the pitch of anchored CLCs. The availability of ions in the system stems from the choice of initiator, additives, and curing/photopolymerization conditions. Recent efforts in this area have focused on the development of switchable glasses, which may have applications for smart window technology [64–67].

2.2. Reflection Notch Tuning

Another dynamic mechanism of PSCLCs is responsive tuning of the selective reflection color through the application of external stimuli. Electrical, mechanical, optical, and thermal inputs have been used to tune the selective reflection peak. Generally, the use of a small electrical field to induce color change is preferred. Nematic LCs with a negative $\Delta\varepsilon$ are utilized for tuning behavior in PSCLCs, as the application of a DC field will not reorient the LC mesogens. This enables tuning of the reflection notch instead of switching behavior.

Previous methods used to electrically induce color changes have employed cell architectures containing interdigitated electrodes [68]. Interdigitated electrodes lead to non-uniform coloration across the cell due to the complex geometry of the electrodes [69,70]. In these systems, focal conic domains, which readily scatter light, form. However, polymer stabilization of these systems has largely corrected this issue regarding the complex geometry of the electrodes and the scattering of the focal conic texture [40,44,54,60]. Another downside to interdigitated electrodes is the need for large electric fields to induce the color tuning of the CLC.

The complexity of cell architectures containing interdigitated electrodes paired with large electric fields has driven the use of conventional cell architectures. Examples of small-scale color tuning using conventional cells include annealing defects in a CLC, which enabled the switching of a laser on and off through the application of a DC field [71]. Color tuning was also achieved through tilting of LCs with positive $\Delta\varepsilon$ along their helical axis [72]. Piezoelectric compression of a cell containing CLCs led to a compression of the pitch, which resulted in color tuning [73,74]. Electrostatic deformation of a CLC also enabled pitch compression, resulting in blue tuning of the reflection notch [75]. These approaches lead to small-scale color tuning of the selective reflection of a CLC. In order to achieve a larger tuning range polymer stabilization must be used.

Recently, tunable notch and bandpass filters capable of spanning the entire visible spectrum with simple device architectures have been reported by using the oblique helicoidal cholesteric phase, also referred to as the heliconical cholesteric phase [68]. The heliconical cholesteric phase was attained by mixing the dimeric mesogens CB7CB and CB11CB, nematic liquid crystal 5CB, and left-handed chiral dopant S811. The unique structure of the heliconical cholesteric state endows the material with dielectric torque upon the application of an electric field, the periodicity of the system changes, resulting in tuning of the selective reflection. Tuning of the reflection notch from 465 to 637 nm was observed at low voltages of 1.6–2.1 V μm^{-1}, respectively. Similarly, electrically tunable optical bandpass filters were fabricated. The efficiency of the bandpass filters was 15% due to reflection losses from the cholesteric and attenuation losses of the polarizer.

Though tuning of the selective reflection is attainable using the electro-optic response of bulk CLCs, large-range reflection tuning often requires polymer stabilization. The implementation of polymer stabilization has enabled large magnitude color tuning, greater than 300 nm, while utilizing low to moderate electric fields. In 2005, Yu et al. from Kent Optronics reported a PSCLC system that exhibited large-magnitude reflection notch tuning of 300 nm [76]. A cross-linkable monomer, photoinitiator, and low molecular weight LC with a positive $\Delta\varepsilon$ were employed to generate a highly cross-linked system. Upon the application of an electric field, the polymer network remained unresponsive but provided stabilization, enabling the rotation of the LCs molecules. Rotation of the LCs induced effective refractive index changes, and thus, tuning of the reflection notch. Interestingly, the PSCLC could also be doped with a laser dye, PM-597, to generate a CLC laser with a tunable lasing wavelength. The system exhibited a 33 nm blue shift in the lasing wavelength over a voltage range of 324 V [76].

Large-magnitude color tuning can also be achieved through delocalization of the polymer-stabilizing network to induce pitch modulation. In these systems, the structural chirality of the polymer network is manipulated by applying a DC field through an electrophoretic mechanism. In order to induce deformation of the polymer network, a negative $\Delta\varepsilon$ LC is required. The negative $\Delta\varepsilon$ LCs ensure that the molecular reorientation does not occur. This enables the movement of the polymer network, and therefore, the large magnitude color tuning. This phenomena was first reported by McConney et al. in a PSCLC containing a low molecular weight negative $\Delta\varepsilon$ LC that exhibited red shifting behavior [44]. By applying a DC field of 8 V μm^{-1}, the reflection notch was tuned to nearly 300 nm. This corresponded to a change in the pitch length of 55% the original pitch length [45]. The PSCLC had facile and reproducible color tuning throughout the visible region into the near IR. However, high polymer concentrations up to 20 wt.% were required. The high polymer content resulted in poor transmission values.

To improve the low transmission values of the PSCLC, Lee et al. reduced the overall polymer content to 6–8 wt.%, resulting in increased transmission values [46]. In this system, the reflection notch was tuned from 700 to nearly 2500 nm upon application of 160 V of a DC field. The PSCLC device maintained high out-of-band transmission values across the entire tuning range, as shown in Figure 2a. Tuning throughout the visible spectra was also achieved in the PSCLC by increasing the chiral dopant concentration. However, appreciable creep in the reflection notch position was observed due to the lower polymer content. At lower polymer concentrations, around 5 wt.%, the creep was more pronounced, while samples with 8 wt.% polymer exhibited higher stability. Figure 2b demonstrates the ability of the PSCLC to adjust the notch position of the selective reflections. The electric field was directly applied to 50 V DC and then maintained for 30 min. When the field was turned off, the reflection band was relaxed back to its original position. The same procedure was performed for 200 nm, 300 nm, and 400 nm displacements of the reflection notch. The inset of Figure 2b shows the POM images of the reflection color as the DC voltage was increased from 0 V to 75 V. During the 30 min that the electric field was maintained, the reflection notch continued to move (creep) and the magnitude of the creep depended on the strength of the electric field. However, this is mainly because the degree of deformation of the polymer network is proportional to the length of time of the applied DC field.

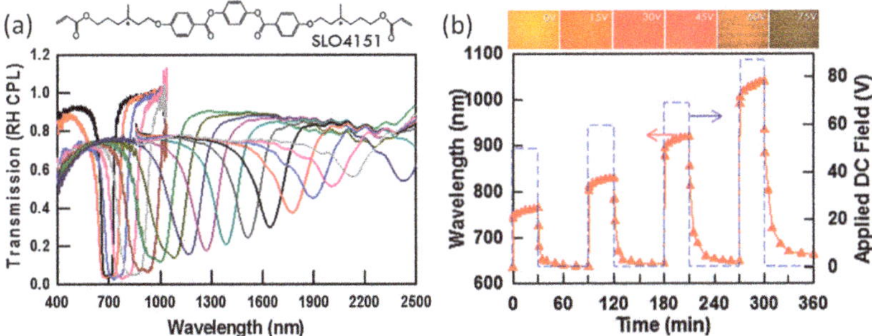

Figure 2. (**a**) Transmission spectra of a PSCLC formulated with 1 wt.% Irgacure 369 and a chiral liquid crystal monomer (SL04151, inset) and cured at 80–100 mW cm^{-2} for 3 min with increasing voltage from 0 V to 160 V. (**b**) Variation in reflection wavelength of a PSCLC with 5 wt.% polymer concentration in an experiment in which DC voltages of 50 V, 60 V, 69 V, and 87 V. The voltages were directly applied to the cells and continuously applied for 30 min. POM images in reflection images increasing DC voltages from 0 V to 75 V. The red triangle indicates the notch position of PSCLC and the blue dashed line represents the applied DC field. Cells with a thickness of 15 μm were used. Adapted from ref. [46].

PSCLCs with tunable reflection notch behavior have been utilized to generate a dynamic bandpass filter that operates in the mid-wave infrared region (MWIR). Here, Worth et al. fabricated a PSCLC containing a 6.5 wt.% diacrylate monomer RM82 and a negative $\Delta\varepsilon$ LC MLC 2079 that exhibited large magnitude tuning across the MWIR [77]. In order to achieve a saturated reflection in the MWIR, a cell thickness of 50 μm was required. Despite the larger cell gap, the PSCLC retained out-of-band transmission values near 70% due to the low polymer content. The PSCLC filter was able to tune over 2 μm from 2.5–4.9 μm by applying a DC field of 110 V. By adjusting the photo-polymerization procedure of the same PSCLC formulation, a bandwidth broadening MWIR filter was able to be generated. Therefore, a dynamic PSCLC exhibiting tunable and broadening behavior in the MWIR was produced by manipulating the structural chirality of the PSCLC polymer network.

Electromechanical distortion of the polymer stabilizing network has largely been utilized for the red shift of the reflection notch upon the application of a DC field. In 2017, the blue shifting of the reflection notch was realized by Lee and coworkers [48]. A PSCLC with a reflection notch centered at 700 nm was blue-shifted to 430 nm by applying a 35 V DC field, the transmission spectra for this sample is shown in Figure 3a. Photographs of the reflection colors in the PSCLC alongside wavelength positions are shown in Figure 3b. The blue shifting behavior was determined to be highly sensitive to the photo-polymerization conditions and the photoinitiator. When the PSCLC was cured at lower intensities of 250 mW cm^{-2} of 365 nm UV irradiation, red shifting was observed after three minutes, and bandwidth broadening was observed between 5–10 min, respectively. Only after the sample was cured for 15 min was blue shifting behavior realized. If the sample was cured at higher intensities of 700 mW cm^{-2} of 365 nm UV irradiation, blue shifting was observed after 10 min of curing. The photoinitiator was also crucial to achieving blue shifting of the reflection notch. The authors determined that morpholino-keto-type initiators, such as Irgacure 369 and 907, were able to induce blue shifting behavior with concentrations as low as 0.5 wt.% after 30 min of irradiation at 250 mW cm^{-2} of 365 nm UV light. When a non-morpholino-keto-type initiator was used, specifically, Irgacure 651, red shifting behavior was observed even at concentrations ranging from 0.5–5 wt.% under similar curing conditions.

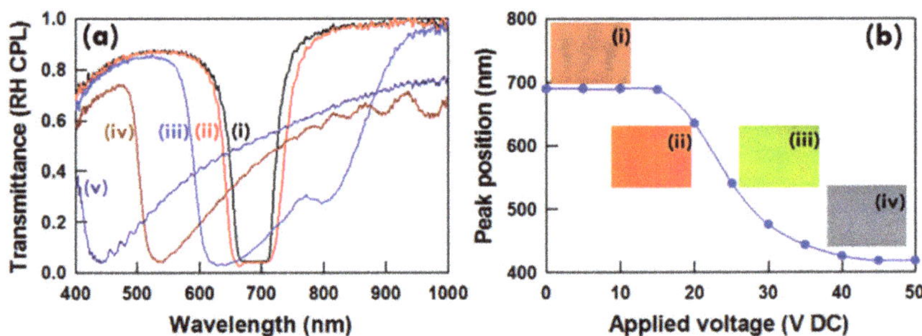

Figure 3. (**a**) Blue shifting of the selective reflection of a 15 µm thick PSCLC at (i) 0 V, (ii) 15 V, (iii) 20 V, (iv) 25 V, and (v) 35 V. (**b**) The wavelength of the selective reflection is plotted as a function of applied DC voltage. Photographs of the reflection at (i) 0 V, (ii) 19 V, (iii) 23 V, and (iv) 36 V are provided as insets. The PSCLC shown was cured at 700 mW cm^{-2} for 30 min. Adapted from ref. [48].

Utilizing the dynamic behavior enabled by electromechanically distorting the polymer network of PSCLCs, Lee et al. produced a PSCLC with complete optical reconfigurability [51]. Figure 4a shows the DC-field-induced large-scale red tuning of the reflection band of a CLC composed of 6 wt.% SL04151 and the positive $\Delta\varepsilon$ LC E7. At appropriate AC field strengths, the orientation of the LC to the polymer network can be overcome, resulting in a switching of the positive $\Delta\varepsilon$ LCs and loss of the reflection notch (Figure 4b). Reversible and repeatable switching was achieved by applying a 150 V (1 kHz) AC field. A DC field was also concurrently applied to the system, yielding an optical element with a tunable and switchable reflection notch. This switchable and tunable behavior is shown in Figure 4c. The high anchoring strength of the LC to the polymer network prohibits the positive $\Delta\varepsilon$ LCs from reorienting upon application of the DC field. This enabled a 1000 nm red shift from 1200 to 2200 nm at 60 V. However, a strong AC field could be applied to overcome the alignment of the positive $\Delta\varepsilon$ LC to the polymer network, resulting in complete removal of the reflection notch. Blue shifting behavior was also attainable in this system by adjusting the curing conditions. The authors were able to fabricate a pixelated display with local control over the reflection color, as shown in Figure 4d–h.

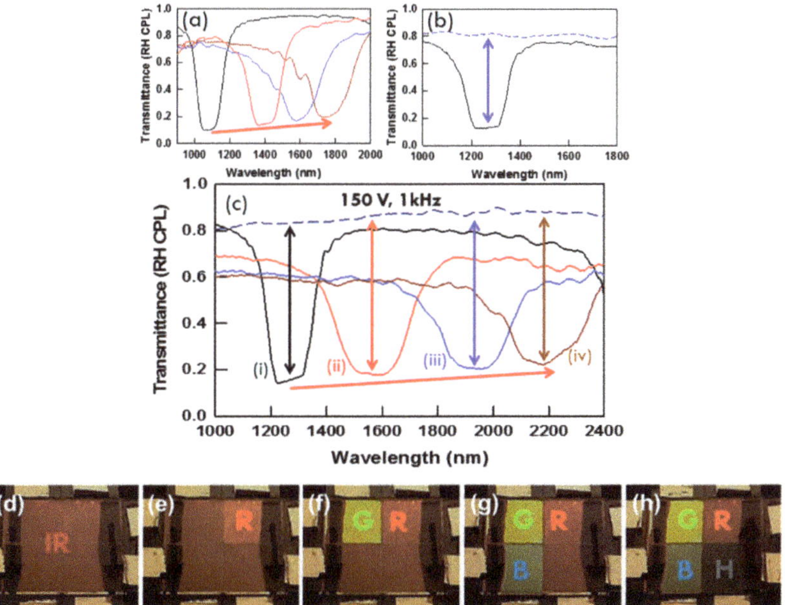

Figure 4. (**a**) Red shifting tuning of the selective reflection of a 15-μm thick PSCLC by the application of DC voltage of 0, 30, 50, and 60 V DC. (**b**) Reflection switching of the PSCLC band gap is induced by the application of 150 V AC. (**c**) Reconfiguration of the selective reflection (tuning and switching) is illustrated in sequential transmission spectra (i) 0, (ii) 40, (iii) 60, and (iv) 70 V DC. Reflection switching of (i–iv) was included with 150 V AC at 1 kHz. The one-sided arrows represent the shift of the reflective band and the vertical double-sided arrows represent the transition between reflective and transparent states. The sample was formulated by mixing 0.4 wt.% I-369, 6 wt.% SL04151, 3 wt.% R1011, and 90.6 wt.% E7 prepared by exposure to a 100 mW cm^{-2} 365 nm wavelength UV light for 3 min. Control of color in PSCLC optical elements separated into four addressable pixels: (**d**) 0 V (IR), (**e**) 3 V of DC field was applied to the upper right pixel (red), (**f**) 6 V of DC field was concurrently applied to the upper left pixel (green), (**g**) 9 V of DC field was concurrently applied to the bottom left pixel (blue), and (**h**) 250 V of AC field at 1 kHz was applied to the bottom right pixel (homeotropic). Adapted from ref. [51].

2.3. Switching Response in PSCLCs

The effects of LC polymer composites can have dramatically different EO properties depending on their relatively low and high molecular weight, and the natural properties of each component [3,22]. In the case of PSCLCs, a bistable switching effect can be attained by using either positive Δε or negative Δε LC hosts, depending on the processing conditions. Bistable switching PSCLCs have two modes: normal mode [5,18,19,36,37] and reverse mode [5,20,21,36–38]. In normal mode, the PSCLC switches from a scattering state to a clear state. In reverse mode, the PSCLC shows a transition from a clear state to a scattering state. By altering the preparation conditions of the PSCLC sample, the two different switching modes can be achieved.

To prepare a PSCLC with positive Δε exhibiting normal mode switching, an electric field is applied during the photopolymerization to align the polymer network in the normal direction of the cell. Removal of the electric field after photopolymerization induces a focal conic scattering state of PSCLCs. This preparation method results in a clear state when an electric field is applied to the sample because the electric field orients the LCs into the homeotropic phase (clear state). In order to prepare a reverse switching mode sample, the PSCLC is prepared without applying an electric field during the photopolymerization. This causes the PSCLC to switch from a clear state to a stable scattering state when the electric field is applied.

Lee and coworkers presented a switchable PSCLC using negative $\Delta\varepsilon$ LCs [49]. Unlike positive $\Delta\varepsilon$ LCs, where the director rotates along the electric field direction, the negative $\Delta\varepsilon$ LCs in the planar geometry are not reoriented by an electric potential. The authors demonstrated the switching response of negative $\Delta\varepsilon$ PSCLCs upon the application of a DC field. The PSCLC samples were prepared with a low polymer concentration (<1.5 wt.%), leading to the generation of a weak polymer network. The polymer network was then detached from the alignment layer by applying a large electric field. The DC field shifts the polymer network towards the negative electrode, resulting in the switching behavior. Figure 5 shows the bistable switching of a PSCLC switched between clear and scatter states by varying the applied electric field to a maximum voltage of 200 V DC.

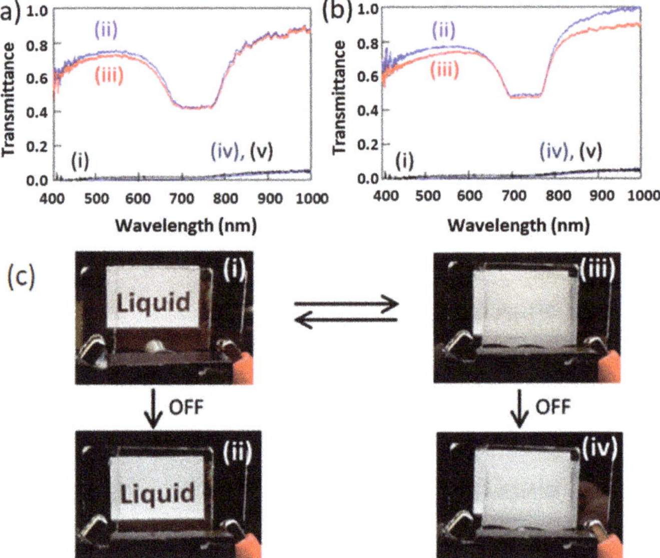

Figure 5. Bistable switching behavior of PSCLCs with (**a**) 30 μm gap thickness: normal mode with (i) 0 V, (ii) 210 V DC, (iii) OFF (0 V), (iv) −45 V DC, and (v) OFF (0 V), (**b**) (i) 0 V, (ii) 100 V AC, (iii) OFF (0 V), (iv) 45 V DC, and (v) 0 V, and (**c**) photographs of samples with 15 μm thickness at (i) 50 V AC, (ii) 0 V, (iii) 100 V DC, and (iv) 0 V. Adapted from ref. [49].

Figure 6 shows the response times of switchable PSCLCs with a thickness of 30 μm. Figure 6a shows the optical response of the sample to an applied DC and AC field. The initial scattering sample (Figure 6a(i)) became reflective when 210 V of DC voltage (forward direction) was applied (Figure 6a(ii)). The reflective state remains stable when the applied DC voltage is removed. A scattering state is then induced from the reflective state by the application of −45 V DC (reverse direction) (Figure 6a(iv)); the scattering state remains stable when the electric stimuli is removed. Interestingly, the reflective state could also be achieved by applying 100 V of an AC voltage at a frequency of 1 kHz (Figure 6(a(vi)). As shown in Figure 6a, the scattering and reflective states remain unchanged when the applied electric field is removed (Figure 6a(iii,v,vii)). In the scattering mode, the PSCLC has a transmittance less than 1%. In reflectance mode, the PSCLC exhibits transmittance values near 60% at a wavelength of 500 nm and a reflection notch in the infrared region.

When the DC voltage is removed, the reflective mode persists and the scattering mode can be recovered upon the application of AC voltage. The average response time for switching from scattering mode to reflective mode was measured to be 200 ms, and switching from the reflective to scattering mode was measured to be 350 ms (Figure 6b).

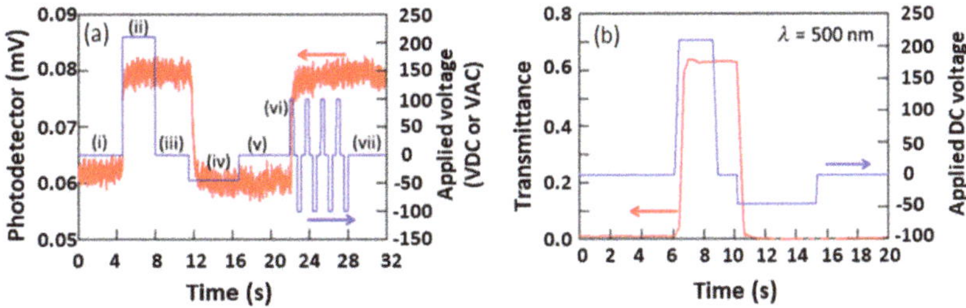

Figure 6. Response time of PSCLCs with 30 μm thickness. An optical setup employing a photodetector connected to an oscilloscope measured the relative intensity of light transmission as the PSCLC was subjected to (**a**) at (**i**) 0 V, (**ii**) 210 V DC, (**iii**) 0 V, (**iv**) −45 V DC, (**v**) 0 V, (**vi**) 100 V AC with 1 kHz, and (**vii**) 0 V over various time intervals. (**b**) The transmission of unpolarized light at 500 nm to measure the response time is illustrated. Adapted from ref. [49].

A new switching mechanism that exploits the reflectance loss of PSCLCs in relatively thin cells using high DC fields was developed by Lee et. al. in 2020 [50]. This mechanism does not rely on the reorientation of positive $\Delta\varepsilon$ LCs or the electrophoretic movement of the polymer network when using negative $\Delta\varepsilon$ LCs. The PSCLCs were able to directly switch from a reflective state to a clear state with complete removal of the reflection notch upon application of a 45 V DC voltage. The PSCLC samples in this study had a much smaller thickness of 5 μm (Figure 7a). Interestingly, this method enabled control over the selective reflection position through moderate changes to the applied DC field. When a lesser voltage of 13 or 23 V was applied to the sample, the selective reflection notch reappeared. However, the reflection notch reappeared at a new position of 670 or 730 nm, respectively (Figure 7b,c). When the applied DC voltage is turned off, it shows a reversible behavior in which the reflection notch relaxes back to the initial notch position (Figure 7d). This study showed that the spectral position and reflectance of the sample could be controlled and turned off through the application of a DC voltage.

The ability to control the red shifting and switching behavior of the PSCLC sample was further refined using a small cell thickness. Figure 8 shows the red tuning and switching response of a thinner 3-μm thick PSCLC sample. The initial notch position of the sample was located at 500 nm; by applying a DC voltage of 23 V the reflection notch red shifts to 730 nm. At a higher applied voltage of 40 V, the reflection notch completely disappears, switching the sample from the reflective state to a clear state. After switching to the clear state, the reflection notch could be recovered by decreasing the applied voltage to 23 V. Interestingly, the position of the reflection was determined by the magnitude of the applied DC voltage (Figure 8a). The voltage-dependent tuning behavior not only led to a change in the reflected wavelength, but also an increase in the transmission of the sample, which is why switching to a clear state was attainable. The controllable tuning behavior at lower applied voltages and switching at higher voltages was also reversible (Figure 8b). Photographs of the PSCLCs reflection color at various DC voltages are shown in Figure 8c. The observed behavior of the thin PSCLCs was attributed to the deformation of the polymer network. As a DC voltage is applied to the cell, the polymer network shifts toward the electrode, which results in a decrease in the number of repeat units throughout the cell. This causes a red shift in the reflection notch as the PSCLCs pitch is elongated. A specific example is the 3-μm PSCLCs appearing transparent at an applied voltage of 40 V, where the limited number of repeated positions along the helical axis leads to loss of the reflection notch.

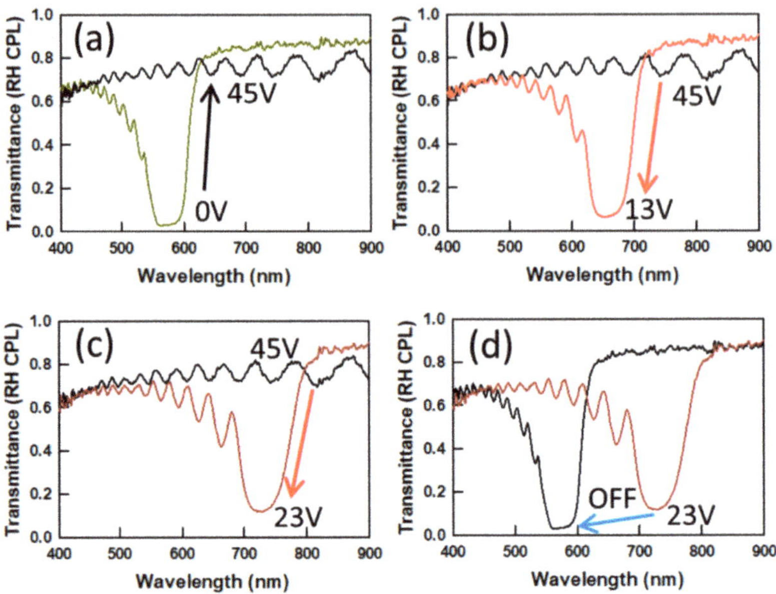

Figure 7. Transmission spectra of a negative Δε PSCLC with 5 ± 0.2 µm thickness showing a (**a**) reflective to transparent or (**b**,**c**) transparent to reflective switching response by application of a DC voltage. (**d**) Spectra for the relaxation to the off state from 23 V DC. Adapted from ref. [50].

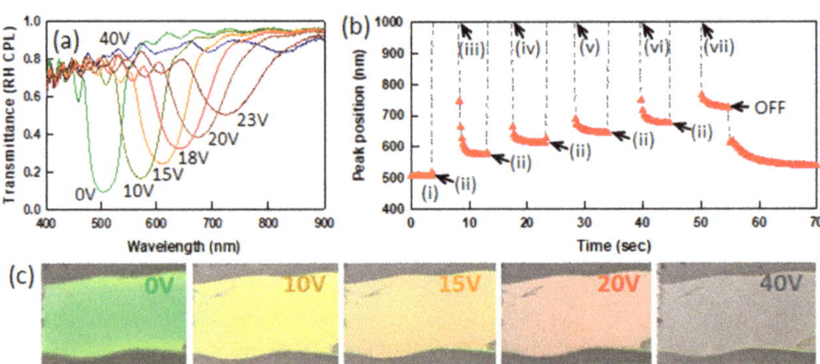

Figure 8. (**a**) Transmission spectra of a negative Δε PSCLC with 3 ± 0.2 µm thickness for DC voltages up to 40 V DC. (**b**) Peak position during the switching of the same PSCLC using various DC voltages. The arrows indicate the times at which the DC voltage was changed to the value indicated. (**c**) Photographs of the reflection color of the PSCLC as a function of the applied DC voltage. Adapted from ref. [50].

Lee et al. also determined that PSCLC samples with a smaller cell thickness had faster response times. Thin cell PSCLC samples exhibited a faster response time when switching from the scattering state to clear state compared to PSCLCs with increased thickness [47,50]. Figure 9 shows the response time for a 3- and 5-µm thick sample, respectively. Response times of 30 ms and 50 ms were observed for the 3- and 5-µm thick PSCLCs when switching from the scattering state to the clear state, respectively. Similarly, response times of ~150 ms and ~200 ms were observed for the 3- and 5-µm thick PSCLCs when switching from the clear state to the scattering state [50]. As the PSCLC thickness was increased to 15 µm, the response times for switching of the sample was determined to be 1.5 s and 3 s for both modes, respectively [47].

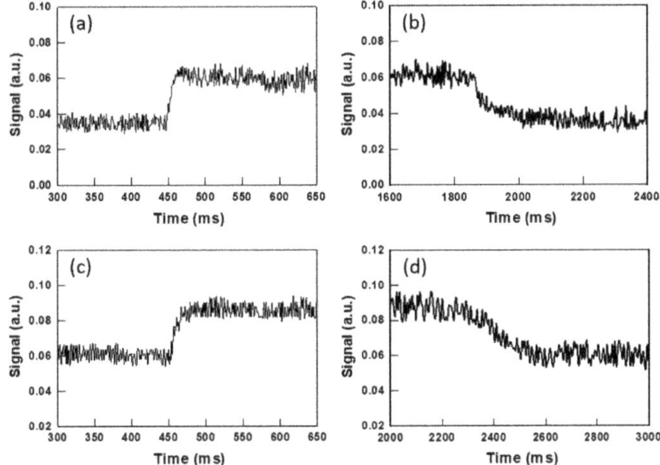

Figure 9. Time evolution of the intensity of the light transmitted by PSCLCs with (**a**,**b**) 3 ± 0.2 μm and (**c**,**d**) 5 ± 0.2 μm thickness when a DC voltage is switched on and off. For the PSCLC with 3 μm, a 40 V DC voltage is (**a**) directly applied at t ~ 450 ms and (**b**) removed at t ~ 1850 ms. Rise and fall times are ~30 ms and ~150 ms. For the PSCLC with 5-μm thickness, 45 V DC are (**c**) applied at t ~450 ms and (**d**) removed at t ~2300 ms. Rise and fall times are ~50 ms and ~200 ms. Adapted from ref. [50].

2.4. Other Effects on Dynamic Behavior of PSCLCs

A variety of factors influence the responsive behavior of PSCLCs and their specific applications. In addition to the components in the PSCLC mixture, mainly the chiral dopant concentration and mesogen type, the processing parameters of boundary conditions and purity influence the electro-optic response. This section will discuss the effects of cell thickness and ion density on the dynamic response of PSCLCs. In addition, the high reflectivity (>50%) of a PSCLC at oblique incidence is discussed.

2.4.1. Cell Thickness

Modifying the cell thickness of PSCLCs has a dramatic effect on their reversible electro-optic response, including the reflection notch tuning range, in-band transmittance, and response time [50,55]. The pitch of the PSCLC samples prepared at various cell thicknesses was ~0.40 μm, as shown in Figure 10. Each PSCLC examined exhibited an electrically induced red shift of the reflection notch, but the tuning range was dependent on the thickness of the cell. The PSCLCs with larger cell thicknesses of 7.6~14.1 μm showed a red shift of ~400 nm at high DC voltages (55–90 V DC), as shown in Figure 10a–c. The transmittance of the reflection notch with a notch position of ~620 nm was <5% over the entire tuning range when the right-handed circularly polarized light (RH-CLP) was used as the probe beam. However, the reflection notch of PSCLCs with cell thicknesses ≤5 μm exhibited a narrower tuning range and higher transmittance, as shown in Figure 10d–f. At higher voltages, the PSCLCs with thinner cell thicknesses appear transparent due to the limited number of repeated positions along the helical axis, as the pitch of the PSCLC is deformed, resulting in a significant decrease in the reflection efficiency. This electric field increases the transparency of the cell because the number of repeat units for the extended pitch is insufficient to attain good reflection efficiency in the cell thickness. However, the thinner PSCLCs show a faster response of ~30 ms and ~50 ms rise times and ~150 ms and ~200 ms fall times for 3-μm and 5-μm thick samples, respectively, than those of thick samples, which were observed to be 1.5–3 s for a 15-μm thick sample.

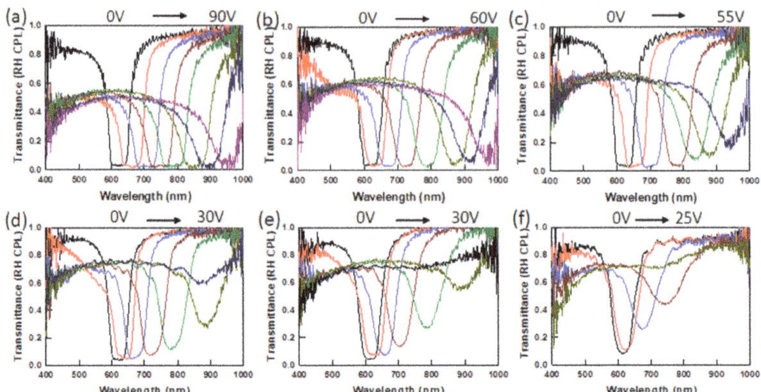

Figure 10. Transmission spectra of negative Δε PSCLCs with various cell thicknesses and DC voltages in the ranges shown above each graph: (**a**) 14.1 ± 0.3 μm thickness, 0 V to 90 V DC, (**b**) 10.3 ± 0.3 μm, 0 V to 60 V DC, (**c**) 7.6 ± 0.2 μm, 0 V to 55 V DC, (**d**) 4.9 ± 0.2 μm, 0 V to 30 V DC, (**e**) 3.6 ± 0.2 μm, 0 V to 30 V DC and (**f**) 2.5 ± 0.2 μm, 0 V to 25 V DC. Composition: 0.4 wt.% I-369, 6 wt.% SL04151, 5 wt.% R1011, 4 wt.% R811, and 84.6 wt.% MCL 2079. Adapted from ref. [50].

2.4.2. Ion Effect

Since the electro-optic response in PSCLCs is commonly attributed to an ion-facilitated electro-mechanical deformation [48,78], changes in the ion density have been found to influence the optical response. During photopolymerization of the polymer-stabilizing network, ions are formed from impurities in the LC monomers and the excess photoinitiator added to the mixture [78,79]. The residual ions that are formed act as impurities, enabling the deformation of the polymer network, subsequently altering the electro-optic response in PSCLCs. The effect of ion concentration on the dynamic response in PSCLCs is further discussed.

It has been documented that polymer networks can be formed from LC mixtures without the presence of photoinitiators [79,80]. The PSCLCs prepared without photoinitiators show lower ion concentrations and different electro-optic responses than PSCLCs prepared in the presence of photoinitiators. Lee and coworkers reported that the LC monomers C3M and C6M were first recrystallized in methanol and separated into two fractions; a purified fraction containing low impurities and low ion concentrations, and a residual fraction containing impurities and high ion concentration [79]. The PSCLCs prepared using the as-received LC monomers had initial ion densities of 7.6×10^{13} ions cm^{-3}; these samples exhibited a bandwidth broadening response as the applied DC voltage was increased (Figure 11a). Interestingly, the PSCLCs prepared with recrystallized LC monomers had lower ion concentrations than the as-received LC monomers and exhibited a smaller bandwidth-broadening response (Figure 11b). While the PSCLCs prepared with the fully purified LC monomers showed a red tuning of the reflection notch instead of bandwidth broadening (Figure 11c). This behavior is attributed to the polymer network entrapping the ions after photopolymerization, the ions can then lead to deformations in the polymer network as the electric stimuli is applied and increased.

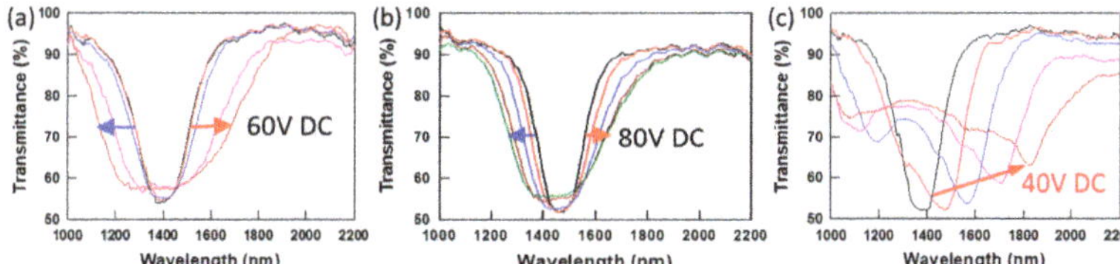

Figure 11. Transmission spectra of PSCLCs formulated with 3.5 wt.% R1011, 90.5 wt.% MLC-2079, and (**a**) 6 wt.% as-received C3M with an initial ion density of 7.6×10^{13} ions cm^{-3}, (**b**) 6 wt.% purified C3M with an initial ion density of 1.6×10^{13} ions cm^{-3}, and (**c**) 6 wt.% residual C3M with an initial ion density of 1.0×10^{14} ions cm^{-3}. Adapted from ref. [79]. The arrows in (**a**,**b**) represent the bandwidth broadening and the arrow in (**c**) indicates the shift of the reflection band in CLC.

Further evidence that ion concentration directly impacts the electro-optic response of PSCLCs was discovered by Lee et al., where samples were exposed to UV light during the application of an electric stimuli. Here, PSCLCs were prepared using 1 wt.% of the photoinitiator Irgacure 369; the PSCLCs were then exposed to irradiation with a UV lamp during the application of a DC voltage. A red-orange reflection color was observed for the initial sample, which became a deep red color when the DC voltage was increased to 20 V. The color became a darker red, ultimately shifting into the infrared region when UV light was exposed during the application of the 20 V. Images of the PSCLCs at different conditions and the transmission spectra of the observed behavior are shown in Figure 12a,b. The observed behavior was attributed to the generation and activation of more ions upon exposure to UV light during the application of the electric field. The increased ions enabled a further redshift of the reflection notch and the resulting color change. Interestingly, UV exposure induces a reversible response.

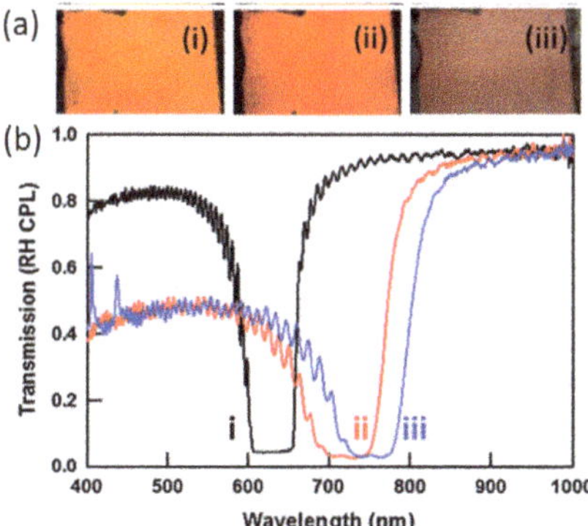

Figure 12. (**a**) Reflection images of PSCLC prepared with 1 wt.% Irgacure 369 at (i) 0 V, (ii) during application of 20 V DC and (iii) UV exposure during application of 20 V DC. (**b**) Transmission spectra at (i) 0 V, (ii) during application of 40 V DC, and (iii) during application of 40 V DC in the presence of 50 mW cm^{-2} of UV light. Adapted from ref. [78].

The driving force for the electro-optic response in PSCLCs is thought to come from the movement of ionic impurities, which will be further discussed in Section 2.5. Upon the application of an electric field, the ionic impurities trapped within the polymer network migrate to the electrodes, inducing a polymer deformation. The deformation of the polymer-stabilizing network results in the tuning of the reflection band of the PSCLC. The reliance on ionic impurities to induce the reflection notch tuning could be a potential issue for utilizing PSCLCs in display technology, as it could induce image sticking. Image sticking is a phenomenon in which a previously displayed image is observed after the applied voltage has been removed and the display has been refreshed [81]. To alleviate the potential issue of image sticking, the choice of negative Δε LC and alignment layer would need to be optimized to reduce the affinity for the ions to adsorb to the alignment layer [81]. Another potential solution that has been shown to reduce image sticking is through doping of either the LC phase or alignment layer with nanoparticles such as γ-Fe2O3 [82] and TiO2 [83]. The doping method would need to be further explored in PSCLC systems, as the proposed mechanism for reducing image sticking relies on ion impurities being absorbed by the nanoparticles, which could reduce tuning ability.

2.4.3. High Reflectivity of a PSCLC at Oblique Incidence

This type of structure results in unpolarized light transmitting one handedness through the CLC while reflecting the opposite handedness of light. In order to achieve total reflectance of unpolarized light, a CLC system must be designed to have CLC layers with different handedness, either creating physically distinct right-handed and left-handed regions within a single CLC sample [31,34,35,84–88] or stacking CLCs with different handedness [42,89,90]. The design and fabrication of these CLC systems often result in a myriad of problems limiting the effectiveness of the system. In order to minimize the problems arising from the difficult fabrication, Tondiglia et al. demonstrated an electrically controlled total reflection within a single-layer PSCLC by observing the reflection at oblique incidence angles [91]. The researchers observed a reflectance notch with full reflection of unpolarized light when the PSCLC device was tilted at an angle of 60° (Figure 13). The increased reflectance was attributed to an increased path length through the PSCLC; this also resulted in an effective refractive index change within the PSCLC layer, resulting in scattering losses. In addition to achieving full reflectance, the pitch of the PSCLC and bandwidth of the reflectance peak could be controlled by the application of DC fields. Figure 14 shows the bandwidth broadening and red tuning response of PSCLC samples at normal incidence and at an incidence angle of 60° as a function of the applied DC field. At a 60° incidence angle, both the bandwidth broadening and red tuning samples showed large electro-optic responses with the total reflection being much less than the 50% reflectance of the sample measured with a 0° incidence angle. The total reflectance behavior was achieved using a simple fabrication method through the alteration of the incidence angle, which enabled an electrically controlled total reflection in a single-layer PSCLC.

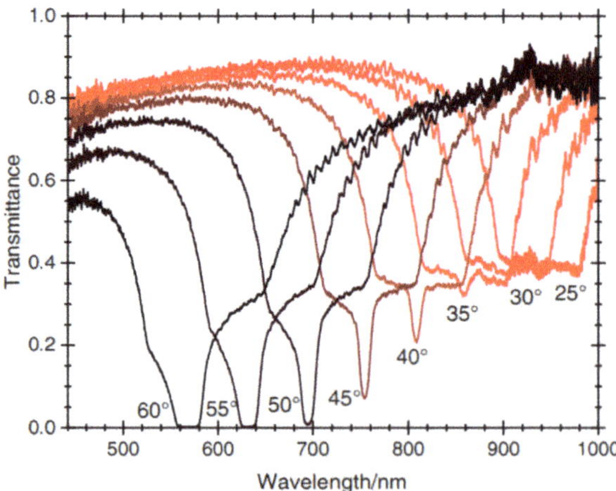

Figure 13. Transmission spectra (unpolarized white light probe) collected from sample CLC with a thickness of 30 μm as a function of incidence angle. The initial notch position of the sample is 1080 nm. Adapted from ref. [91].

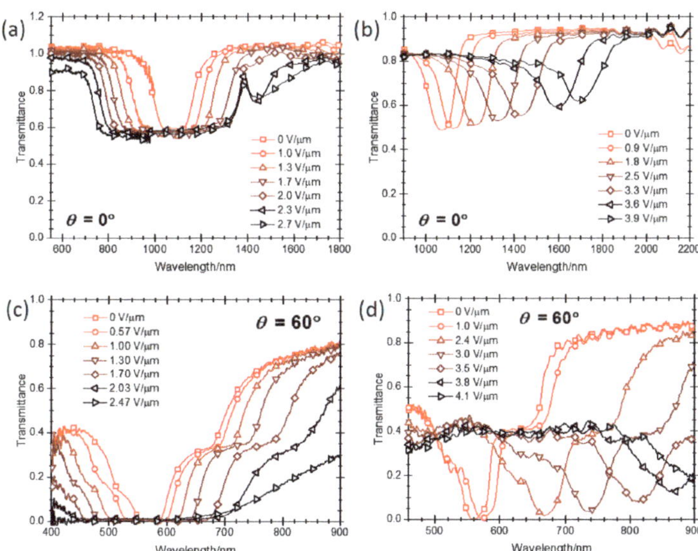

Figure 14. Variation of the optical properties of PSCLC with DC field: (**a**,**b**) transmission spectra at θ = 0° for unpolarized light and (**c**,**d**) transmission spectra at θ = 60° for unpolarized light. Adapted from ref. [91]. A 30-μm thick PSCLC was used for bandwidth broadening of the reflection band (**a**,**c**), and a PSCLC with a thickness of 15 μm was used for red tuning response (**b**,**d**).

2.5. Potential Mechanism

The electrically induced optical response of PSCLCs is attributed to the deformation of the polymer network stabilizing in the CLC medium [39,41,46,50,79]. Electro-optic experiments reveal a strong correlation between the magnitude of the optical response and the concentration of ionic impurities in the LC host. In LCs or LC mixtures, ions exist in the range of 10^9–10^{14} ions/cm^3 and are residual impurities generated from initiators, catalysts, salts, moisture used during synthesis and purification processes [78]. Further studies have also determined that the UV curing process used to form the polymer stabilizing network can further increase the ion density due to degradation of the liquid crystals [92–95] and alignment layers [63]. Three types of ionic impurities can be present in LC mixtures: positive, negative and neutral ions. The positive ions in the mixture can be bonded to the ester group of the liquid crystal monomer and the polar group of the photoinitiator. During photopolymerization a fraction of the positively charged ions are trapped on or in the polymer network, represented by green charges, while some exist as free ions, represented as red positive and blue negative ions, as shown in Figure 15a. Application of a DC electric field causes charge screening, in which free ions present in the system migrate toward the oppositely charged electrodes. An increase in the DC field can then move the trapped ions in the polymer, leading to the deformation of the polymer network across the cell thickness direction, as can be seen in Figure 15b. If the polymer network is attached to the substrates while applying a DC field, the number of pitches must remain constant and the deformation of the polymer network induces a pitch variation across the cell; i.e., pitch compression and expansion near the negative electrode and the positive electrode, respectively. The degree of deformation of the polymer network is mainly influenced by two factors: the viscoelastic properties of the polymer network [46,47] and the type and concentration of ions trapped in the polymer network [48,79,80]. Such field-induced deformation and related pitch distribution depend on the nature of the liquid crystals, the polymers, the type and concentration of ionic impurities, and the chemical and physical properties of the crosslinked polymer network.

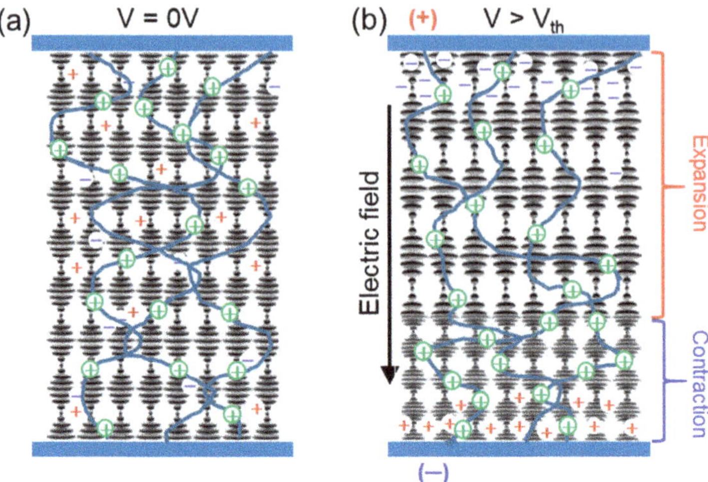

Figure 15. Schematic of the polymer network in a deformable PSCLC with Δε < 0: (**a**) with no applied field, (**b**) with a DC field applied between the top and bottom substrates. The blue lines represent the polymer network, the grey horizontal bars are the low-molecular weight CLC molecules, (+) and (−) are the free cationic and anionic impurities and ⊕ are trapped cationic impurities. Adapted from ref. [50].

The mechanism of the responsive electro-optic behavior of PSCLCs being driven by the pitch modulation of the CLC medium through polymer deformation has been further experimentally supported. Pitch changes can be probed using confocal or multiphoton fluorescence polarization microscopy, which enables visualization of the cross-sectional image of PSCLCs [41,46]. Due to the LC molecules anchoring to the polymer network, the pitch of the CLC is deformed. The change in the pitch is directly observable using fluorescence confocal microscopy. Figures 16 and 17 exhibit the fluorescence confocal microscopy images of PSCLCs, showing bandwidth broadening and red-tuning, respectively [41,46]. In both cases, applying a DC field results in non-uniform pitch distortion. In Figure 16, Nemati and colleagues were able to determine the helical pitch change of PSCLCs exhibiting bandwidth broadening [41] using an LC mixture consisting of 91.48% negative Δε LC HCCH, 2% chiral dopant R811, 6.5% monomer RM257, and 0.02% fluorescence dye BTBP, *N,N′*-bis(2,5-di-*tert*-butylphenyl)-3,4,9,10-perylenedicarboximide). The pitch of the resulting PSCLC was determined to be ~5 μm. Figure 16a shows the cross-sectional image of the florescence radiation intensity profile of the sample before the application of an electric field, and shows a layered structure having a uniform half pitch (P/2). When a DC field of 1.1 V μm^{-1} was applied, a non-uniform pitch change was observed. The CLC layers expanded near the positive (top) electrode and compressed near the negative (bottom) electrode, as shown in Figure 16b. When the polarity of the applied DC voltage was reversed, the CLC layers near the negative (top) electrode were compressed, while the CLC layers near the positive (bottom) electrode were expanded. However, it was observed that a large fraction of the CLCs in the bulk maintained their initial pitch. This observation indicates that, in the case of symmetrical broadening, most of the CLCs maintain their initial pitch, while a fraction of pitches are compressed near the negative electrode and expanded near the positive electrode.

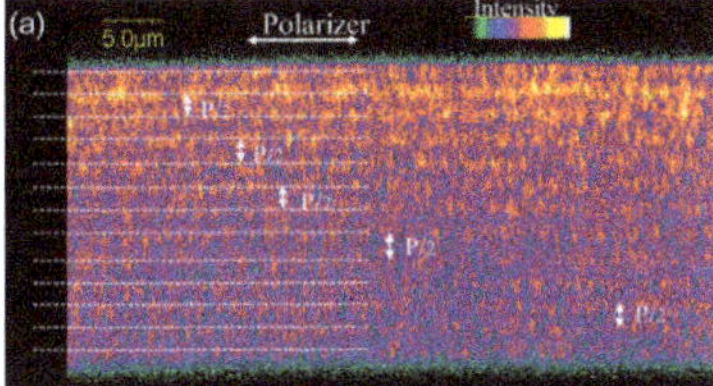
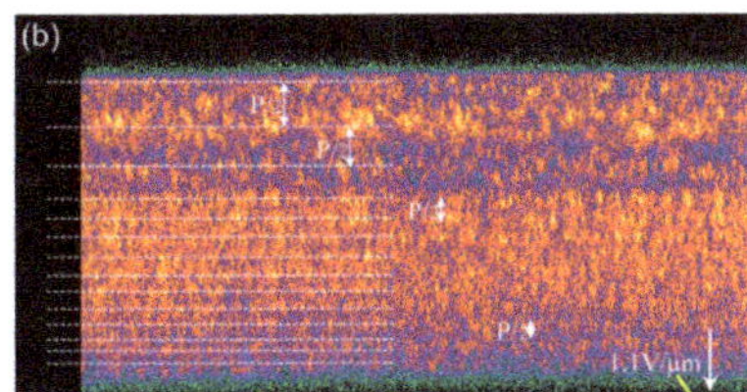

Figure 16. Polarizing confocal microphotographs of the polymer-stabilized cholesteric liquid crystal in the absence and presence of a DC electrical field. (**a**) 0 V μm^{-1} and (**b**) 1.1 V μm^{-1}. The dashed lines are a guide to the eye. Adapted from ref. [41].

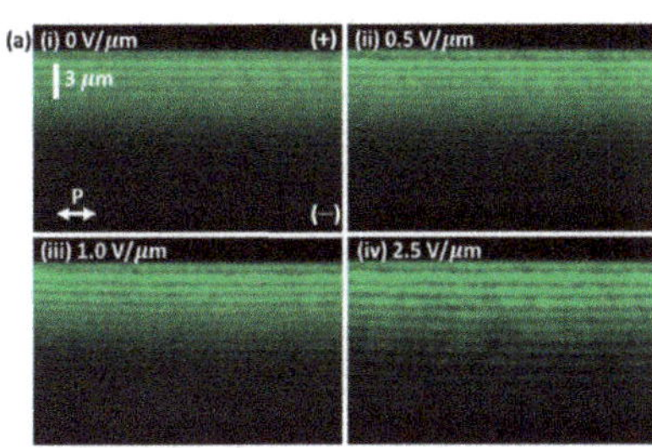

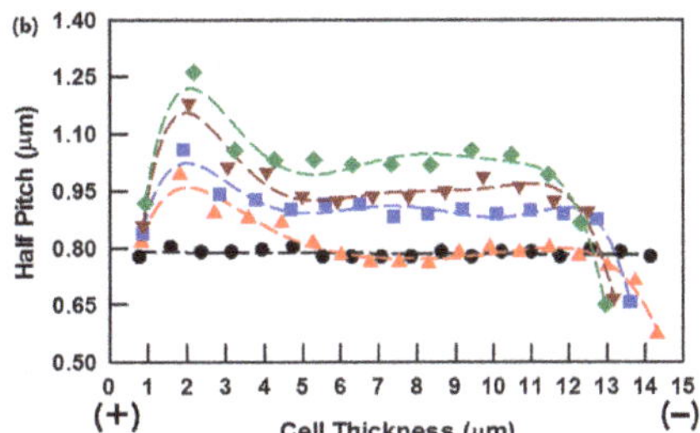

Figure 17. (**a**) Map of the fluorescence intensity by three-photon excitation along a 2D cross-section of a PSCLC without or with an applied DC field (reflection notch at 2.45 μm at zero field and ~15 μm cell thickness): (i) 0 V μm^{-1}, (ii) 0.5 V μm^{-1}, (iii) 1.0 V μm^{-1}, and (iv) 2.5 V μm^{-1}. The vertical direction is along the cell normal and the helical axis, the horizontal direction is parallel to the cell substrates. The white arrow labeled "P" in (**i**) is the polarization direction of the excitation light. The polarity of the DC voltage is indicated by the "(+)"and "(−)" signs. (**b**) Magnitude of the half pitch through the cell thickness as obtained from the fluorescence images in (**a**) for the following values of the electric field: (•) 0 V μm^{-1}, (▲) 0.5 V μm^{-1}, (■) 1.0 V μm^{-1}, (▼) 2.0 V μm^{-1}, and (♦) 2.5 V μm^{-1}. Composition of the CLC mixture: 0.2 wt.% I-369, 4 wt.% SL04151, 2 wt.% RM23, 2 wt.% CB15, and 91.8 wt.% MLC 2079. The PSCLC was formed by curing the mixture with exposure to 365 nm UV light at 100 mW cm^{-2} for 5 min. Adapted from ref. [46].

Similar to responsive PSCLCs exhibiting bandwidth broadening, tunable PSCLCs are also driven by pitch modulation. The pitch of a PSCLC exhibiting red tuning is shown in Figure 17a; the period of the half-pitch starts to change near the top and bottom substrates by increasing the DC field [46]. Each line represents the half pitch of the CLC. At this low DC field, a pitch change similar to the bandwidth broadening sample is observed. As the electric field strength increases, most of the pitches in the bulk CLC expand, except for a small number of pitches that are compressed near the negative electrode. The corresponding half pitch variation in cell thickness is shown in Figure 17b. At a low DC field of 0.5 V μm^{-1}, the pitch change occurs near the top and bottom of the substrate, with one side expanding and the other side contracting. At larger DC fields greater than 1.0 V μm^{-1}, the pitch deviates from the original pitch value across the cell thickness and increases with the applied DC field. The sample highlighted in Figure 17 and similarly formulated PSCLCs show a red shift of the Bragg reflection peak with increasing the DC field, but no significant change in bandwidth [44,46].

3. Conclusions

In this review, the electrically induced optical response of PSCLCs fabricated with positive or negative Δε LCs was summarized. Polymer stabilization of the CLC phase enabled responsive behaviors not observed in bulk CLCs; specifically, the tuning of the selective reflection across the electromagnetic spectrum and bandwidth broadening. Large-magnitude dynamic bandwidth broadening of the reflection notch of PSCLCs was observed in PSCLCs with relatively low crosslink density of the polymer network, whereas a red shift or blue shift of the reflection band was observed in PSCLCs with higher crosslink densities. The bistable switching between reflective and scattering states of PSCLCs was attributed to an electromechanical polymer displacement mechanism. In thin cells, a reversible switching between reflective and transparent states was observed because the number of effective repeat units in the cholesteric pitch became too small to achieve measurable reflection efficiency. The potential mechanism for the dynamic responses of PSCLCs was attributed to the electric-field-induced polymer deformation in a CLC medium. Increasing the electric field controls the degree of polymer deformation, inducing a pitch variation across the cell thickness. The dynamic control of optical properties enables the PSCLCs to be used in several optical applications, including displays, optical filters, smart windows, and mirrors. The use of PSCLCs in these technologies requires addressing several potential challenges, such as high DC driving voltages, image sticking problems, and modest optical properties. In addition, the deformation of the polymer network to induce the dynamic response of the PSCLC results in a relatively slow response time in the order of seconds, lower transmittance, and increased haze. These practical issues of PSCLCs require further investigation and optimization in future studies.

Author Contributions: Early conceptualization of the studies: K.M.L. and N.P.G.; writing and original draft preparation: K.M.L., Z.M.M., E.P.C., U.N.T., C.P.A., S.M.W. and N.P.G.; writing, review and editing: K.M.L., Z.M.M., E.P.C., U.N.T., C.P.A., S.M.W., K.J.C., H.N.L., M.E.M. and N.P.G. All authors have read and agreed to the published version of the manuscript.

Funding: This research was funded by Materials and Manufacturing Directorate of the Air Force Research Laboratory, grant number #FA8650-16-D-5404.

Data Availability Statement: The data presented in this study are available on request from the corresponding author.

Acknowledgments: The authors acknowledge funding from the Materials and Manufacturing Directorate of the Air Force Research Laboratory under contract #FA8650-16-D-5404.

Conflicts of Interest: The authors declare no conflict of interest.

References

1. Geelhaar, T.; Griesar, K.; Reckmann, B. 125 Years of Liquid Crystals—A Scientific Revolution in the Home. *Angew. Chem. Int. Ed.* **2013**, *52*, 8798–8809. [CrossRef] [PubMed]
2. Kitzerow, H.-S.; Bahr, C. *Chirality in Liquid Crystals*; Springer: New York, NY, USA, 2001.
3. Wu, S.T.; Yang, D.-K. *Reflective Liquid Crystal Displays*; Wiley: Chichester, UK, 2001.
4. Blinov, L.M. *Electro-Optical and Magneto-Optical Properties of Liquid Crystals*; Wiley: New York, BY, USA, 1983.
5. Yang, D.-K.; Chien, C.-C.; Donna, J.W. Cholesteric liquid crystal/ polymer dispersion for hazy-free light shutters. *Appl. Phys. Lett.* **1992**, *60*, 3102–3104. [CrossRef]
6. Yang, D.-K.; Doane, J.W.; Yaniv, Z.; Glasser, J. Cholesteric reflective display: Drive scheme and contrast. *Appl. Phys. Lett.* **1994**, *64*, 1905. [CrossRef]
7. Yang, D.-K.; West, J.L.; Chien, L.C.; Doane, J.W. Control of reflectivity and bistability in displays using cholesteric liquid crystals. *J. Appl. Phys.* **1994**, *76*, 1331–1333. [CrossRef]
8. Lu, M.-H. Bistable reflective cholesteric liquid crystal display. *J. Appl. Phys.* **1997**, *81*, 1063–1066. [CrossRef]
9. Taheri, B.; Doane, J.W.; Davis, D.; St. John, D. Optical properties of bistable cholesteric reflective displays. *SID Int. Symp. Dig. Tec.* **1996**, *27*, 39–42.
10. Anderson, J.; Watson, P.; Ruth, J.; Sergan, V.; Bos, P. Fast frame rate bistable cholesteric texture reflective displays. *SID Int. Symp. Dig. Tec.* **1998**, *29*, 806–809. [CrossRef]
11. Yang, D.-K. Flexible bistable cholesteric reflective displays. *J. Disp. Technol.* **2006**, *2*, 32. [CrossRef]
12. Li, C.-C.; Tseng, H.-Y.; Pai, T.-W.; Wu, Y.-C.; Hsu, W.-H.; Jau, H.-C.; Chen, C.-W.; Lin, T.H. Bistable cholesteric liquid crystal light shutter with multielectrode driving. *Appl. Opt.* **2014**, *53*, E33–E37. [CrossRef]
13. Huang, C.-Y.; Fu, K.-Y.; Lo, K.-Y.; Tsai, M.-S. Bistable transflective cho lesteric light shutters. *Opt. Express* **2003**, *11*, 560–565. [CrossRef]

14. Lin, F.-C.; Wei, L. Color-reflective dual frequency cholesteric liquid crystal displays and their drive schemes. *Appl. Phys. Express* **2011**, *4*, 112201. [CrossRef]
15. Kumar, P.; Kang, S.-W.; Lee, S.H. Advanced bistable cholesteric light shutter with dual frequency nematic liquid crystal. *Opt. Mater. Express* **2012**, *2*, 1121–1134. [CrossRef]
16. Xu, M.; Yang, D.-K. Dual frequency cholesteric liquid crystals. *Appl. Phys. Lett.* **1997**, *70*, 720. [CrossRef]
17. Wen, C.-H.; Wu, S.T. Dielectric heating effects of dual-frequency liquid crystals. *Appl. Phys. Lett.* **2005**, *86*, 231104. [CrossRef]
18. Zhang, F.; Yang, D.-K. Polymer stabilized cholesteric dichroic dye displays. *SID Int. Symp. Digest Tec.* **2012**, *33*, 469–471. [CrossRef]
19. Lin, Y.-H.; Ren, H.; Fan, Y.-H.; Wu, Y.-H.; Wu, S.T. Polarization-independent and fast-response phase modulation using a normal-mode polymer-stabilized cholesteric texture. *J. Appl. Phys.* **2005**, *98*, 043112. [CrossRef]
20. Ren, H.; Wu, S.-T. Reflective reversed-mode polymer stabilized cholesteric texture light switches. *J. Appl. Phys.* **2002**, *92*, 797–800. [CrossRef]
21. Yin, Y.; Li, W.; Cao, H.; Guo, J.; Li, B.; He, S.; Ouyang, C.; Cao, M.; Huang, H.; Yang, H. Effects of monomer structure on the morphology of polymer network and the electro-optical property of reverse-mode polymer stabilized cholesteric texture. *J. Appl. Polym. Sci.* **2009**, *111*, 1353–1357. [CrossRef]
22. Dierking, I. *Polymer-Modified Liquid Crystals*; Royal Society of Chemistry: Cambridge, UK, 2019.
23. Fung, Y.K.; Yang, D.-K.; Ying, S.; Chien, L.C.; Zumer, S.; Doane, J.W. Polymer networks formed in liquid crystals. *Liq. Cryst.* **1995**, *19*, 797–801. [CrossRef]
24. Hikmet, R.A.M.; Lub, J. Anisotropic networks and gels obtained by photopolymerization in the liquid crystalline state: Synthesis and applications. *Prog. Polym. Sci.* **1996**, *21*, 1165–1209. [CrossRef]
25. Hikmet, R.A.M. Electrically induced light scattering from anisotropic gels. *J. Appl. Phys.* **1990**, *68*, 4406–4412. [CrossRef]
26. Hikmet, R.A.M.; Howard, R. Structure and properties of anisotropic gels and plasticized networks containing molecules with a smectic-A phase. *Phys. Rev. E* **1993**, *48*, 2752–2759. [CrossRef] [PubMed]
27. Dierking, I. Polymer Network-Stabilized Liquid Crystals. *Adv. Mater.* **2000**, *12*, 167–181. [CrossRef]
28. Dierking, I. Recent development in polymer stabilized liquid crystals. *Polym. Chem.* **2010**, *1*, 1153–1159. [CrossRef]
29. Hikmet, R.A.M.; Boots, H.M.J. Domain structure and switching behavior of anisotropic gels. *Phys. Rev. E* **1995**, *51*, 5824–5831. [CrossRef]
30. Hikmet, R.A.M.; Kemperman, H. Electrically switchable mirrors and optical components made from liquid-crystal gels. *Nature* **1998**, *392*, 476–479. [CrossRef]
31. Guo, J.; Cao, J.; Wei, J.; Zhang, D.; Liu, F.; Pan, G.; Zhao, D.; He, W.; Yang, H. Polymer stabilized liquid crystal films reflecting both right- and left-circularly polarized light. *Appl. Phys. Lett.* **2008**, *93*, 201901. [CrossRef]
32. Robbie, K.; Broer, D.J.; Brett, M.J. Chiral nematic order in liquid crystals imposed by an engineered inorganic nanostructure. *Nature* **1999**, *399*, 764–766. [CrossRef]
33. Guo, J.; Liu, F.; Chen, F.; Wei, J.; Yang, H. Realisation of cholesteric liquid-crystalline materials reflecting both right- and left-circularly polarised light using the wash-out/refill technique. *Liq. Cryst.* **2010**, *37*, 171–178. [CrossRef]
34. McConney, M.E.; Tondiglia, V.P.; Hurtubise, J.M.; White, T.J.; Bunning, T.J. Photoinduced hyper-reflective cholesteric liquid crystals enabled via surface initiated photopolymerization. *Chem. Commun.* **2011**, *47*, 505–507. [CrossRef]
35. McConney, M.E.; Tondiglia, V.P.; Hurtubise, J.M.; Natarajan, L.V.; White, T.J.; Bunning, T.J. Thermally Induced, Multicolored Hyper-Reflective Cholesteric Liquid Crystals. *Adv. Mater.* **2011**, *23*, 1453–1457. [CrossRef] [PubMed]
36. He, Z.; Zeng, J.; Zhu, S.; Zhang, D.; Ma, C.; Zhang, C.; Yu, P.; Miao, Z. A bistable light shutter based on polymer stabilized cholesteric liquid crystals. *Opt. Mater.* **2023**, *136*, 113426. [CrossRef]
37. Wang, H.; Wang, L.; Chen, M.; Li, T.; Cao, H.; Yang, D.-K.; Yang, Z.; Yang, H.; Zhu, S. Bistable polymer-dispersed cholesteric liquid crystal thin film enabled by a stepwise polymerization. *RSC Adv.* **2015**, *5*, 58959–58965. [CrossRef]
38. Li, X.; Guo, Y.; Zhang, M.; Zhang, C.; Niu, R.; Ma, H.; Sun, Y. Colorable Light-Scattering Device Based on Polymer-Stabilized Ion-Doped Cholesteric Liquid Crystal and an Electrochromatic Layer. *ACS Appl. Mater. Interfaces* **2023**, *15*, 7184–7195. [CrossRef] [PubMed]
39. Tondiglia, V.P.; Natarajan, L.V.; Bailey, C.A.; McConney, M.E.; Lee, K.M.; Bunning, T.J.; Zola, R.; Nemati, H.; Yang, D.-K.; White, T.J. Bandwidth broadening induced by ionic interactions in polymer stabilized cholesteric liquid crystals. *Opt. Mater. Express* **2014**, *4*, 1465. [CrossRef]
40. Tondiglia, V.P.; Natarajan, L.V.; Bailey, C.A.; Duning, M.M.; Sutherland, R.L.; Yang, D.-K.; Voevodin, A.; White, T.J.; Bunning, T.J. Electrically induced bandwidth broadening in polymer stabilized cholesteric liquid crystals. *J. Appl. Phys.* **2011**, *110*, 053109. [CrossRef]
41. Nemati, H.; Liu, S.; Zola, R.S.; Tondiglia, V.P.; Lee, K.M.; White, T.J.; Bunning, T.J.; Yang, D.-K. Mechanism of electrically induced photonic band gap broadening in polymer stabilized cholesteric liquid crystals with negative dielectric anisotropies. *Soft Matter* **2015**, *11*, 1208. [CrossRef]
42. Lee, K.M.; Tondiglia, V.P.; McConney, M.E.; Natarajan, L.V.; Bunning, T.J.; White, T.J. Color-tunable mirrors based on electrically regulated bandwidth broadening in polymer-stabilized cholesteric liquid crystals. *ACS Photonics* **2014**, *1*, 1033. [CrossRef]
43. Radka, B.; Lee, K.M.; Godman, N.P.; White, T.J. Electro-optic characteristics of stabilized cholesteric liquid crystals with non-liquid crystalline polymer networks. *Soft Matter* **2022**, *18*, 3013–3018. [CrossRef]

44. McConney, M.E.; Tondiglia, V.P.; Natarajan, L.V.; Lee, K.M.; White, T.J.; Bunning, T.J. Electrically induced color changes in polymer-stabilized cholesteric liquid crystals. *Adv. Opt. Mater.* **2013**, *1*, 417–421. [CrossRef]
45. White, T.J.; Lee, K.M.; McConney, M.E.; Tondiglia, V.P.; Natarajan, L.V.; Bunning, T.J. Stimuli-Responsive Cholesteric Liquid Crystal Composites for Optics and Photonics. *SID Int. Symp. Dig. Tec.* **2014**, *45*, 555. [CrossRef]
46. Lee, K.M.; Tondiglia, V.P.; Lee, T.; Smalyukh, I.I.; White, T.J. Large range electrically-induced reflection notch tuning in polymer stabilized cholesteric liquid crystals. *J. Mater. Chem. C* **2015**, *3*, 8788–8793. [CrossRef]
47. Lee, K.M.; Tondiglia, V.P.; Rumi, M.; White, T.J. Time-dependent deformation of structurally chiral polymer networks in stabilized cholesteric liquid crystals. *J. Polym. Sci. Part B* **2018**, *56*, 1087–1093. [CrossRef]
48. Lee, K.M.; Tondiglia, V.P.; Godman, N.P.; Middleton, C.M.; White, T.J. Blue-shifting Tuning of the Selective Reflection of Polymer Stabilized Cholesteric Liquid Crystals. *Soft Matter* **2017**, *13*, 5842–5848.
49. Lee, K.M.; Tondiglia, V.P.; White, T.J. Bistable switching of polymer stabilized cholesteric liquid crystals between transparent and scattering modes. *MRS Commun.* **2015**, *5*, 223–227. [CrossRef]
50. Lee, K.M.; Crenshaw, E.P.; Rumi, M.; White, T.J.; Bunning, T.J.; McConney, M.E. Effect of Cell Thickness on the Electro-optic Response of Polymer Stabilized Cholesteric Liquid Crystals with Negative Dielectric Anisotropy. *Materials* **2020**, *13*, 746. [CrossRef]
51. Lee, K.M.; Tondiglia, V.P.; White, T.J. Electrically Reconfigurable Liquid Crystalline Mirrors. *ACS Omega* **2018**, *3*, 4453. [CrossRef]
52. Broer, D.J.; Lub, J.; Mol, G.N. Wide-band reflective polarizers from cholesteric polymer networks with a pitch gradient. *Nature* **1995**, *378*, 467–469. [CrossRef]
53. White, T.J.; McConney, M.E.; Bunning, T.J. Dynamic color in stimuli-responsive cholesteric liquid crystals. *J. Mater. Chem.* **2010**, *20*, 9832–9847. [CrossRef]
54. Zhang, L.; Nie, Q.; Jiang, X.-F.; Zhao, W.; Hu, X.; Shui, L.; Zhou, G. Enhanced Bandwidth Broadening of Infrared Reflector Based on Polymer Stabilized Cholesteric Liquid Crystals with Poly(N-vinylcarbazole) Used as Alignment Layer. *Polymers* **2021**, *13*, 2238. [CrossRef]
55. Hu, X.; de Haan, L.T.; Khandelwal, H.; Schenning, A.P.H.J.; Nian, L.; Zhou, G. Cell thickness dependence of electrically tunable infrared reflectors based on polymer stabilized cholesteric liquid crystals. *Sci. China Mater.* **2018**, *61*, 745–751. [CrossRef]
56. Khandelwal, H.; Debije, M.G.; White, T.J.; Schenning, A.P.H.J. Electrically tunable infrared reflector with adjustable bandwidth broadening up to 1100 nm. *J. Mater. Chem. A* **2016**, *4*, 6064–6069. [CrossRef]
57. Hu, X.; Zeng, W.; Yang, W.; Xiao, L.; De Haan, L.T.; Zhao, W.; Li, N.; Shui, L.; Zhou, G. Effective electrically tunable infrared reflectors based on polymer stabilised cholesteric liquid crystals. *Liq. Cryst.* **2019**, *46*, 185–192. [CrossRef]
58. Yu, M.; Wang, L.; Nemati, H.; Yang, H.; Bunning, T.J.; Yang, D.-K. Effects of polymer network on electrically induced reflection band broadening of cholesteric liquid crystals. *J. Polym. Sci. Part B* **2017**, *55*, 835–846. [CrossRef]
59. Kim, J.; Kim, H.; Kim, S.; Choi, S.; Jang, W.; Kim, J.; Lee, J.-H. Broadening the reflection bandwidth of polymer-stabilized cholesteric liquid crystal via a reactive surface coating layer. *Appl. Opt.* **2017**, *56*, 5731–5735. [CrossRef]
60. Zhou, H.; Wang, H.; He, W.; Yang, Z.; Cao, H.; Wang, D.; Li, Y. Research Progress of Cholesteric Liquid Crystals with Broadband Reflection. *Molecules* **2022**, *27*, 4427. [CrossRef]
61. Nemati, H.; Liu, S.; Moheghi, A.; Tondiglia, V.P.; Lee, K.M.; Bunning, T.J.; Yang, D.-K. Enhanced reflection band broadening in polymer stabilized cholesteric liquid crystals with negative dielectric anisotropy. *J. Mol. Liq.* **2018**, *267*, 120–126. [CrossRef]
62. Duan, M.-y.; Cao, H.; Wu, Y.; Li, E.-l.; Wang, H.H.; Wang, D.; Yang, Z.; He, W.-l.; Yang, H. Broadband reflection in polymer stabilized cholesteric liquid crystal films with stepwise photo-polymerization. *Phys. Chem. Chem. Phys.* **2017**, *19*, 2353–2358. [CrossRef]
63. Lu, L.; Sergan, V.; Bos, P.J. Mechanism of electric-field-induced segregation of additives in a liquid-crystal host. *Phys. Rev. E* **2012**, *86*, 051706. [CrossRef]
64. Li, C.-C.; Tseng, H.-Y.; Chen, C.-W.; Wang, C.-T.; Jau, H.-C.; Wu, Y.-C.; Hsu, W.-H.; Lin, T.-H. Versatile Energy-Saving Smart Glass Based on Tristable Cholesteric Liquid Crystals. *ACS Appl. Energy Mater.* **2020**, *3*, 7601–7609. [CrossRef]
65. Liang, X.; Chen, M.; Guo, S.; Zhang, L.; Li, F.; Yang, H. Dual-Band Modulation of Visible and Near-Infrared Light Transmittance in an All-Solution-Processed Hybrid Micro-Nano Composite Film. *ACS Appl. Mater. Interfaces* **2017**, *9*, 40810–40819. [CrossRef]
66. Liang, X.; Guo, S.; Chen, M.; Li, C.; Wang, Q.; Zou, C.; Zhang, C.; Zhang, L.; Guo, S.; Yang, H. A temperature and electric field-responsive flexible smart film with a full broadband optical modulation. *Mater. Horiz.* **2017**, *4*, 878–884. [CrossRef]
67. Khandelwal, H.; Loonen, R.C.G.M.; Hensen, J.L.M.; Debije, M.G.; Schenning, A.P.H.J. Electrically switchable polymer stabilised broadband infrared reflectors and their potential as smart windows for energy saving in buildings. *Sci. Rep.* **2015**, *5*, 11773. [CrossRef] [PubMed]
68. Nava, G.; Ciciulla, F.; Simoni, F.; Iadlovska, O.; Lavrentovich, O.D.; Lucchetti, L. Heliconical cholesteric liquid crystals as electrically tunable optical filters in notch and bandpass configurations. *Liquid Cryst.* **2021**, *48*, 1534–1543. [CrossRef]
69. Xianyu, H.; Faris, S.; Crawford, G.P. In-plane switching of cholesteric liquid crystals for visible and near-infrared applications. *Appl. Opt.* **2004**, *43*, 5006–5015. [CrossRef] [PubMed]
70. Escuti, M.J.; Bowley, C.C.; Crawford, G.P.; Žumer, S. Enhanced dynamic response of the in-plane switching liquid crystal display mode through polymer stabilization. *Appl. Phys. Lett.* **1999**, *75*, 3264–3266. [CrossRef]
71. Lin, T.H.; Jau, H.-C.; Chen, C.H.; Chen, Y.J.; Wei, T.H.; Chen, C.-W.; Fuh, A.Y.G. Electrically controllable laser based on cholesteric liquid crystal with negative dielectric anisotropy. *Appl. Phys. Lett.* **2006**, *88*, 86–89. [CrossRef]

72. Scaramuzza, N.; Ferrero, C.; Carbone, V.; Versace, C. Dynamics of selective reflections of cholesteric liquid crystals subject to electric fields. *J. Appl. Phys.* **1995**, *77*, 572–576. [CrossRef]
73. Umeton, C.; Bartolino, R.; Cipparone, G.; Simoni, F. Liquid crystal laser tuner. *Appl. Opt.* **1988**, *27*, 210–211. [CrossRef]
74. Umeton, C.; Bartolino, R.; Cipparrone, G.; Xu, F.; Ye, F.; Simoni, F. Wavelength Modulation Using Cholesteric Liquid Crystals. *Mol. Cryst. Liq. Cryst.* **1987**, *144*, 323–336. [CrossRef]
75. Bailey, C.A.; Tondiglia, V.P.; Natarajan, L.V.; Duning, M.M.; Bricker, R.L.; Sutherland, R.L.; White, T.J.; Durstock, M.F.; Bunning, T.J. Electromechanical tuning of cholesteric liquid crystals. *J. Appl. Phys.* **2010**, *107*, 013105. [CrossRef]
76. Yu, H.; Tang, B.Y.; Li, J.; Li, L. Electrically tunable lasers made from electro-optically active photonics band gap materials. *Opt. Express* **2005**, *13*, 7243–7249. [CrossRef] [PubMed]
77. Worth, B.; Lee, K.M.; Tondiglia, V.P.; Myers, J.; Mou, S.; White, T.J. Dynamic, infrared bandpass filters prepared from polymer-stabilized cholesteric liquid crystals. *Appl. Opt.* **2016**, *55*, 7134–7137. [CrossRef] [PubMed]
78. Lee, K.M.; Tondiglia, V.P.; White, T.J. Photosensitivity of reflection notch tuning and broadening in polymer stabilized cholesteric liquid crystals. *Soft Matter* **2016**, *12*, 1256–1261. [CrossRef] [PubMed]
79. Lee, K.M.; Bunning, T.J.; White, T.J.; McConney, M.E.; Godman, N.P. Effect of Ion Concentration on the Electro-Optic Response in Polymer-Stabilized Cholesteric Liquid Crystals. *Crystals* **2021**, *11*, 7. [CrossRef]
80. Lee, K.M.; Ware, T.H.; Tondiglia, V.P.; Mcbride, M.K.; Zhang, X.; Bowman, C.N.; White, T.J. Initiatorless Photopolymerization of Liquid Crystal Monomers. *ACS Appl. Mater. Interfaces* **2016**, *8*, 28040–28046. [CrossRef] [PubMed]
81. Xu, D.; Peng, F.; Chen, H.; Yuan, J.; Wu, S.-T.; Li, M.-C.; Lee, S.-L.; Tsai, W.-C. Image sticking in liquid crystal displays with lateral electric fields. *J. Appl. Phys.* **2014**, *116*, 193102. [CrossRef]
82. Liu, N.; Wang, M.; Tang, Z.; Gao, L.; Xing, H.; Meng, X.; He, Z.; Li, J.; Cai, M.; Wang, X.; et al. Influence of γ-Fe_2O_3 nanoparticles doping on the image sticking in VAN-LCD. *Chin. Opt. Lett.* **2020**, *18*, 033501. [CrossRef]
83. Son, S.-R.; An, J.; Choi, J.-W.; Lee, J.H. Fabrication of TiO2-Embedded Polyimide Layer with High Transmittance and Improved Reliability for Liquid Crystal Displays. *Polymers* **2021**, *13*, 376. [CrossRef]
84. Mitov, M.; Dessaud, N. Cholesteric liquid crystalline materials reflecting more than 50% of unpolarized incident light intensity. *Liq. Cryst.* **2007**, *34*, 183. [CrossRef]
85. Wu, T.; Li, J.; Li, J.; Ye, S.; Wei, J.; Guo, J. A bio-inspired cellulose nanocrystal-based nanocomposite photonic film with hyper-reflection and humidity-responsive actuator properties. *J. Mater. Chem. C* **2016**, *4*, 9687. [CrossRef]
86. Agez, G.; Mitov, M. Cholesteric Liquid Crystalline Materials with a Dual Circularly Polarized Light Reflection Band Fixed at Room Temperature. *J. Phys. Chem. B* **2011**, *115*, 6421–6426. [CrossRef] [PubMed]
87. Li, Y.; Liu, Y.J.; Dai, H.T.; Zhang, X.H.; Luo, D.; Sun, X.W. Flexible cholesteric films with super-reflectivity and high stability based on a multi-layer helical structure. *J. Mater. Chem. C* **2017**, *5*, 10828. [CrossRef]
88. Guo, J.; Chen, F.; Qu, Z.; Yang, H.; Wei, J. Electrothermal Switching Characteristics from a Hydrogen-Bonded Polymer Network Structure in Cholesteric Liquid Crystals with a Double-Handed Circularly Polarized Light Reflection Band. *J. Phys. Chem. B* **2011**, *115*, 861–868. [CrossRef] [PubMed]
89. Makov, D.M. Peak reflectance and color gamut of superimposed left- and right-handed cholesteric liquid crystals. *Appl. Opt.* **1980**, *19*, 1274–1277.
90. Schadt, M.; Fünfschilling, J. New Liquid Crystal Polarized Color Projection Principle. *Jpn. J. Appl. Phys.* **1990**, *29*, 1974. [CrossRef]
91. Tondiglia, V.P.; Rumi, M.; Idehenre, I.U.; Lee, K.M.; Binzer, J.F.; Banerjee, P.P.; Evans, D.R.; McConney, M.E.; Bunning, T.J.; White, T.J. Electrical Control of Unpolarized Reflectivity in Polymer Stabilized Cholesteric Liquid Crystals at Oblique Incidence. *Adv. Opt. Mater.* **2018**, *6*, 1800957. [CrossRef]
92. Lee, W.; Wang, C.-T.; Lin, C.-H. Recovery of the electrically resistive properties of a degraded liquid crystal. *Displays* **2010**, *31*, 160–163. [CrossRef]
93. Gosse, B.; Gosse, J.P. Degradation of liquid crystal devices under d.c. excitation and their electrochemistry. *J. Appl. Electrochem.* **1976**, *6*, 515–519. [CrossRef]
94. Wen, C.-H.; Gauza, S.; Wu, S.T. Ultraviolet stability of liquid crystals containing cyano and isothiocyanato terminal groups. *Liq. Cryst.* **2004**, *31*, 1479–1485. [CrossRef]
95. Lin, P.-T.; Wu, S.T.; Shang, C.Y.; Hsu, C.S. UV Stability of High Birefirngence Liquid Crystals. *Mol. Cryst. Liq. Cryst.* **2004**, *411*, 243–253. [CrossRef]

 materials

Article

Research on Vibration Control Technology of Robot Motion Based on Magnetorheological Elastomer

Xuegong Huang *, Yutong Zhai and Guisong He

School of Mechanical Engineering, Nanjing University of Science & Technology, Nanjing 210094, China
* Correspondence: huangxg@njust.edu.cn

Abstract: The vibration and impact of a humanoid bipedal robot during movements such as walking, running and jumping may cause potential damage to the robot's mechanical joints and electrical systems. In this paper, a composite bidirectional vibration isolator based on magnetorheological elastomer (MRE) is designed for the cushioning and damping of a humanoid bipedal robot under foot contact forces. In addition, the vibration isolation performance of the vibration isolator was tested experimentally, and then, a vibration isolator dynamics model was developed. For the bipedal robot foot impact, based on the vibration isolator model, three vibration reduction control algorithms are simulated, and the results show that the vibration damping effect can reach 85%. Finally, the MRE vibration isolator hardware-in-the-loop-simulation experiment platform based on dSPACE has been built to verify the vibration reduction control effect of the fuzzy PID algorithm. The result shows the vibration amplitude attenuates significantly, and this verifies the effectiveness of the fuzzy PID damping control algorithm.

Keywords: bipedal robot; dynamics simulation; MRE isolator; vibration reduction control algorithms

Citation: Huang, X.; Zhai, Y.; He, G. Research on Vibration Control Technology of Robot Motion Based on Magnetorheological Elastomer. *Materials* **2022**, *15*, 6479. https://doi.org/10.3390/ma15186479

Academic Editors: Yang Qin, Bing Wang, Tung Lik Lee and Raul D.S.G. Campilho

Received: 24 July 2022
Accepted: 14 September 2022
Published: 18 September 2022

1. Introduction

Most current humanoid robots consist of rigid structures [1]. Although rigid structures obtain substantial strength and are also convenient to assemble, they result in robots that lack flexibility. During walking, running and jumping, the leg will inevitably collide with the ground at the moment of landing, and the contact reaction force from the ground will cause the whole leg and even the torso to be subjected to a certain degree of vibration from the bottom up [2]. Although the foot–ground collision is only momentary, the impact on the robot walking stability is significant. Impacts from the ground can cause sudden changes in the speed of the robot's components, which can lead to problems such as tipping when the sudden change in speed reaches a certain value. It also interferes with the normal operation of the robot's vision and sensing systems, causing serious damage to joints, servo motors and other structurally fragile areas and accelerating the destruction.

Therefore, the vibration damping of humanoid robots has become a hot research topic for scholars at home and abroad in recent years. This problem has been addressed by passive vibration isolation methods, such as installing cushioned support brackets at important parts [3] and adding impact-resistant materials to the bottom of the foot [4]. Arvind et al. at the MIT Bionic Robotics Laboratory designed a bionic robotic leg that mimics a musculoskeletal structure, using elastic bands to enhance flexibility [5]. Grizzle et al. at Michigan State University, USA, developed a bipedal robot that takes full advantage of the cushioning and energy-saving effects of springs at specific speeds and environments [6]. Youwang Yin et al. from Shanghai University of Technology proposed a new type of flexible vibration isolation structure, which achieves quasi-zero stiffness characteristics and achieves low-frequency vibration isolation with good effects [7].This type of method does not require external energy input and merely leverages the structural properties of the material itself to achieve damping, but it has a narrow operating band and is less adaptable [8]. Some scholars, at the same time, tried to isolate vibrations by active means. Bjorn Verrelst from the Free University of Brussels designed a bipedal robot with folded pneumatic muscles, and the corresponding active control algorithm controlled the pneumatic muscles to adjust the stiffness, showing an excellent cushioning and vibration damping effect [9]. Jonas Buchli et al. at the University of Southern California's Computational Learning and Motion Control Laboratory utilized a quadruped robot developed by Boston Dynamics as a control object to achieve better robust robot motion in complex and rough terrain [10]. Shoukun Wang et al. from the Beijing Institute of Technology designed an adaptive impedance for wheeled robots based on the Lyapunov method to solve the robot's vibration isolation problem with passing speed bumps [11]. Ji Sun et al. from Anhui University of Science and Technology designed a multidimensional vibration isolation device for aging robots based on the TRIZ theory, which provides an effective method for solving multidimensional vibration problems [12]. Ying Zhang et al. from Beijing Jiaotong University proposed a multidimensional vibration isolation platform for wheel-legged robotic vehicles, supplemented by spring damping modules, combining posture-adjusting and vibration control functions, with vibration isolation effects of over 50% [13]. Yunfeng Wang et al. combined the quasi-zero stiffness isolation system with active damping and investigated the effect of active damping and experimentally verified its good vibration isolation efficiency [14]. However, these methods all depend on continuous external energy input [15], which greatly increases the energy consumption problem of the robot, and the optimization algorithm needs to be improved for different robot structures to be applicable. In this context, the semi-active vibration isolation method is a technique that uses the controlled performance of intelligent materials to change the stiffness and damping of the structure in real time according to the external excitation to achieve vibration isolation, combining the advantages of active and passive vibration isolation. In recent years, many have attempted to apply the method to robot vibration damping problems, and it has become one of the main development directions for solving robot vibration damping problems.

Magnetorheological elastomers are prepared by mixing soft magnetic particles and polymer elastomers, which have the advantages of both liquids and elastomers, and overcoming the drawbacks of magnetorheological liquids, such as easy settling and poor stability. This incredible tunability makes magnetorheological elastomers a favorable application for semi-active vibration control in mechanical systems. Based on the superior stiffness damping controllability of magnetorheological elastomers, domestic and international scholars have made various attempts on magnetorheological elastomer vibration isolators for different practical engineering applications. Fengrong Bi et al. from Tianjin University designed a variable stiffness and damping damper based on MRE, which has obvious variable characteristics [16]. Fu et al. from Chongqing University developed a shear-compression composite mode MRE vibration isolator, demonstrating better performance in the MRE composite operating mode [17]. M.D. Christie at the University of Wollongong, Australia, developed an MRE vibration isolator for C-leg joints and measured that the robot leg can achieve up to 48% change in stiffness through a proprietary experimental platform built [18]. Weijia Ma et al. from Nanjing University of Science and Technology proposed an on–off control strategy for sweeping excitations and demonstrated good vibration isolation effects through experimental studies based on the frequency-shifting characteristics of magnetorheological elastomeric vibration isolator [19]. All of the above MRE isolators have excellent performance in transverse or longitudinal isolation or have been optimized for specific areas. It has also been demonstrated that great variable stiffness and variable damping characteristics can be achieved with composite directional vibration isolators.

The work proposed a composite bidirectional MRE vibration isolator designed for the vibration and impact problem of the robot after contact with the ground, using a magnetorheological elastomer, an intelligent material, to be installed on the foot of the robot. The stiffness and damping of the vibration isolators were adjusted in real time according to the vibration and shock they were subjected to, improving the humanoid robot's impact resistance to complex environments. Performance tests, dynamics modelling and semi-active control of the MRE isolator were carried out in this paper. In order to simulate the actual situation of the robot walking, this paper completed a virtual prototype-based dynamics simulation to analyze the vibration and impact on the robot foot. It is adapted to the semi-active vibration isolator designed in this paper to provide a new solution to the robot vibration damping problem.

2. Simulation of Bipedal Robot Dynamics

The inverted pendulum model is a common model-based approach to gait planning for bipedal robots [20]. Bipedal robots generally have hip and knee joints, as well as ankle joints. In order to obtain the vibration and impact signals of the foot of the bipedal robot, it is modeled in a simplified way, and the gait planning of the bipedal robot is based on the inverted pendulum model to access the corner curves of its joints. The dynamics simulation in Adams was finished, following adding constraints, drives, sensors and other steps. The scaling parameters for the dynamics simulation are shown in Table 1. Setting the simulation time for the humanoid robot model, we can obtain the key frames for a single step cycle of the gait simulation, as shown in Figure 1 [21].

Table 1. Kinetic scaling parameters in Adams.

Parameters	Values and Units
Stiffness factor	10^5 (N/mm)
Force Index	2.2
Damping factor	10 (Ns/mm)
Penetration depth	0.1
Coefficient of static friction	0.5
Coefficient of dynamic friction	0.3
Size of each foot	$10 \times 5 \times 3$ cm^3
Mass of each foot	100 g
Simulation time	9.6 s

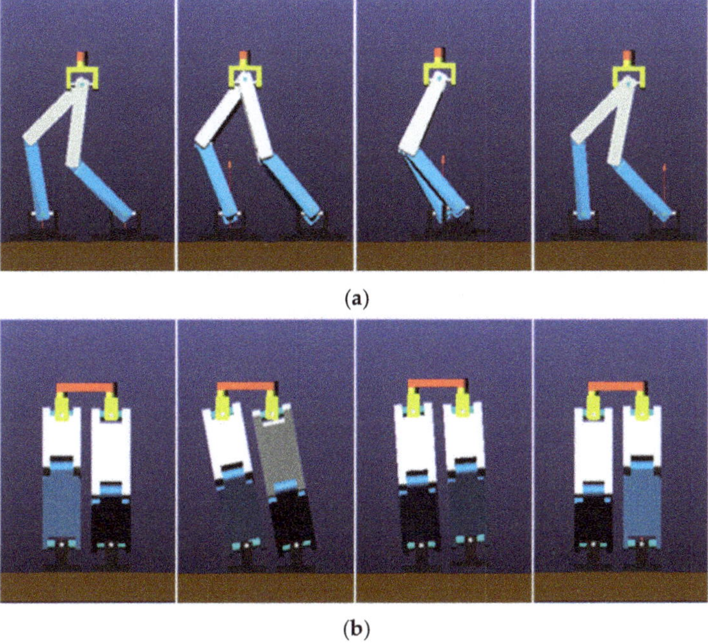

Figure 1. Kinetic simulation keyframes: (**a**) forward plane and (**b**) lateral plane.

The data curves for the contact forces between the robot's foot and the ground can be calculated through the Adams simulation analysis in Figure 2. It can be concluded from the figure that the humanoid robot has a periodic contact force with the ground during its movement and that the soles of its feet are mainly subjected to forces in the vertical direction. There are sharp peaks in the horizontal and vertical curves due to the gravity on the bipedal robot itself, relative speed of the foot to the ground and the friction between the foot and the ground when the robots first enter the unipedal support phase [22]. The walking dynamics simulation in this paper was carried out for a single step cycle, so during walking, the movement pattern of the left and right foot is symmetrical, and therefore, the curves of the left and right foot are similar in shape and have the same sharp peaks. Only the curves of the left and right foot have a phase difference in time. In the vertical direction, the contact reaction force on the left foot is the same as that on the right foot during the bipedal support period, while during the unipedal support period, the contact reaction force on the swing leg drops to zero, and the contact reaction force on the support leg, which bears the entire mass of the humanoid robot, is approximately doubled. The force on the foot in the horizontal direction may be related to frictional forces. In the simulation environment, there are vector marks representing the forces. These vector marks appear in approximately opposite directions. Set one of them in the positive direction and the other in the negative direction. The sharp peaks in Figure 2b are not strictly symmetrical, with some deviations owing to positive and negative directions artificially defined. Taking the foot–ground contact force as the input signal, we can get plantar vibration and impact signals, as shown in Figure 3.

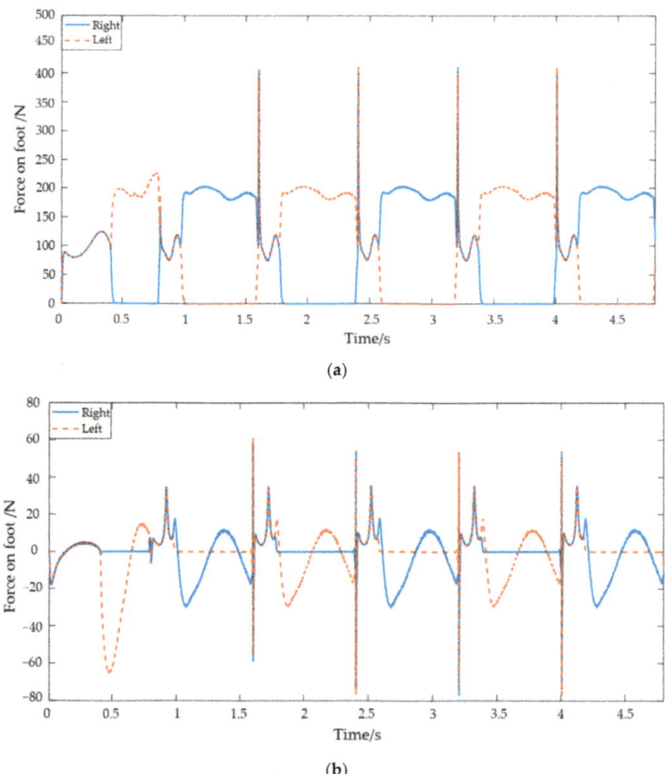

Figure 2. Humanoid robot with ground contact forces: (**a**) vertical direction and (**b**) horizontal direction.

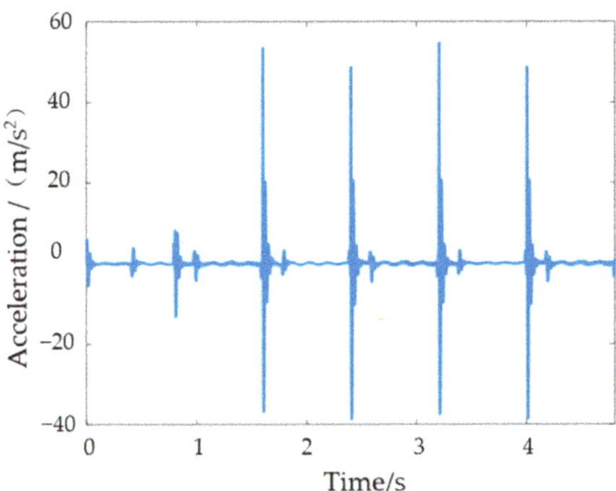

Figure 3. Plantar vibration and impact signals.

3. Design and Experimentation of MRE Vibration Isolators for Robot Feet

3.1. Magnetorheological Elastomer-Based Vibration Isolator Design and Vibration Damping Performance Testing

According to the vibration situation of the robot's feet, a vibration isolator based on the magnetorheological elastomer is designed. Given the relationship between the direction of motion of the magnetorheological elastomer in the vibration isolation support and the direction of the applied magnetic field, the two main modes of operation are the shear mode and pull–press mode, the principle of which is shown in Figure 4. As relevant studies show, the modulus of the magnetorheological elastomers in the pull–compression mode in the zero-field case is twice that in the shear mode [23,24], while the amount of change in the compressive and shear module is essentially the same for the same magnetic field strength. The MRE in shear mode of operation allows for greater magnetorheological effects. Meanwhile, the MRE in the pull–pressure mode has a greater load-bearing capacity.

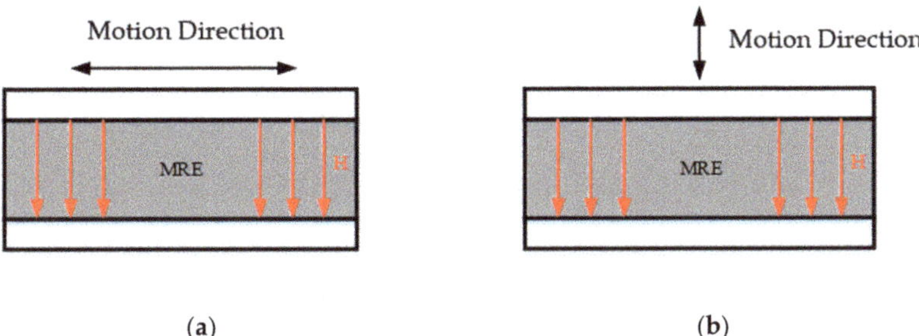

Figure 4. Operating modes of the magnetorheological elastomers: (**a**) shear mode and (**b**) pull–press mode.

In this manuscript, carbonyl iron powder (type: MPS-MRF-15) is chosen as the magnetic particle. Silicone rubber (type: RTV-704) with good adhesion and aging resistance is chosen as the elastic matrix, and dimethyl silicone oil is chosen as the auxiliary additive. The carbonyl iron powder, dimethyl silicone oil and 704 silicone rubber were first mixed in a beaker at a mass ratio of 8:1:1 and dispersed well using a mixer. The mixture was then placed in a vacuum drying oven to extract the air inside from the stirring process in three stages, followed by pouring the defoamed mixture into a mold and placing it in the drying oven again to extract the air three times and, finally, covering the mold with a lid and placing it in a magnetizing device at 800 mT for 2 h to pre-structure the anisotropic MRE material.

The force on the foot during robot walking is a multidirectional force, which inevitably leads to vibration and shock in multiple directions. Conventional MRE isolators for one direction only are not a reasonable solution to this problem. This manuscript therefore attempts to design a composite bidirectional MRE isolator, which is capable of isolating both vertical and horizontal directions. The working principle of the designed composite bidirectional MRE isolator is expressed in Figure 5. The vertical vibration isolation of this composite bidirectional MRE is accomplished by a hollow circular MRE and a disc-shaped MRE (i.e., disc MRE1). A physical diagram of the MRE material is shown in Figure 6. When the vibration isolator is subjected to forces in the vertical direction, the disc MRE1 will operate in the pull-press mode, which gives the vibration isolator a greater load carrying capacity in the vertical direction, while the circular MRE, which is fixed between the upper and lower coils by hexagonal fastening screws, will operate in the shear mode in a similar way, which has a greater magnetorheological effect than the disc MRE1. In other words, the range of variation in stiffness and damping will be more notable. These two magnetorheological elastomers enable the vibration isolator to possess a large load-bearing capacity in the main vibration isolation direction and a large range of stiffness and damping variation, making it well-suited to large load-bearing and large impact conditions. Meanwhile, the horizontal vibration isolation, which is relatively weak in terms of vibration and shock, is completed by another disc-shaped MRE (i.e., disc MRE2). Its operation in shear mode allows it to solve the problem of vibration isolation in the horizontal direction to a certain extent. It is worth noting that MRE1 can hardly isolate horizontal vibrations, such as MRE2.

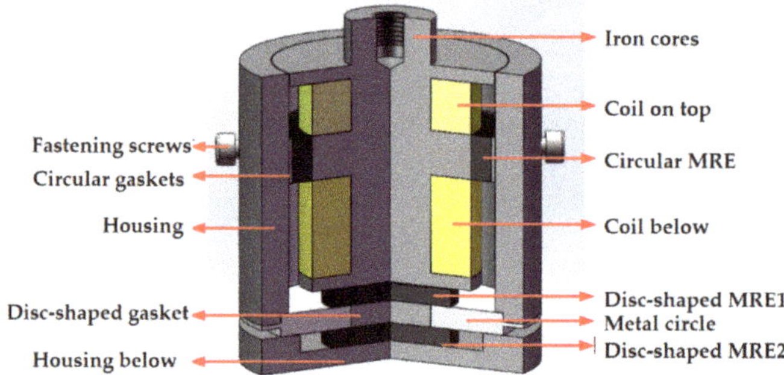

Figure 5. Composite bidirectional MRE vibration isolators.

Figure 6. Physical view of the MRE.

The bidirectional MRE isolator designed in this paper is a strictly cylindrical axisymmetric structure, so the simulation analysis is carried out by a one-half model in the X-Z plane. The material properties of each part are first defined via Maxwell's material library, followed by the selection of the appropriate air boundaries as boundary conditions. Then, we define the loading current density and the number of winding turns for the coil. Finally, the division of the mesh and the solution options setting are made, and then, the simulation can be carried out. The magnetic field distribution of the MRE vibration isolator the one-half model with a 2-A current applied is shown in Figure 7. In Figure 7, the magnetic circuit in the MREs meets the design expectations. In the one-half simulation diagram, the MREs are all represented by black rectangles. The center of the rectangles is selected as the location where the magnetic induction intensity is recorded, varying the current (0–2 A) for multiple simulations. According to the recording, the simulations in Maxwell-2D give the relationship between the coil current and the magnetic induction strength in the three MREs, as shown in Figure 8.

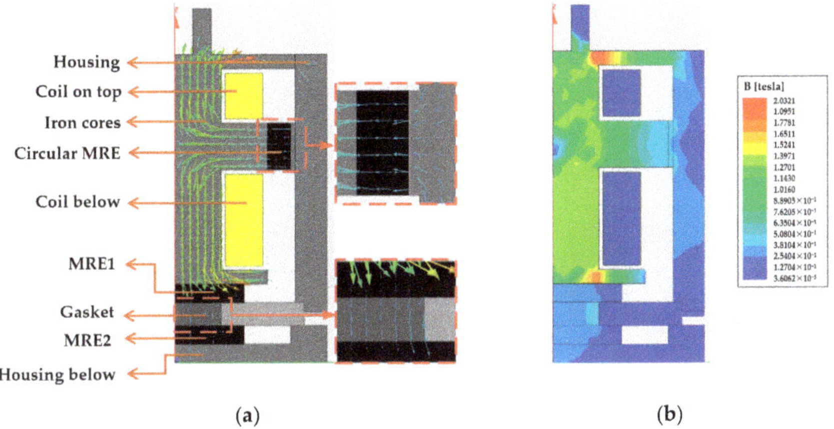

Figure 7. Simulation of the magnetic field of a vibration isolator under a 2-A applied current: (**a**) vector image and (**b**) magnitude image.

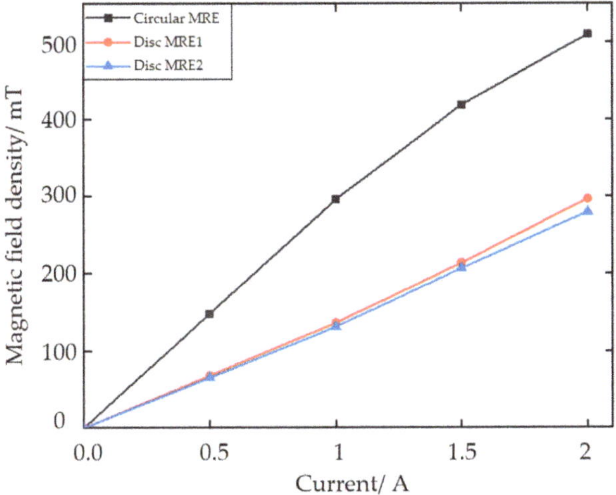

Figure 8. Trend of the magnetic induction strength with the current.

MREs are a type of viscoelastic material, which means both viscous and elastic deformation mechanisms exist. The mechanical properties of viscoelastic materials are shown in Figure 9 [25].

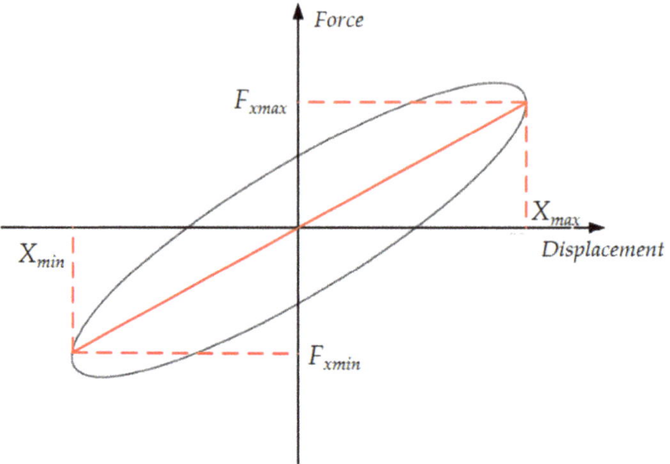

Figure 9. Force–displacement hysteresis loops.

Based on the mechanical properties, the following equations for equivalent stiffness and equivalent damping are deduced [26]:

$$K_{eff} = \frac{F_{xmax} - F_{xmin}}{X_{max} - X_{min}} \tag{1}$$

$$C_{eq} = \frac{S}{2\pi^2 f X^2} \tag{2}$$

where K_{eff} denotes the equivalent stiffness of the isolator, X_{max} and X_{min} denote the maximum and minimum displacement values of the MRE isolator and F_{xmax} and F_{xmin} denote the force output of the MRE isolator at the maximum and minimum displacement values, respectively. C_{eq} denotes the equivalent damping of the isolator, S denotes the area of the force–displacement hysteresis loop, f denotes the excitation frequency and X denotes the excitation amplitude.

In order to further establish the mechanical performance of the designed MRE vibration isolator, a performance test rig was set up for testing in both the vertical and horizontal directions. The sinusoidal excitation was used to perform the experiments in order to verify the vibration isolator' s damping and vibration isolation performance. Simulation experiments for random signals will be carried out in our future research. In addition, both the sinusoidal excitation and the foot–ground contact force are periodic signals. Therefore, experiments with periodic signals are more in line with the scenario of robots. During the test experiments in the vertical direction, the signal generator was to generate different sinusoidal excitations, while the current provided by the DC supply was increased from 0 A to 2 A. The slope of the long axis of the ellipse represents the equivalent stiffness of the isolator, and the area of the ellipse represents the equivalent damping of the isolator as the applied current increases, as shown in Figure 10 for the vertical direction. The equivalent stiffness and damping of the MRE isolator are therefore positively related to the applied current.

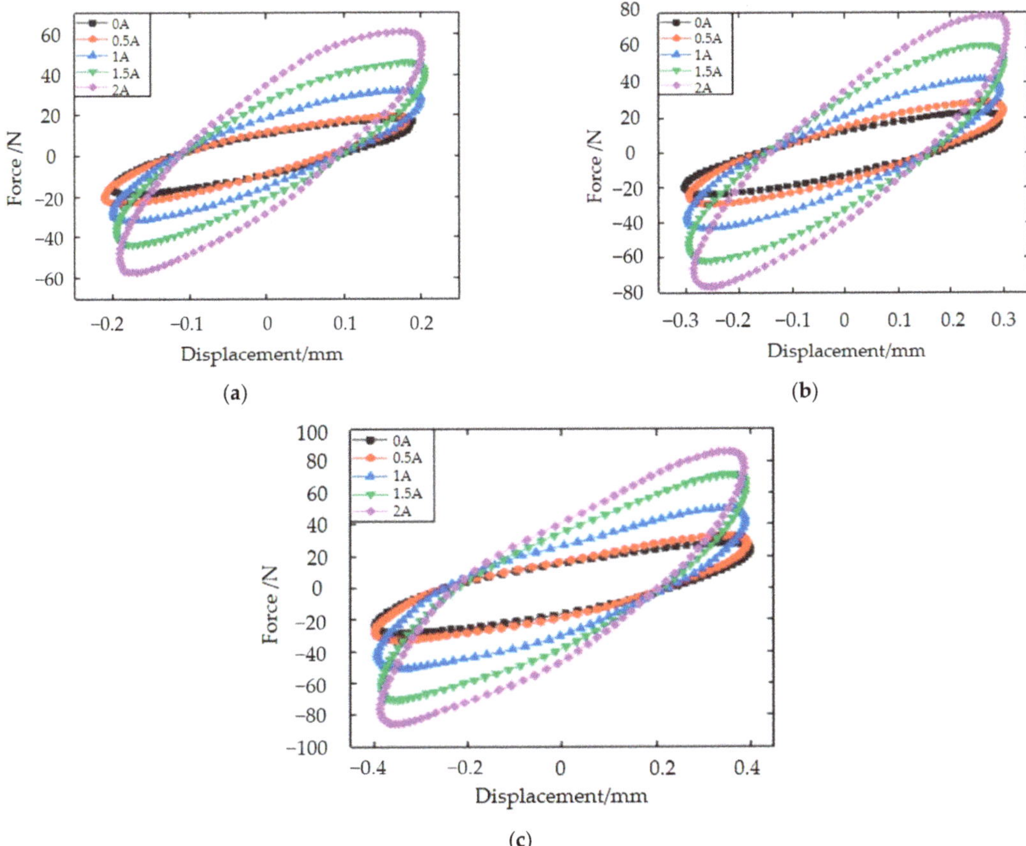

Figure 10. Force–displacement hysteresis loops in the vertical direction of the MRE isolator for 8-Hz excitation: (**a**) 8 Hz, 0.2 mm; (**b**) 8 Hz, 0.3 mm and (**c**) 8 Hz, 0.4 mm.

In the horizontal direction, aiming to test the performance of the MRE vibration isolator under more working conditions, the signal generator was to generate sinusoidal excitation frequency of 8 Hz, and the displacement amplitudes at each excitation frequency were 0.4 mm, 0.6 mm and 0.8 mm, respectively. The current supplied by the DC supply is from 0 A to 2 A. The 8-Hz sinusoidal excitation experiments are shown in Figure 11. Compared to the force–displacement hysteresis loop in the vertical direction, the overall trend is the same, with the equivalent stiffness and damping of the MRE isolator being positively correlated with the applied current, and the higher the current, the higher the peak under the same conditions. In a word, the results of this experiment are consistent with the theory. In the previous section on the design principles of the vibration isolator, a total of three MREs operate in the vibration isolator. The circular MRE and MRE1 are responsible for causing the stiffness and damping changes in the axial direction (vertical direction), and MRE2 is responsible for causing the stiffness and damping changes in the radial direction (horizontal direction). Therefore, there are two MREs operating in the axial direction and only one MRE in the radial direction. That is why the theoretical dependence of the axial direction on the current is more significant. This explains the much less current dependence in Figure 11. However, as can be seen from the two figures, the trend in the hysteresis loop is similar.

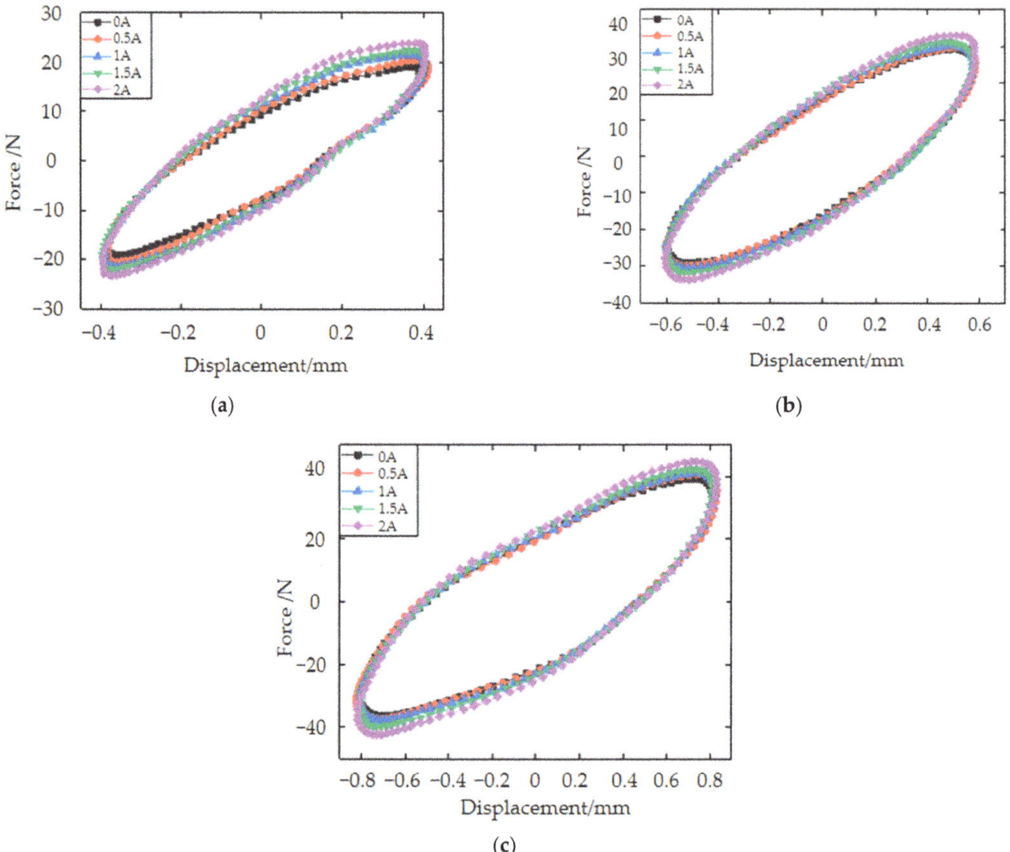

Figure 11. Force–displacement hysteresis loops in the horizontal direction of the MRE isolator for 8-Hz excitation: (**a**) 8 Hz, 0.4 mm; (**b**) 8 Hz, 0.6 mm and (**c**) 8 Hz, 0.8 mm.

Liu Tao et al. [27,28] proposed a shear mode MRE isolator. The isolator is applied to a unidirectional seismic isolation system and has almost no variable stiffness characteristics in the vertical direction of the isolation direction, although it has a good dynamic response in the isolation direction. Compared to that isolator, our isolator has a dynamic response in both directions and can be controlled by the appropriate algorithm to achieve variable stiffness and variable damping in both directions. In terms of the vibration isolation direction, the originality is conspicuous.

3.2. Dynamical Modeling of the Magnetorheological Elastomeric Vibration Isolator

Both simulation and semi-physical experiments require a dynamic model of the vibration isolator. In this paper, the Kelvin model is selected to describe the mechanical properties of the MRE vibration isolator, which mainly equates the viscoelastic material as a combination of spring and viscous pot units. In conjunction with the foot–ground collision in the previous gait simulation of robot walking, the MRE vibration isolation system is modeled as shown in Figure 12.

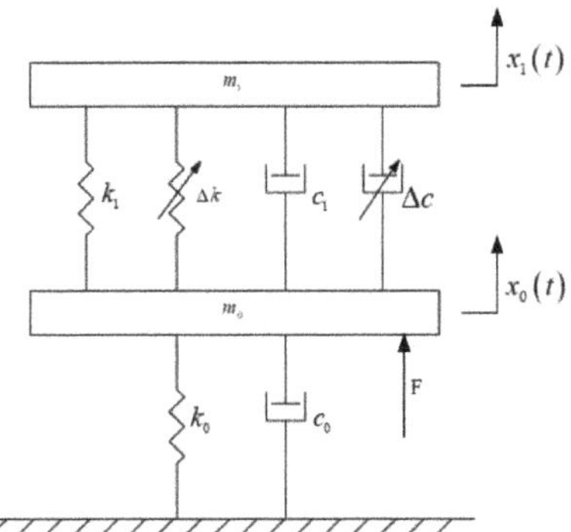

Figure 12. Dynamical model of the MRE vibration isolation system.

From the dynamics model, the equations of motion for this MRE vibration isolation system can be obtained as follows:

$$\begin{cases} m_1\ddot{x}_1 = k_1(x_0 - x_1) + c_1(\dot{x}_0 - \dot{x}_1) + F_c \\ m_0\ddot{x}_0 = F - F_c - k_0x_0 - c_0\dot{x}_0 - k_1(x_0 - x_1) - c_1(\dot{x}_0 - \dot{x}_1) \\ F_c = \Delta k(x_0 - x_1) + \Delta c(\dot{x}_0 - \dot{x}_1) \end{cases} \tag{3}$$

where m_0 denotes the unsprung mass, i.e., the part of the robot below the MRE isolator, and m_1 denotes the sprung mass, i.e., the sum of the mass of the robot body and the mass of the core in the MRE isolator. k_0 and c_0 denote the stiffness and damping coefficients set during the collision between the robot's foot and the ground in the gait simulation, respectively. k_1 and c_1 denote the initial equivalent stiffness and damping of the vibration isolator without an applied magnetic field. Δk and Δc denote the equivalent stiffness and damping of the vibration isolator as the magnetic field changes after a magnetic field is applied. $x_0(t)$ denotes the excitation displacement, i.e., the vibration displacement of the robot after vibration and shock, and $x_0(t)$ denotes the response displacement, i.e., the vibration displacement transmitted to the robot body through the MRE isolator. F denotes the foot–ground contact force obtained in the previous simulation, and F_c denotes the magnetic field control force generated when the isolator is energized, which is obtained by controlling the equivalent stiffness and damping of the MRE isolator.

4. Simulation Experiments of Control Algorithms Based on Foot–Ground Signals

The vibration isolator model is known to be in contact with the robot's foot–ground signal. Simulated vibration isolation experiments are accomplished in Simulink for three algorithms and compared to the vibration isolator output for the no-isolation case.

4.1. Fuzzy Control

Fuzzy control is robust and well-suited to the control study of MRE vibration isolators. The design flow of the fuzzy control focuses on the design of its controller, which is built in Simulink in a block diagram, as shown in Figure 13. The input variables are divided into seven classes: NB, NM, NS, ZO, PS, PM and PB, while the applied current as an output variable is divided into only four fuzzy classes: ZO, PS, PM and PB, as there are no negative values. The input affiliation function is of Gaussmf type, with the domain set to $[-2, 2]$, and the output affiliation function is of trimf type, with the domain set to $[0, 2]$. As both input variables have seven fuzzy levels, the fuzzy rules should be designed as 49 levels. The design process is based on the corresponding theoretical relationships analyzed from the experience and experimental results of various scholars in the field of vibration control [29].

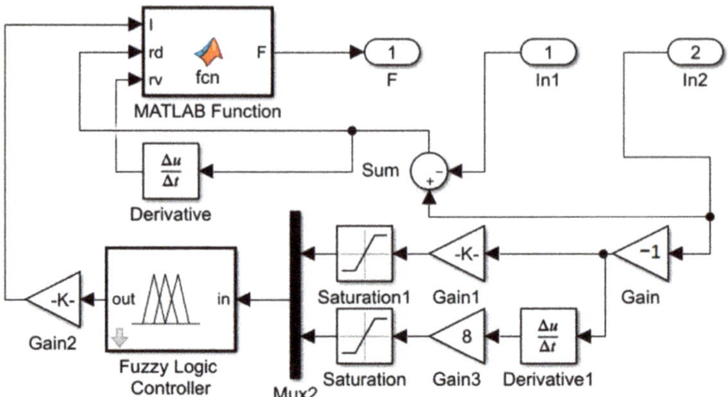

Figure 13. Fuzzy controller block diagram.

4.2. PID Control

The PID control law adopted for the vibration isolators in this paper is as follows:

$$u = k_P \left[e(t) + \frac{1}{T_I} \int_0^t e(t) + \frac{T_D de(t)}{dt} \right] \tag{4}$$

where *e(t)* denotes the deviation, i.e., the difference between the actual output and the theoretical value. k_p, k_i and k_d denote the scale factor, the integral time constant and the differential time constant, respectively. The magnetically induced control force of the MRE isolator cannot meet the control force output by the PID controller at all times, so the control force output by the PID must be limited, as shown in Equation (5):

$$u = \begin{cases} 0 & u \cdot F \leq 0 \\ u & u \cdot F > 0 \,\&\, abs(u) \leq abs(F_{\max}) \\ F_{\max} & u \cdot F > 0 \,\&\, abs(u) > abs(F_{\max}) \end{cases} \tag{5}$$

where u denotes the control force output from the PID controller, and F_{max} denotes the maximum magnetically induced control force of the vibration isolator.

4.3. Fuzzy PID Control

The fuzzy PID control algorithm adjusts the parameters of the PID in real time through the inference capability of the fuzzy controller. Fuzzy PID control algorithms come in various forms, but their basic operating principle remains the same. The fuzzy PID algorithm uses deviation e and the deviation change rate *de/dt* as the input of the fuzzy controller, output k_p, k_i and k_d and reasoning using fuzzy control rules to rectify the three parameters of the PID in real time, and eventually, the PID controller completes the real-time control of the controlled object. The setup of this control algorithm is similar in principle to the first two, where the fuzzy inference rules draw on the practical experience of relevant experts, and the rules for the three parameters are similar. A block diagram is therefore established in Simulink as follows in Figure 14 [30]:

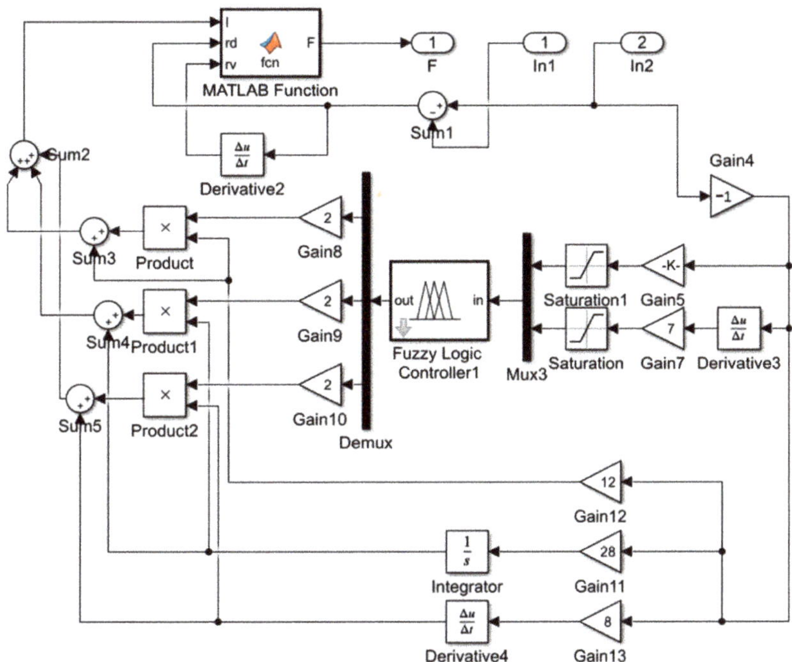

Figure 14. Fuzzy PID controller block diagram.

4.4. Semi-Active Control Simulation Results and Analysis

Based on the foot contact force curve and the MRE vibration isolator dynamics model, a block diagram of the integrated MRE vibration isolation control system was constructed, as shown in Figure 15. The results after the four control methods are shown in Figure 16, and their specific details are shown in Table 2.

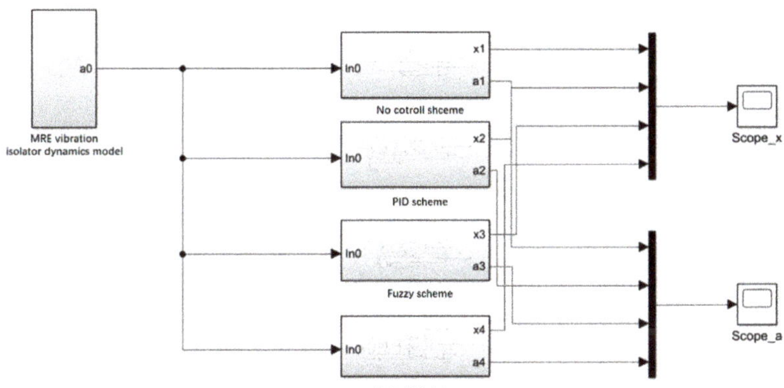

Figure 15. Block diagram of the MRE vibration isolation control system.

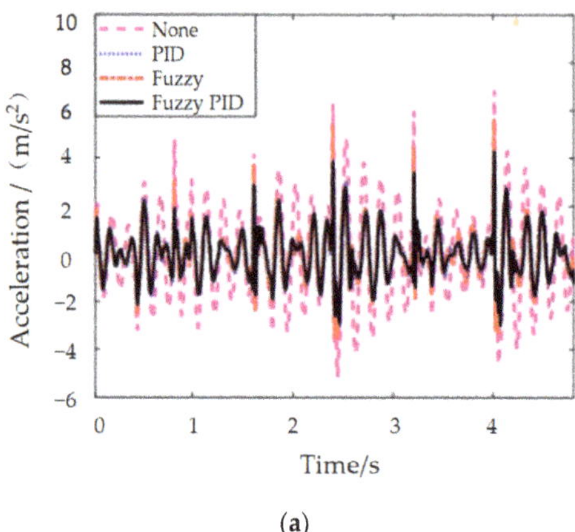

(a)

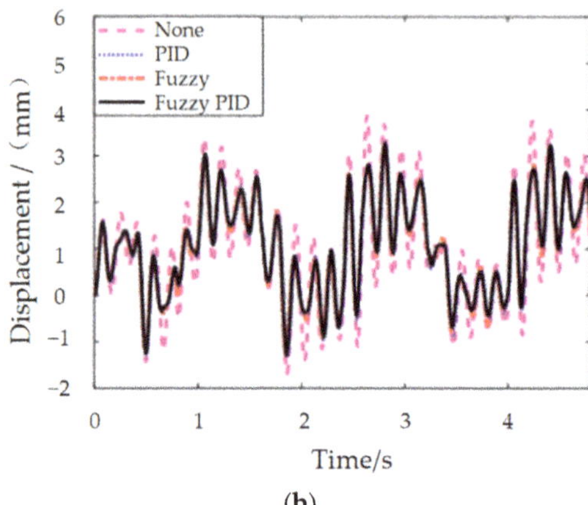

(b)

Figure 16. MRE vibration isolator response signal: (**a**) acceleration and (**b**) displacement;.

Table 2. Response of the vibration isolators under vibration signals.

	None		PID		Fuzzy		Fuzzy PID	
	Peak	**RMS**	**Peak**	**RMS**	**Peak**	**RMS**	**Peak**	**RMS**
displacement	3.871	1.604	3.214	1.4452	3.259	1.4398	3.289	1.4536
acceleration	6.872	1.978	4.266	1.0076	5.691	1.0623	4.267	0.997

The analysis of the results in conjunction with the above leads to the following conclusions:

(1) Overall, the MRE vibration isolator seems to have a fabulous damping effect in the robot walking process. The peak acceleration and root mean square (RMS) values of the robot foot vibration signal are reduced by more than 85% under the MRE vibration isolator's processing. The vibration displacement amplitude is within 4 mm, which means it has a small impact on the robot control accuracy.

(2) From the acceleration control effect, comparing several control algorithms, PID control reveals the best control effect on the acceleration peak, followed by fuzzy PID control, while fuzzy control has the third-place effect. The acceleration RMS value is the best effect of fuzzy PID control algorithm, followed by PID control, and the third-place is still fuzzy control.

(3) In terms of displacement control, the PID control algorithm is the least effective, the fuzzy control is the second-most effective, and the fuzzy PID has the highest peak displacement. This is because, in buffered damping, in order to cope with the impact of higher acceleration, a certain amount of vibration amplitude is demanded to better offset the impact effect, i.e., by increasing the displacement in exchange for a tinier acceleration. Therefore, fuzzy and PID control of the displacement control is more expressive than fuzzy PID.

5. Hardware-in-the-Loop-Simulation of Vibration Isolator Damping Control Based on Foot–Ground Signal

To investigate the rationality of the control algorithm, a test bench for vibration isolation systems based on dSPACE is built, as shown in Figure 17. dSPACE is a semi-physical real-time simulation system that can directly interact with MATLAB/Simulink modules. The software part is mainly responsible for the controller design, automatic code generation and download, including the visualization and management of experiments; the hardware part mainly consists of a high-performance processor board, a multi-channel high-speed A/D converter board and a high-resolution D/A converter board.

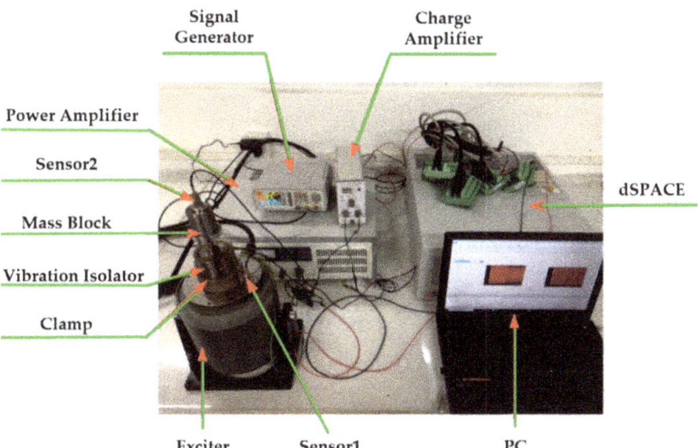

Figure 17. Experimental platform.

The experimental procedure for the whole vibration isolation control system is as follows: Firstly, the control system is set up in a computer using the MATLAB/Simulink R2018a module, as shown in Figure 18. The signal generator emits a sinusoidal waveform with a frequency corresponding to the foot–ground contact, which is amplified by a power amplifier to drive the exciter for excitation. The input acceleration to the controller is obtained by collecting the signal from sensor 1, while the signal from sensor 2 is collected by another ADC module, which is connected to a charge amplifier for conversion and then connected to the board for acquisition. The current–voltage conversion is based on the coil resistance of the MRE isolator. The output of the board is connected to the winding coil of the MRE isolator to complete the control of the applied current. The software module completes the code-related task, and the simulation results are obtained in controldesk, as shown in Figure 19.

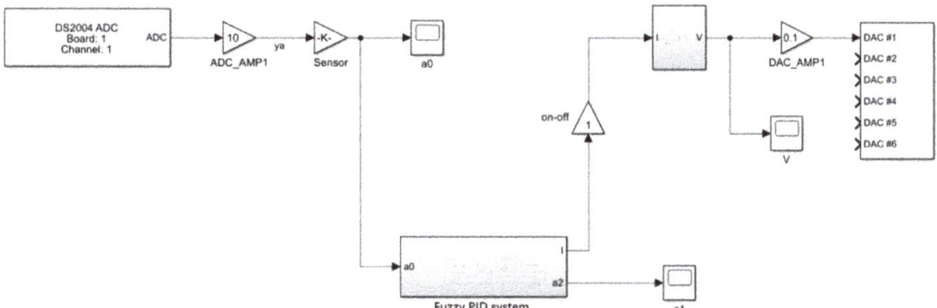

Figure 18. Block diagram of the control system.

Figure 18 shows the acceleration response curve obtained by the experiment. As can be seen from the figure, the response acceleration amplitude decreases to a certain extent at 10 s after the fuzzy PID controller control, and from the overall peak value, the acceleration peak value decreases from 0.5736 m/s^2 without the control to 0.3539 m/s^2 with the fuzzy PID control applied, which decreases by 38.3% and has a better control effect. The effectiveness and reasonableness of the fuzzy PID algorithm were confirmed.

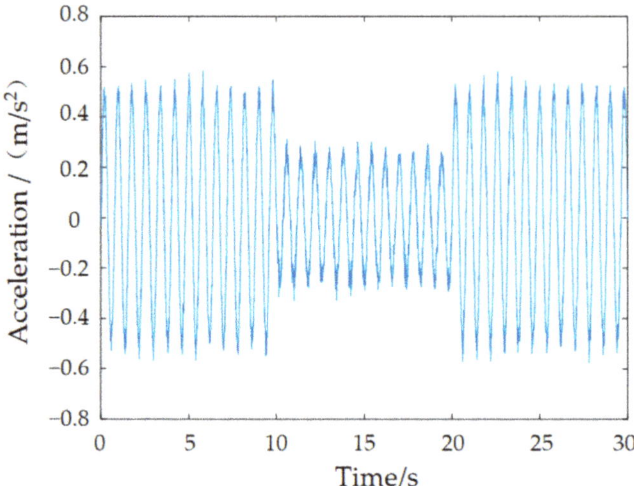

Figure 19. Response acceleration curve.

6. Conclusions

This paper presents a composite bidirectional MRE vibration isolator solution for the cushioning and vibration damping of a robot foot: a composite bidirectional MRE vibration isolator is designed using magnetorheological elastomers with high magnetorheological effects. Based on the vibration and impact signals to which the foot is subjected during walking, three controllers: fuzzy, PID and fuzzy PID, are designed, and the damping effects of the control algorithms are compared with a simulation analysis. The experimental platform of the vibration isolation control system is finally built by means of a dSPACE hardware-in-the-loop-simulation system, and the effectiveness of the MRE vibration isolator and the designed control algorithm is verified for sinusoidal excitation signals. The significances and impacts of this work are as follows:

Firstly, this manuscript proposes a dynamical approach to the gait analysis of a bipedal robot, providing an important vibration isolation perspective where the forces on the robot's foot cause vibrations. Secondly, this paper designed a bidirectional vibration isolator using MREs with high magnetorheological effects and conducted experiments to demonstrate its good bidirectional dynamic performance, offering a solution for scenarios requiring multi-directional engineering vibration isolation. Finally, three intelligent control algorithms with good results were designed in this paper, enriching the ideas of intelligent algorithm control.

Author Contributions: Conceptualization, X.H. and Y.Z.; methodology, G.H.; software, G.H.; validation, X.H. and Y.Z.; formal analysis, X.H.; data curation, Y.Z.; writing—original draft preparation, Y.Z.; writing—review and editing, X.H. and project administration, X.H. All authors have read and agreed to the published version of the manuscript.

Funding: This research was funded by the Natural Science Foundation of China, grant number 52075264.

Institutional Review Board Statement: Not applicable.

Informed Consent Statement: Not applicable.

Data Availability Statement: The data used to support the findings of this study are available from the corresponding author upon request.

Conflicts of Interest: The authors declare no conflict of interest.

References

1. Li, M. Research on Cushioning and Vibration Damping Mechanism of Goat Hoof Box and Design of Bionic Cushioned Foot for Footed Robot. Master's Thesis, Jilin University, Jilin, China, 2017.
2. Shen, J.H.; Ding, E. A study of the walking stability of a humanoid robot. *J. Harbin Eng. Univ.* **2004**, *04*, 536–539.
3. Jing, L.; Wang, Z.H.; Zhao, L.M. Progress in the study of the mechanical properties of porous metals and their sandwich structures. *Mech. Pract.* **2015**, *37*, 1–24.
4. Miao, H.B. Multivariate Coupled Cushioning Mechanism of the Canine Leg-Foot System and Its Bionic Study. Ph.D. Thesis, Jilin University, Jilin, China, 2020.
5. Ananthanarayanan, A.; Azadi, M.; Kim, S. Towards a bio-inspired leg design for high-speed running. *Bioinspiration Biomim.* **2012**, *7*, 46005. [CrossRef] [PubMed]
6. Grizzle, J.; Hurst, J.; Morris, B.; Park, H.W.; Sreenath, K. Mabel, a new robotic bipedal walker and runner. In Proceedings of the 2009 American Control Conference, St. Louis, MO, USA, 10–12 June 2009.
7. Yin, Y.W.; Zheng, P. Design and dynamic analysis of a flexible vibration isolation structure. *J. Dyn. Control.* **2021**, *19*, 16–24.
8. Cunningham, D.; Davis, P. A multiaxis passive isolation system for a magnetic bearing reaction wheel. *Adv. Astronaut. Sci.* **1993**, *95*, 80426.
9. Verrelst, B.; Ham, R.V.; Vanderborght, B.; Daerden, F.; Lefeber, D. The Pneumatic Biped "Lucy" Actuated with Pleated Pneumatic Artificial Muscles. *Auton. Robot.* **2005**, *18*, 201–213. [CrossRef]
10. Buchli, J.; Kalakrishnan, M.; Mistry, M.; Pastor, P.; Schaal, S. Compliant quadruped locomotion over rough terrain. In Proceedings of the 2009 IEEE/RSJ International Conference on Intelligent Robots and Systems, St. Louis, MO, USA, 10–15 October 2009.
11. Wang, S.K.; Shi, M.X.; Yue, B.K.; Xu, K.; Wang, J.Z. Adaptive impedance control based vibration isolation control for wheeled and footed robots. *Trans. Beijing Inst. Technol.* **2020**, *40*, 888–893.
12. Sun, J.; Wang, C.J. Design of multi-dimensional vibration isolator for vibration stress relief robot based on TRIZ theory. *South Agric. Mach.* **2022**, *53*, 1–4.
13. Sun, Y.; Sun, H.; Ma, S.S. Design of a kind of multi-dimensional attitude adjustment and vibration isolation platform based on 4-UPS/CPC parallel mechanisms. *China Mech. Eng.* **2021**, *32*, 1513–1522.
14. Wang, Y.F.; Wu, S.; Li, Z.X.; Xu, S. A study on vibration isolation based on a buckled beam quasi-zero-stiffness isolator and active damping. *J. Vib. Shock.* **2021**, *40*, 79–84.
15. Hou, Z.G.; Zhao, X.G.; Cheng, L.; Wang, Q.M.; Wang, W.Q. Research advances in rehabilitation robotics and intelligent assistive systems. *J. Autom.* **2016**, *42*, 1765–1779.
16. Bi, F.R.; Cao, R.K.; Wang, X.; Wang, J.; Ma, T. MRE-based variable stiffness variable damping damper design study. *Vib. Shock.* **2019**, *42*, 192–198.
17. Fu, J.; Li, P.D.; Liao, G.Y.; Lai, J.J.; Yu, M. Development and Dynamic Characterization of a Mixed Mode Magnetorheological Elastomer Isolator. *IEEE Trans. Magn.* **2017**, *53*, 1–4. [CrossRef]
18. Christie, M.D.; Sun, S.S.; Ning, D.H.; Du, H.; Zhang, S.W.; Li, W.H. A torsional MRE joint for a C-shaped robotic leg. *Smart Mater. Struct.* **2017**, *26*, 015002. [CrossRef]
19. Ma, W.J.; Huang, X.G.; Wang, H.X.; Zhang, G.; Wang, J. Vibration isolation control and an experimental study of magnetorheological elastomer isolators. *J. Vib. Shock.* **2020**, *39*, 118–122.
20. Zhang, W.; Du, J.H. Gait planning for bipedal walking robots. *Comput. Eng. Appl.* **2002**, *38*, 214–216.
21. Chen, N.; Lu, W. Gait smulation and results analysis of multi-articular biped walking robot in ADAMS environment. *South Agric. Mach.* **2021**, *52*, 1–6.
22. Lu, L. Bipedal Robot Modelling and Gait Planning Analysis. Master's Thesis, Shenyang University of Architecture, Shenyang, China, 2014.
23. Lu, C. Design Study of Magnetorheological Elastomers and Their Intelligent Vibration Isolators for Composite Service Conditions. Master's Thesis, Nanjing University of Science and Technology, Nanjing, China, 2016.
24. Du, G.L. Design of Composite Magnetorheological Elastomeric Vibration Isolators and Study of Their Control Systems. Master's Thesis, Nanjing University of Science and Technology, Nanjing, China, 2018.
25. Zhu, M. Research on Magnetorheological Elastomer Materials and Devices for Lateral Vibration Control. Master's Thesis, Chongqing University, Chongqing, China, 2016.
26. Przybylski, M.; Sun, S.S.; Li, W.H. Development and characterization of a multi-layer magnetorheological elastomer isolator based on a Halbach array. *Smart Mater. Struct.* **2016**, *25*, 105015. [CrossRef]
27. Liu, T.; Wang, H.X.; Ma, W.J. Dynamic model identification of MRE vibration isolators based on genetic algorithm. *Noise Vib. Control.* **2021**, *41*, 50–57.
28. Wang, H.X.; Liu, T.; Ma, W.J.; Zhang, G. Analysis of variable stiffness semi-active vibration isolation system based on MRE isolator. *J. Hunan Univ.* **2021**, *48*, 27–36.
29. Cheng, Z.L.; Pan, D.Y.; Xiao, P.; Shi, P.C. Analysis of vibration isolation performance of magnetorheological mounting system based on fuzzy control. *J. Chongqing Inst. Technol.* **2020**, *34*, 15–21.
30. Zhang, S.; Shen, C.W.; Liu, X.B.; Xie, M.J. Active vibration isolation technology based on fuzzy PID. *J. Chang. Univ. Technol.* **2020**, *41*, 67–72.

Article

Improving Aeroelastic Stability of Bladed Disks with Topologically Optimized Piezoelectric Materials and Intentionally Mistuned Shunt Capacitance

Xin Liu [1,2], Yu Fan [1,2,*], Lin Li [1,2] and Xiaoping Yu [3]

[1] School of Energy and Power Engineering, Beihang University, Beijing 100191, China; liu_xin_flying@buaa.edu.cn (X.L.); feililin@buaa.edu.cn (L.L.)
[2] Beijing Key Laboratory of Aero-Engine Structure and Strength, Beijing 100191, China
[3] China Academy of Aerospace Aerodynamics, Beijing 100074, China; xiaopingyu@buaa.edu.cn
* Correspondence: fanyu04@buaa.edu.cn

Abstract: It is well known that bladed disks with certain patterns of mistuning can have higher aeroelastic stability than their tuned counterparts. This requires small but accurate deviation of the mechanical properties on each blade sector, and currently it is difficult to realize by mechanical manufacturing. In this paper, we propose an adaptive strategy to realize the intentional mistuning for the improvement of aeroelastic stability. The basic idea is to bond or embed piezoelectric materials to each blade and use different shunt capacitance on each blade as the source of mistuning. When the shunt capacitance varies from zero (open-circuit, OC) to infinity (short-circuit, SC), the stiffness of each blade changes within a relatively small interval. In this way, the required small difference of stiffness among blades is altered into a relatively larger difference of the shunt capacitance. This provides a more feasible and robust way to implement the intentional mistuning, provided that the variation interval of blade stiffness between OC and SC contains the limits of required mistuning. Thus, it is critical to maximize the ability of changing the blade stiffness by shunt capacitance with limited amount of piezoelectric materials. To do so, a straightforward approach is proposed to get the best distribution of piezoelectric materials on the blade for the targeting mode. This approach is based on the FE model of the bladed disc, and the piezoelectric materials are introduced by replacing elements (if they are embedded) or adding an extra layer of elements (if they are bonded). An empirical balded disc with NASA-ROTOR37 profile is used as the example. With a proper design of the mistuning pattern and replace use piezoelectric materials of only 10% the blade mass, the proposed method can significantly improve the aeroelastic stability of bladed disks.

Keywords: aeroelastic stability; bladed disk; intentional mistuning; piezoelectric material; topological optimization

Citation: Liu, X.; Fan, Y.; Li, L.; Yu, X. Improving Aeroelastic Stability of Bladed Disks with Topologically Optimized Piezoelectric Materials and Intentionally Mistuned Shunt Capacitance. *Materials* **2022**, *15*, 1309. https://doi.org/10.3390/ma15041309

Academic Editor: Georgios C. Psarras

Received: 6 January 2022
Accepted: 2 February 2022
Published: 10 February 2022

1. Introduction

Improving aero-elastic stability to avoid flutter is an essential task for bladed disks in modern high-performance aero-engines. Previous studies [1,2] have proved the beneficial effects of mistuning on aeroelastic stability of cascades. It was frequently reported that mistuning can increase the minimum aerodynamic damping of the cascades while decreasing the maximum one [3,4]. Mistuning refers to slight mechanical difference among blade sectors, and it is unavoidable due to manufacturing tolerances and in-service wear [5,6]. Such intrinsic mistuning is randomly distributed and thus uncontrollable. As the frequency migration caused by mistuning is identified as the crucial factor influencing the aeroelastic stability [7,8], researchers have to intentionally impose certain mistuning pattern to maximum frequency migrations so as to achieve the best beneficial effects. Researchers have studied alternate patterns [9–13], sinusoidal patterns [14,15] and others [16].

Extensive experimental evidences have been reported regarding this topic, and the implementation of mistuning pattern is among the key techniques. Groth et al. [11] milled grooves the shroud to realize the intentional mistuning. Their results show that a well-designed mistuning can push the unstable boundary away from the turbine operating envelope. Figaschewsky et al. [17] applied heavy paint to the blades to realize the intentional mistuning. This is a non-destructive approach, but the passage of fluid may be narrowed. Adding mass to the tips of blades is also adopted to realize the intentional mistuning by some researchers [12,18]. Note that intentional mistuning requires small but accurate deviation of the mechanical properties on each blade sector. It can be a difficult task for mechanical manufacturing because it may prone to additional errors for small amount of manufacturing. Moreover, the best mistuning pattern may vary for different vibration modes, or at different working conditions. Pure mechanical implementation lacks the capability to adjust.

With the two-way energy transfer capability between the mechanical and electric fields, piezoelectric materials can be used as an adaptive way to tailor the mechanical properties by shunting different external circuits. For example, connecting a resistor to the electrodes is equivalent to adding an external viscous damper to the host structure. Likewise, a resistor-inductor circuit is equivalent to an external oscillator [19]. Finally, a capacitor is equivalent to an external grounded spring, thus changing the capacitance can adjust the natural frequencies of the host structure.

On this basis, we propose an alternative and more robust approach to implement mistuning. The basic idea is to bond or embed piezoelectric materials to each blade, and use different shunt capacitance on each blade as the source of mistuning. When the shunt capacitance varies from zero (open-circuit, OC) to infinity (short-circuit, SC), the stiffness of each blade changes within a relatively small interval. In this way, the required small difference of stiffness among blades is altered into a relatively larger difference of the shunt capacitance. This provides a more feasible and robust way to implement the intentional mistuning, provided that the variation interval of blade stiffness between OC and SC contains the limits of required mistuning.

The change in mechanical properties is limited by the model electromechanical coupling factor (MEMCF), which quantifies the energy exchange capability of a given piezoelectric structure. MEMCF is defined as the fraction of eigenvalue (square of natural frequencies) deviation from OC to SC. It can also be demonstrated that an external (positive) capacitance can only changes a modal frequency of the host structure from the values with OC to it with SC. Thus, maximizing the ability of changing the blade natural frequencies by shunt capacitance is equivalent to maximizing the MEMCF. This is critical especially when limited amount of piezoelectric materials are allowed to use in future engineering practice.

Existing studies [20–22] have pointed out that MEMCF (for a given mode) only depends on the number, shape, size, and location of the attached piezoelectric materials. Therefore, maximizing MEMCF is leading us to designing the geometrics of piezoelectric materials, as reported in the applications of modal sensing [23], actuating [24], energy harvesting [25], vibration mitigation [26–29], and so on. In these studies, the host structures to place piezoelectric materials are relatively simple and the researchers mainly takes position and direction of the given shape piezoelectric patches. For more complex structures like bladed disk, more factors such usage, shape, connected ways, arrangement method, etc., should be considered in the optimization process. Moreover, these methods are based on certain optimization processes such as the genetic algorithms, therefore modal analysis will be repeatedly conducted with updated design variables. Despite that modal analysis is much faster than forced response, for complex structures like bladed disks (the FE model can have millions of DOFs), it is still a heavy task.

In this work, an alternative approach is proposed and it only requires a single modal analysis in prior. First, we point out that MEMCF is also the fraction of electric energy over elastic potential energy associate with the structural mode. Consequently, piezoelectric materials should be placed in priority to the places with higher modal electric energy to achieve highest possible MEMCF. In this way, the shape of piezoelectric material is determined only by a single modal analysis, and we do not need to invoke any standard topological optimization algorithm. Iteration is avoided and the proposed approach has added very little computing load. This can be the major advantage of the proposed approach.

The claimed originality of the work is two-fold. The first one is using capacitance variation of piezoelectric shunt to modify the mechanical properties of each blade. To the authors knowledge, there is no other open literature exploring this idea than a preliminary work done by the authors [30]. This former publication has three shortages: (1) no unstable modes, namely, a stable working condition of the fluid field is selected; (2) the shape of piezoelectric materials is not optimized; and (3) capacitance is directly following the harmonic mistuning pattern rather than being determined by the harmonic mistuning pattern of mechanical properties. To further demonstrate the feasibility of such an original idea with more convincing data, we conduct the work reported in this manuscript and resolved all three aforementioned shortages. In particular, the topological design approach with just a single modal analysis can be attributed to the second original contribution of the conducted work. As mentioned in the previous review and the following detailed presentation, this design approach is significantly different from the existing method in current literature and it is especially suitable for the problem raised in this manuscript.

In later parts of this paper, an empirical bladed disk with NASA-ROTOR37 profile (Section 1) is introduced, and it is used as an example to illustrate and validate the proposed approach. The theory and procedure of the topological optimization approach are enclosed in Sections 3.1 and 3.2. Bladed disks with optimized distribution of piezoelectric materials are given at Section 3.3. The performance of intentional mistuning realized by the mistuning of external capacitance is presented in Sections 3.1 and 3.2. Eventually, the robust of this adaptive method in the presence of random mistuning is also examined (Section 3.3).

2. Problem Formulation

The bladed disk shown in Figure 1 is considered in this work. It consists of 36 blades with NASA-Rotor37 profile and a dummy disk. The material parameters of the bladed disk are as follows: modulus of elasticity 2.8×10^5 MPa, density 7.8×10^{-9} t/mm^3 and Poisson ratio 0.3. In the following analysis, the displacement at inner diameter of disk is constrained, in order to simulate the actual installation conditions.

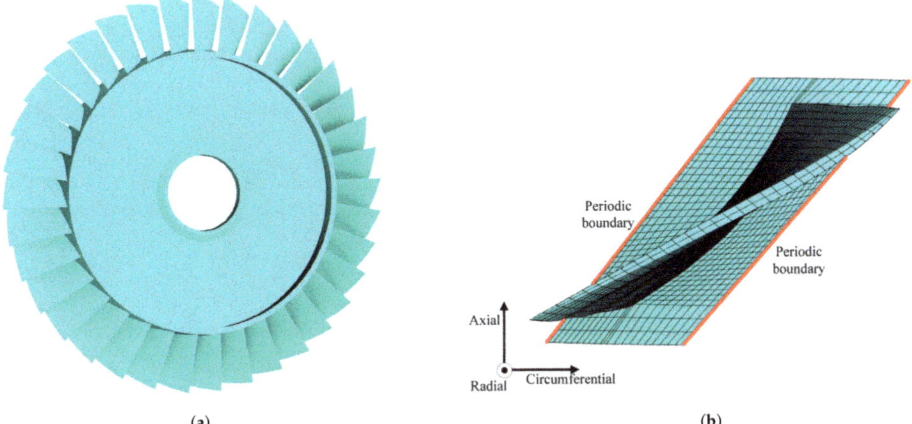

Figure 1. The FEM model of the Rotor37 bladed disk. (**a**) Overall model. (**b**) Sector model.

First, we conduct modal analysis of the tuned bladed disk with no fluid–structure interaction, as shown in Figure 2a. We have checked the mesh density and a finer mesh can only provide very minor improvements of the results. The modes with similar blade deformation but different engine-order (the number of nodal lines when the bladed disk vibrates) are classified to the same modal family, and they are link by the same line in Figure 2a. In this work, we use the aeroelastic stability of the first modal family to illustrate the proposed approaches. The modes in this family have similar frequency for they are all dominated by the blade bending deformation as shown in Figure 2b.

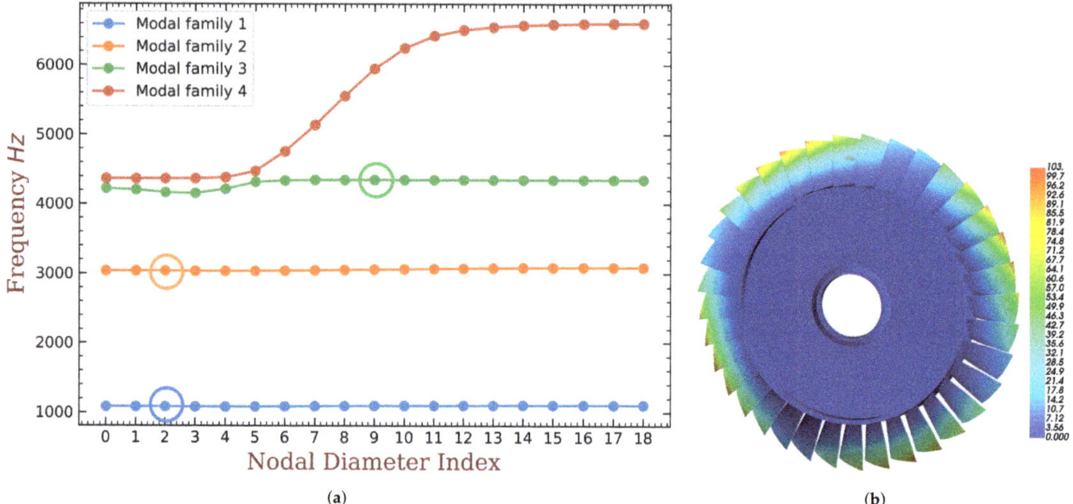

Figure 2. Modal analysis results of the empirical bladed disk. (**a**) The frequency versus nodal diameter index. (**b**) Modal shape marked with blue circle.

Second, we access the aeroelastic stability of the tuned bladed disk as a reference. To do this, aerodynamic influence coefficient (AIC) [31] is employed to model the aeroelastic force caused by the movement of blades, leading to a linearized dynamic equation of the bladed disk for free vibration:

$$\hat{\mathbf{M}}\ddot{\mathbf{y}} + (\hat{\mathbf{K}} + \mathbf{L})\mathbf{y} = \mathbf{0} \quad (1)$$

where $\hat{\mathbf{M}}$ and $\hat{\mathbf{K}}$ are mass and stiffness matrices of the bladed disk; $\mathbf{y}$ is the displacement vector. Matrix $\mathbf{L}$ is constructed by the AICs and it contains complex numbers. Analyzing the modal characteristics in this case will lead to a complex eigenvalue problem:

$$(-\hat{\omega}_j^2\hat{\mathbf{M}} + \hat{\mathbf{K}} + \mathbf{L})\mathbf{y} = \mathbf{0} \quad (2)$$

and the natural frequencies $\hat{\omega}_j$ may become complex values. Aerodynamic damping ratio ξ_j of the jth mode can be obtained by

$$\xi_j = -\frac{\mathrm{Im}(\hat{\omega}_j)}{|\hat{\omega}_j|} \quad (3)$$

The system become unstable if there are negative values of aerodynamic damping ratio. Therefore, we will use the minimum value of aerodynamic damping ratio among all the modes, denoted by ξ_{min}, as an indicator for the aeroelastic stability of the system. If $\xi_{\mathrm{min}} < 0$ then the system is unstable, and if ξ_{min} increases after some treatment we can conclude that the aeroelastic stability is improved.

This method is relatively mature and details can be found in the literature [32]. For the sake of brevity, we do not repeat the basics of this method but only presents the key results. Specifically, a numerical simulation of the flow field should be carried out to determine AICs, and the calculation domain is shown in Figure 3a. The performance characteristics of the bladed disk are analyzed and shown in Figure 3b. The lines refer to different operational rotation speed, and the speed is proportional to the design speed of the rotor. When the bladed disk operates under specific rotation speed, the pressure ratio would drastically decrease as the mass flow increase to the block margin. In addition, the pressure ratio would keep almost constant as the mass flow decreases to the unstable region. When the rotor operates close to the unstable region, the bladed disk is more prone to flutter.

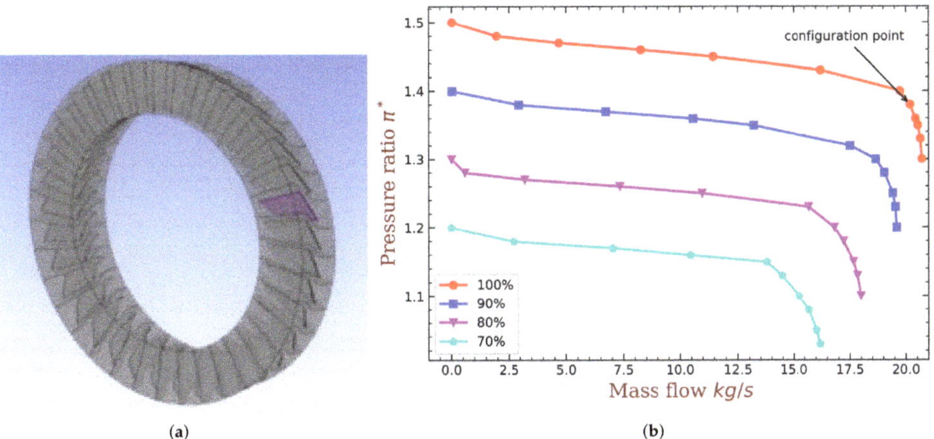

Figure 3. Outline of the CFD analysis of the bladed disk. (**a**) Flow field calculation domain. (**b**) The performance curves.

We choose the case (marked in Figure 3b) close to the unstable region as the working point to illustrate the feasibility of the adaptive method in improving the aeroelastic stability of the bladed disk. The boundary conditions and more details about the flow field settings can be found in our previous publication [33]. The aerodynamic influence coefficients for each blades when only the 1st blade is vibrating with the first bending mode are shown in Figure 4a. Thus, each AIC represents the aeroelastic force acted on each blade, generated by the movement of the 1st blade, and projected to the same modal coordinate (the first bending mode). The AICs presented in Figure 4a can be regarded as a row of matrix **L** in Equation (2). Changing between different moving blades can yield other rows of matrix **L**. Due to cyclic symmetry, all the remaining rows of matrix **L** can be generated by shifting the order of AICs presented in Figure 4a. The blade with larger distance to the reference (1st) blade has smaller aerodynamic force, and this is expected. Because of the blade torsion, the AIC values are not symmetric with respect to blade index 1, namely, AIC of blade 2 does not equal to blade 36. This means that **L** is not a symmetric matrix and the eigenvalues of Equation (2) are no longer double roots. The aerodynamic damping is computed and shown in Figure 4b. In this working condition, it is notable that there exists unstable modes with negative the aeroelastic damping. In our later investigation, we will use the proposed method to alleviate the unstable modes.

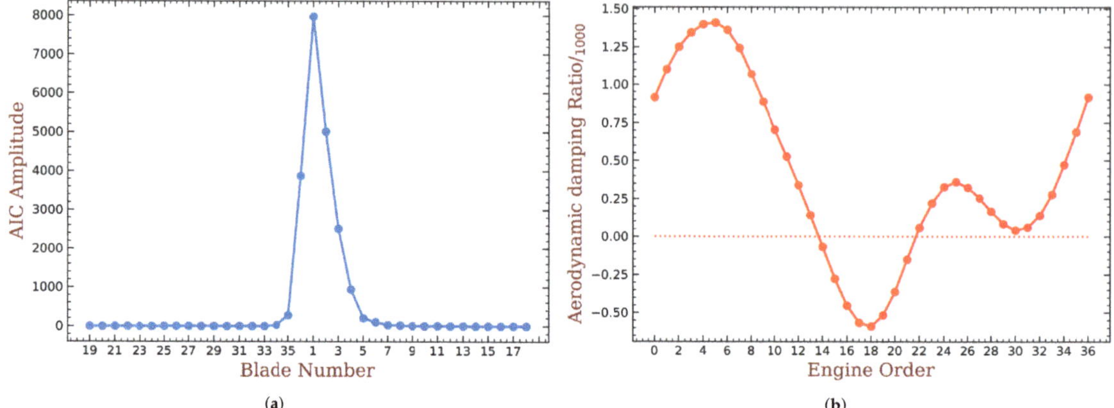

Figure 4. The aeroelastic stability analysis of the tuned bladed disk. (**a**) Aerodynamic influence coefficients. (**b**) Aerodynamic damping.

3. Topology Optimization for the Piezoelectric Materials

3.1. Theoretical Basis

A structure with piezoelectric materials has two coupled physical fields: the mechanical field and the electric field. The coupling strength between these two fields for the jth mode is quantify by modal electromechanical coupling factors (MEMCF), and it is defined as [22,34–36]

$$k_j^2 = \frac{(\omega_j^{\mathrm{OC}})^2 - (\omega_j^{\mathrm{SC}})^2}{(\omega_j^{\mathrm{SC}})^2} \tag{4}$$

where ω_j^{OC} and ω_j^{SC} are angular frequencies with open-circuit and short-circuit, respectively. We will reveal (1) how is this factor related to the maximum ability of external capacitance to change the natural frequencies and (2) how is this factor related to the geometric design of the piezoelectric materials.

Let us recall the dynamic equation of the piezoelectric structure with two electrodes (one voltage DOF):

$$\begin{aligned} \mathbf{M}\ddot{\mathbf{x}} + \mathbf{K}\mathbf{x} - \boldsymbol{\eta} V &= \mathbf{f}(t) \\ C_{\mathrm{p}} V + \boldsymbol{\eta}^{\mathrm{T}}\mathbf{x} &= Q(t) \end{aligned} \tag{5}$$

where **M** and **K** are the mass and stiffness matrices, respectively; η is the piezoelectric matrix; **x** is the displacement vector; V is the voltage between the electrodes; $\mathbf{f}(t)$ is external force vector; $Q(t)$ is the electric quantity of the circuit connected to the piezoelectric patch; and C_{p} is the intrinsic capacitance. Shunting an external capacitance C_{e} gives an additional equation:

$$Q(t) = -C_{\mathrm{e}} V \tag{6}$$

and thus $Q(t)$ in the second equation of (5) can be eliminated, and the equation becomes

$$(C_{\mathrm{p}} + C_{\mathrm{e}})V + \boldsymbol{\eta}^{\mathrm{T}}\mathbf{x} = 0 \tag{7}$$

If the external capacitance $C_{\mathrm{e}} = +\infty$, the voltage between the piezoelectric electrodes is zero and this makes the piezoelectric patch short circuit. Accordingly, Equation (5) becomes

$$\mathbf{M}\ddot{\mathbf{x}} + \mathbf{K}\mathbf{x} = \mathbf{f}(t) \tag{8}$$

Otherwise, if the external capacitance $C_{\mathrm{e}} = 0$, the piezoelectric patch is open circuit and Equation (5) becomes

$$\mathbf{M}\ddot{\mathbf{x}} + (\mathbf{K} + C_{\mathrm{p}}^{-1}\boldsymbol{\eta}^{\mathrm{T}}\boldsymbol{\eta})\mathbf{x} = \mathbf{f}(t) \tag{9}$$

It is clear that when external C_e varies in interval $[0, +\infty)$, it can only induce limited change to the stiffness and the natural frequencies can only vary in a limited zone. MEMCF is defined upon the maximum fraction of frequency changing.

Solving eigenvalue problem of Equations (8) and (9), respectively, we can get the angular frequencies ω_j^{OC} and ω_j^{SC}. Owing to the orthogonality of the piezoelectric structure modal shape in open-circuit and short-circuit, we can get the following expression:

$$\begin{gathered}\boldsymbol{\phi}_{\mathrm{OC},j}^{\mathrm{T}}\mathbf{K}\boldsymbol{\phi}_{\mathrm{OC},j} + V_j C_{\mathrm{p}} V_j = \omega_{oc,j}^2 \boldsymbol{\phi}_{\mathrm{OC},j}^{\mathrm{T}}\mathbf{M}\boldsymbol{\phi}_{\mathrm{OC},j} \\ [2ex]\boldsymbol{\phi}_{\mathrm{SC},j}^{\mathrm{T}}\mathbf{K}\boldsymbol{\phi}_{\mathrm{SC},j} = \omega_{\mathrm{SC},j}^2 \boldsymbol{\phi}_{\mathrm{SC},j}^{\mathrm{T}}\mathbf{M}\boldsymbol{\phi}_{\mathrm{SC},j}\end{gathered} \tag{10}$$

where $\boldsymbol{\phi}_{\mathrm{OC},j}$, $\boldsymbol{\phi}_{\mathrm{SC},j}$ are the jth modal shapes with open circuit and close circuit respectively. We assume that the changing of electrodes status does not result in significant structural deformation difference, namely,

$$\boldsymbol{\phi}_{oc,j} \approx \boldsymbol{\phi}_{sc,j} \tag{11}$$

This assumption is reasonable when the amount of piezoelectric materials are minor and it is the case in this paper.

Combining Equations (4) and (10), we can get

$$\frac{V_j C_{\mathrm{p}} V_j}{\boldsymbol{\phi}_{\mathrm{SC},j}^{\mathrm{T}}\mathbf{K}\boldsymbol{\phi}_{\mathrm{SC},j}} \approx \frac{(\omega_j^{\mathrm{OC}})^2 - (\omega_j^{\mathrm{SC}})^2}{(\omega_j^{\mathrm{SC}})^2} = k_j^2 \tag{12}$$

This indicates that MEMCF k_j^2 also represents the proportion of the electric energy in the mechanical energy when the piezoelectric structure vibrates in the form of $\boldsymbol{\phi}_{sc,j}$. In limited usage of the piezoelectric material, the geometry parameters of the piezoelectric materials, such as position and shape, do not change the mechanical energy of the structure significantly, but play a dominated role of the electric energy stored in the piezoelectric patches. The relation between the voltage, V, and the electric field intensity, E, is

$$V = Ed \tag{13}$$

where d is the distance between the electrodes of the piezoelectric patch. Based on the constitutive relationship of the piezoelectric material, the electric field intensity, E, is the linear summation of the strain in every direction:

$$E_i = \sum_{j=1}^{6} h_{ij} S_j \tag{14}$$

where h_{ij} is piezoelectric constant and i refers to the direction of the electric field. The local coordinate system defined by the polarization direction of the piezoelectric material is shown in Figure 5a, the constitutive relationship of the piezoelectric material is expressed in the local coordinate system. For instance, when the polarization direction is z (3), the direction of the electric field is set up in the direction z (3), which means the electric field intensity E in Equation (13) is E_3.

Therefore, based on the strain distribution of the structure, we can determine the place to arrange the piezoelectric material. Namely, piezoelectric materials should be placed at the area with large $|E_3|$ of the blade to achieve the highest possible modal electromechanical coupling factor. In addition, we assume that the modal strain distribution would not be significantly altered by the introduction of piezoelectric materials. Therefore, $|E_3|$ can be estimated by the modal strain field of the structure before the installation of piezoelectric materials. In this way, the priority places for the installation of piezoelectric materials is determined by a single modal analysis and minor additional computing (Equation (14)). The detailed procedure will be enclosed in the next section.

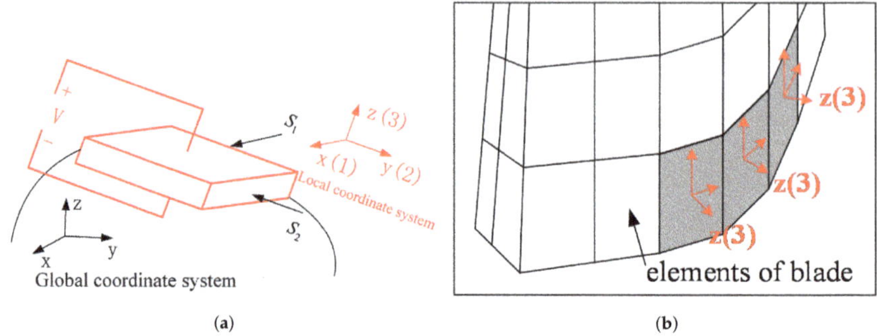

(a) (b)

Figure 5. Illustration of the local coordinates on which the constitutive equation of piezoelectric materials is defined. (**a**) Local coordinates and the polarization direction. (**b**) Local coordinates on the blade.

3.2. The Optimization Procedure

The topology optimization for the piezoelectric material (PZT-5H is used, as shown in Appendix A) on the bladed disk can be done based on FEM model. The piezoelectric materials are introduced by replacing elements (if they are embedded) or adding an extra layer of elements (if they are bonded). In order to not weaken the strength of the blades or not add too much weight to the whole structure, we set the ratio, R_m, of the piezoelectric material mass on a single sector, m_pzt, and the blade mass on a single sector (does not include disk), m_blade, as our design constrain:

$$R_\mathrm{m} = \frac{m_\mathrm{pzt}}{m_\mathrm{pzt} + m_\mathrm{blade}} \tag{15}$$

When the blades vibrate, the induced strain mainly happens along the blade surface. In the normal direction of the blade, the strain is small. Therefore, the working mode of the piezoelectric elements is '3-1' mode, so the electric field intensity E in Equation (13) is E_3. Moreover, the polarization direction of the piezoelectric elements is along the normal direction of the blade aera, and the normal direction is set to the outward of the blade surface (as shown in Figure 5b).

The objective function of the optimization is

$$\mathrm{Obj} : \max(\sum_{\mathrm{elem}} |E_3|) \tag{16}$$

As discussed, this can maximize the MEMCF and endow the largest possible capability for the an external capacitance to tailor the natural frequencies of the structure.

Based on the FEM model of the bladed disk, the position to place the piezoelectric elements is decided by the element strain. The optimization procedure is given as follows (shown in Figure 6):

1. Conduct modal analysis of the tuned bladed disk and obtain its modal information. This can be done with the sector model plus periodic conditions as shown in Figure 1b. Our target modes are those dominated by the blade 1st bending deformation (1st modal family).
2. Extract the blade surface elements strain. It should be noted that the strain need to be calculated at the element local coordinate system, and the way to set local coordinate system is shown in Figure 5b. Use the strain to calculate the electric field intensity $|E_3|$ of each element based on Equation (13). Then, sort the element according to the electric field intensity $|E_3|$. Note that element strain is an averaged value from the distributed strain in the element. In principle, it could even generate a null field (for example in a cantilever beam, if the element is located at the neutral axis in a pure

bending). To avoid this abnormal situation, it is suggested to have a finer mesh so that the strain does not vary significantly inside a single element.

3. Choose the way to place the piezoelectric material.
    (a) If the piezoelectric materials are embedded to the blades. First, set the elements coordinate system of the blade surface elements. The z direction of the local coordinate system is along the outward normal direction of the blade surface. Then, replace the blade surface elements with the piezoelectric material elements. In ANSYS, this means changing element type from SOLID185 to SOLID226. The criterion is replacing the blade element based on the electric field intensity $|E_3|$ from the largest ones until the mass ratio R_{m} meets the condition.
    (b) If the piezoelectric materials are bonded to the blades. We generate new elements along the norm direction of the blade elements. Namely, we will create a new layer of elements on the installation place. The remaining operations are the same as when we embed the piezoelectric materials.
4. Simulate the electrodes. The electrodes applied on a continuous area of piezoelectric materials are modeled by coupling the voltage DOFs on the top and bottom surfaces.
5. Interconnect discontinued piezoelectric materials on the blade. After the optimization, there may be several discontinued areas of piezoelectric materials. We will interconnect their electrodes so there is only one port on each blade, as shown in Figure 7. To do this, the poling direction of some areas should be reversed so that all the areas have the same signs of charges. Based on the sign of electric field intensity E_3, we can judge the poling direction.

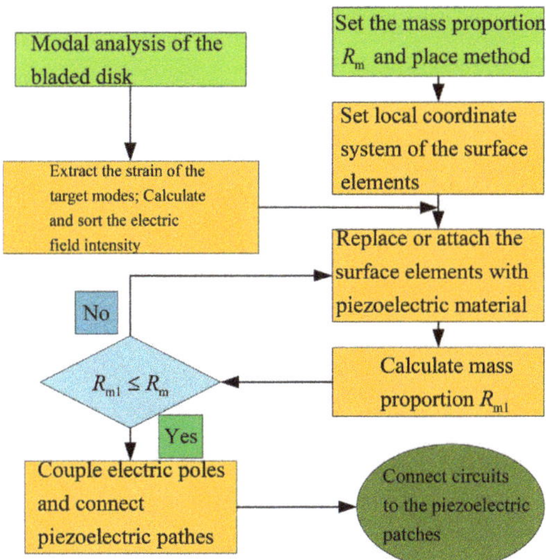

Figure 6. Optimization procedure of the piezoelectric materials distribution on the blades.

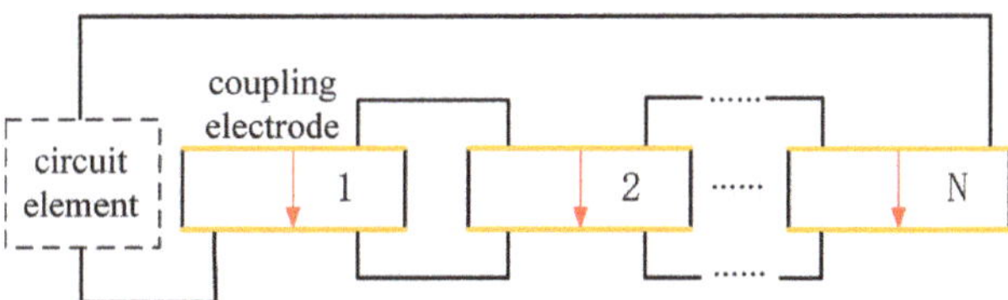

Figure 7. The electric connection of several distributed piezoelectric patches to the same circuits.

3.3. Optimized Distributions of Piezoelectric Materials on the Blade

Although our primary object is the first modal family (1st blade bending), the proposed algorithm can be used to any modes. To illustrate this, we optimize the distribution of piezoelectric materials for the first three modal families. The deformations shown in Figures 8a and 9a are the modal shapes in the second and third modal families, respectively. We can see that the second modal family of the blade disk is dominated by the blade 1st torsional deformation, and the third modal family is dominated by the blade 2nd bending deformation.

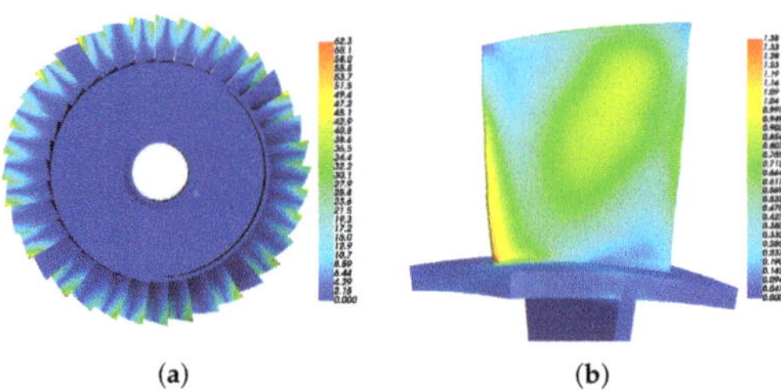

Figure 8. The modal deformation of the bladed disk dominated by the blade 1st torsional mode (marked with orange circle in Figure 2a). (**a**) Overall modal displacement. (**b**) Total strain on a single blade.

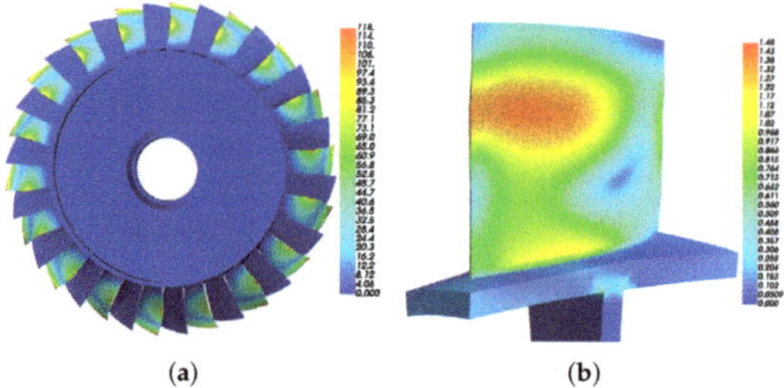

Figure 9. The modal deformation of the bladed disk dominated by the blade 2nd bending mode (marked with green circle in Figure 2a). (**a**) Overall modal displacement. (**b**) Total strain on a single blade.

The modal strain distributions are also given in Figures 8b and 9b. Note that the absolute values have no physical meaning, and we only use the relative values to determine the priority of locations to introduce piezoelectric materials. The priority indicator E_3 are computed according to Equation (16) and the results are shown in Figure 10. The optimized distributions of piezoelectric materials are shown in Figures 11 and 12, where different colors (red and purple) indicate the piezoelectric materials with opposite polarization directions. The best locations results follow the distribution of large E_3 values as imposed by the algorithms. Thus, the agreements between Figures 10 and 11 indicate that the algorithm is performed as expected. According to Equation (16), indicator E_3 can also be regarded as a weighted sum of the strain components. Thus, the distribution of E_3 should also be similar to (but not necessarily the same as) the distribution of strain. This remark can be verified by Figures 8b and 10a.

Eventually we optimize the distribution of piezoelectric materials for the first modal family, which is chosen to illustrate the performance of the proposed method in the later sections. The results are shown in Figure 13b for the embedded case. We can see that the suggested area is not completely located at the root area of the blade, but is expands from the root to center of blade. Such results can be explained by E_3 shown in Figure 13a. The suggested distribution for the bonded case are given in Figure 13c,d. We can also see that the placement method would not significantly change the distribution area of the piezoelectric materials.

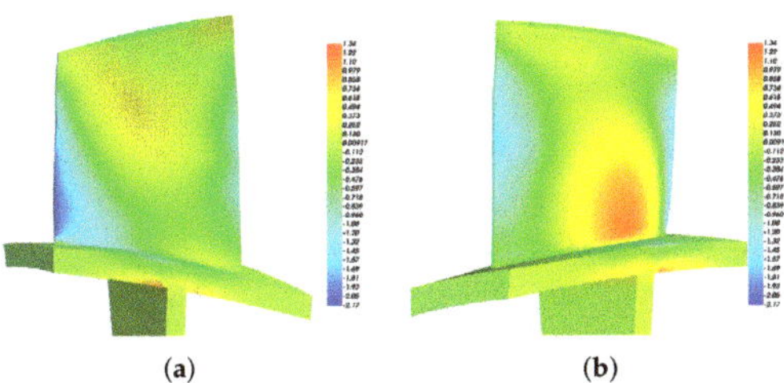

Figure 10. The electric intensity E_3 for the blade 1st torsional mode. (**a**) Pressure side. (**b**) Suction side.

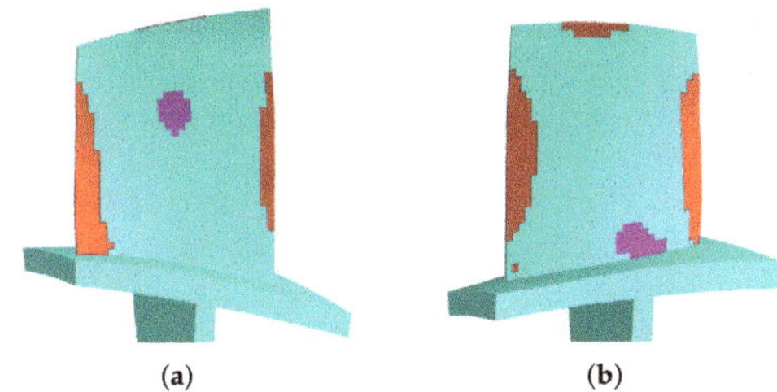

Figure 11. The optimized distribution of the embedded piezoelectric material for the 2nd modal family dominated by the blade 1st torsional mode (mass ratio = 10%). (**a**) Pressure side. (**b**) Suction side.

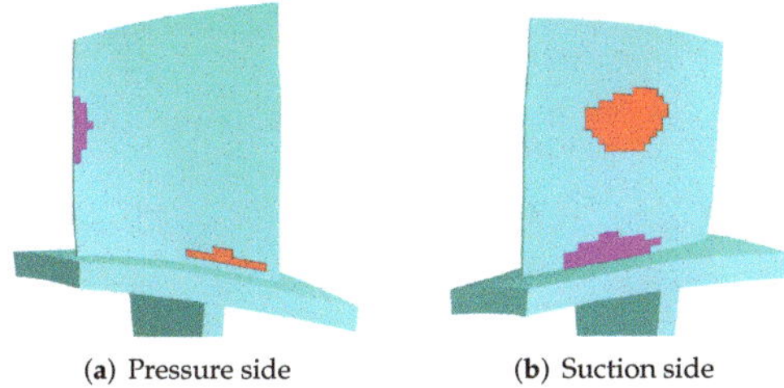

Figure 12. The optimized distribution of the embedded piezoelectric material for the 3rd modal family dominated by the blade 2nd bending mode (mass ratio = 10%). (**a**) Pressure side. (**b**) Suction side.

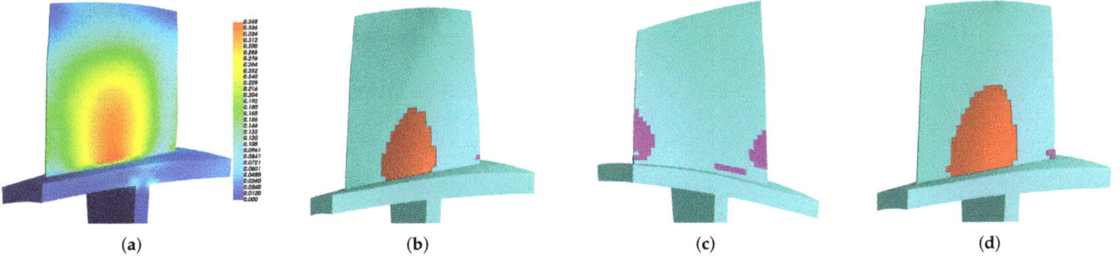

Figure 13. The optimized distribution of piezoelectric materials for the 1st modal family, mass ratio = 10%. (**a**) E_3, (**b**) Embedded Pzt (Suction side), (**c**) bonded Pzt (Pressure side), (**d**) bonded Pzt (Suction side).

The modal shapes with piezoelectric materials being open-circuit and short-circuit are compared with the original blades in Figure 14. Little visual difference can be found among them, and we can use modal assurance criterion (MAC) to quantify their similarity:

$$\mathrm{MAC}(\mathbf{a},\mathbf{b}) = \frac{|\mathbf{a}^{\mathrm{H}}\mathbf{b}|}{|\mathbf{a}||\mathbf{b}|} \tag{17}$$

where **a** and **b** are vectors whose similarity to be evaluated, superscript H refers to conjugate transpose. MAC varies from 0 to 1, and a closer value to 1 indicates a stronger similarity between **a** and **b**. Figure 15 summaries MAC values among the original, open-circuit and short-circuit modal shapes for the first three modal groups. We can see their MAC values are very close to 1. These results support the assumption that introduction of piezoelectric materials and the status of electrodes do not significantly change the modal shapes (when small amount of piezoelectric materials are used), as expressed in Equation (11).

Figure 16 demonstrates the strain distribution influence by the introduction of piezoelectric materials. The overall distributions of the train field are nearly the same when piezoelectric materials are introduced (bonding or embedding). Note that the edge of the piezoelectric area is determined by the edges of elements, so it is not smooth. Consequently, there will be stress concentration near the edges. This stress concentration only happens in the numerical simulation. In practical, one can choose to use the rounded or rectangular piezoelectric materials that can be purchased with ease, to cover the targeting area obtained by the proposed method. Alternatively, one can choose to customize the piezoelectric materials after smoothing the edges. On one hand, the similarity of the overall distribution before and after the introduction of piezoelectric materials is justifying the design procedure. On the other hand, it also explains why we use the modal strain field of the original blade to build our topological design. If we first introduce a small proportion of piezoelectric materials and the location of the remaining proportions are determined by the strain field of the blade with existing piezoelectric materials, such a 'numerical' stress concentration will mislead the design.

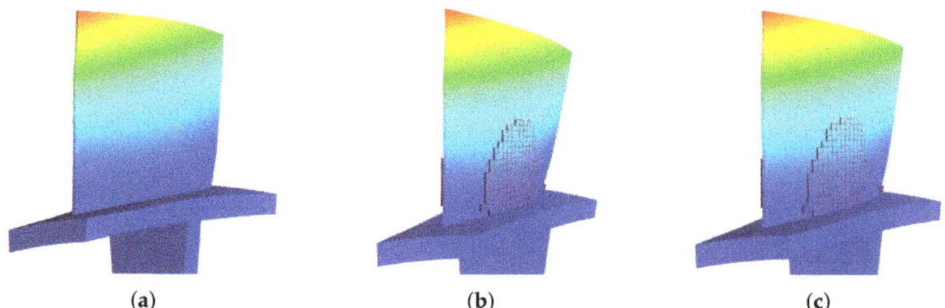

Figure 14. The modal shapes of the 1st modal family at different situations, mass ratio = 10%. (**a**) Original, (**b**) Bonded Pzt, open-circuit, (**c**) Bonded Pzt, short-circuit.

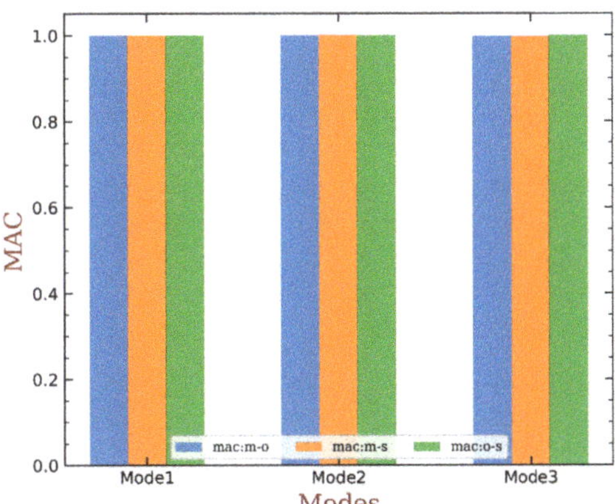

Figure 15. The modal assurance criterion (MAC) between the original (denoted by 'm'), open-circuit ('o') and short-circuit ('s') modal shapes for the first three modal groups.

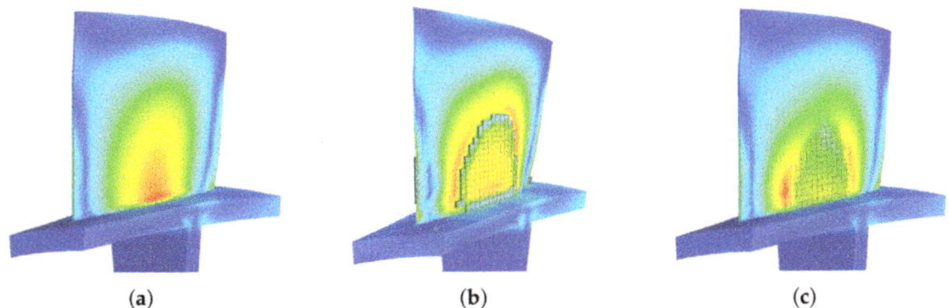

Figure 16. Total strain distribution of the 1st modal family at different situations, mass ratio = 10%. (**a**) Original, (**b**) Bonded Pzt, (**c**) Embedded Pzt.

4. Intentional Capacitance Mistuning to Improve Aeroelastic Stability

4.1. The Relation between Frequency Deviation and Shunt Capacitance

The topology optimization of the piezoelectric materials on the blades allows us to change the natural frequency (or local stiffness) of blades to the maximum extent, under limited amount of piezoelectric materials. The modal frequency of each blade can be varied if we connect different capacitances to the piezoelectric materials on each blade. In this way, we can implement the desired mistuning pattern by capacitance mistuning. To do this, we need to quantify the relationship between the blade frequency with respect to the capacitance. Namely, for the desired mistuning pattern (normally given by the distribution of frequency deviation among sectors, $\Delta\omega_j$ with $j = 1, 2, \ldots$ the sector index), we can implement it by a distribution of capacitance among sectors (C_j with j the sector index) but we must know the relationship between C_j and $\Delta\omega_j$.

The relation between the first modal frequency of the cantilever blade and the capacitance is shown in Figure 17a (embedded pzt) and 17b (bonded pzt). The usage of the piezoelectric material are both 10%. Such results can be obtained by solving the eigenvalue problem associated with Equations (5) and (6) for each given capacitance C_e. The relation between the first modal frequency of the cantilever blade and the capacitance is shown in Figure 17a (embedded pzt) and 17b (bonded pzt). The usage of the piezoelectric material are both 10%. Such results can be obtained by solving the eigenvalue problem associated with Equations (5) and (6) for each given capacitance C_e. The modal frequency of the cantilever blade can be changed from around 924 to 946 Hz by varying the external capacitance for the embedded case. Such a range may not be said wide in general situations, but it is acceptable to create an intentional mistuning. If not satisfied, one can increase this range by using more piezoelectric materials (increasing the value of R_m in the topological optimization), or by using more powerful piezoelectric materials. More importantly, such a varying range of modal frequency is achieved by a much wider range of the capacitance. Please note that the logarithmic scale is used in Figure 17a,b. This means that the value of capacitance can vary within 3 orders of magnitude to achieve such a 20/950 difference in natural frequencies. Therefore, the design capacitance can subject to relatively large uncertainties and would not induce significant error to the desired mistuning pattern. This endows a much higher robustness of the proposed method over the mechanical manufacturing approaches. Similar remarks can be given to the bonded PZT case, where the frequency range is slightly narrower but the overall trend is the same.

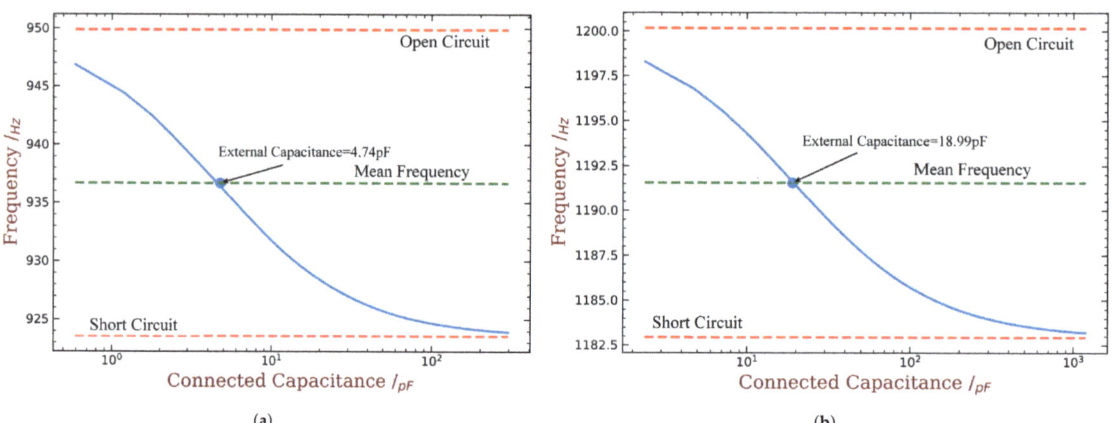

Figure 17. The relation between the blade 1st modal frequency and the shunt capacitance, mass ratio = 10%. (**a**) Embedded PZT and (**b**) bonded PZT.

In summary, we will set a mean frequency that all of the blades on the bladed disk should be first tuned with same external capacitance. In this way, we can additionally modify the connected capacitance to realize the desired intentional mistuning. The capacitance with 4.74 pF and 18.99 pF are chosen as the mean values for the embedded and bonded cases respectively, as shown in Figure 17.

4.2. Mistuning Pattern Design

We follow the literature [14,15] and employ mistuning patterns with harmonic (sinusoidal) forms:

$$\Delta\omega_j = A * \sin(\frac{N}{2\pi}hj + \theta) \tag{18}$$

where N is the overall number of blades and equals to 36 in this paper; $j = 1, 2, \ldots N$ is the sector index; h is the harmonic index and represents the repeating times of pattern along the circumference direction; θ is an arbitrary starting phase and it does not affect the results; A is the amplitude of mistuning. After obtaining the knowledge about modifying the blade frequency with external capacitance, we can implement such an intentional mistuning pattern. As we assume to fully use the adjust range of frequency to maximize the strength of mistuning, A is determined by Figure 17. Then, what remains is only to determine harmonic index h.

The aeroelastic stability of the bladed disk is determined by the minimum aerodynamic damping ratio ξ_{min}. However, there is no general conclusion about the relationship between ξ_{min} and harmonic index h of mistuning. Therefore, we conduct parametric studies to determine the best choice of h for the largest possible ξ_{min}. In our investigation, we also varied the mass ratios of the piezoelectric material from 3% to 10%. The results are summarized in Figure 18 for embedded PZT and Figure 19 for bonded PZT. In each figures, the x-coordinate refers to different mass ratio, and the y-coordinate is the harmonic index for which the capacitance varies along the circumference direction. The color is indicating the magnitude of ξ_{min} when (1) the piezoelectric materials are optimized under the mass ratio shown in the corresponding x-coordinate, and (2) the mistuning pattern is following the harmonic index shown in the corresponding y-coordinate. We have also labeled the values when ξ_{min} is greater than $-0.25‰$, while the values of unlabeled colors can be estimated by the scale shown alongside the figure.

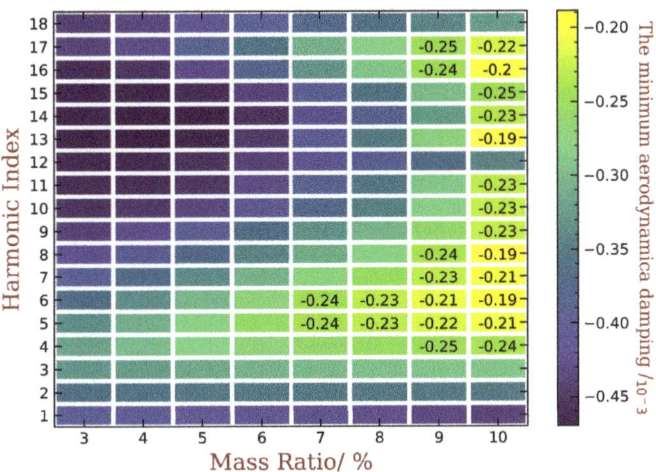

Figure 18. The variation of ξ_{min} with respect to harmonic index h and mass ratios when piezoelectric materials are embedded into the blades.

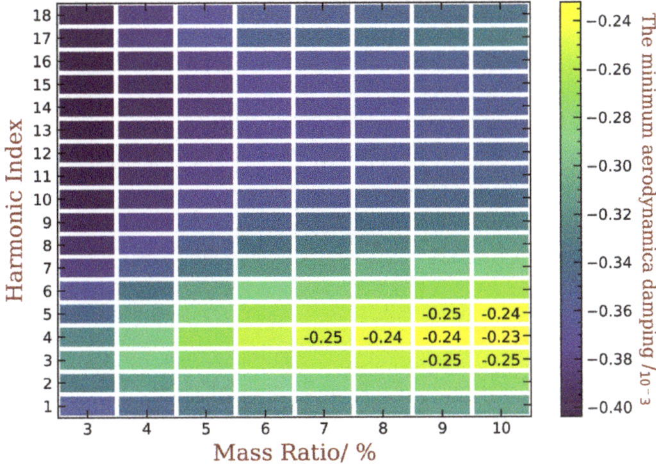

Figure 19. The variation of ξ_{min} with respect to harmonic index h and mass ratios when piezoelectric materials are bonded to the blades.

In both figures, the best results of ξ_{min} are still negative values, this means that the system is still unstable. Note that the original ξ_{min} for the tuned bladed disk is lower than $-0.5‰$, as shown in Figure 4b. Thus, a ξ_{min} in the mistuned cases larger than $-0.5‰$ can be regarded as an improvement of the aeroelastic stability. The remaining negative aeroelastic damping is easier to compensate by material or structural damping. Moreover, we can increase the mass ratio or use more powerful piezoelectric materials. In this paper, we just use NASA-rotor37 blades as example and choose an abnormal working condition to reproduce the flutter and to illustrate the performance of the proposed method. Therefore, the observation of the improved ξ_{min} in such a unstable situation is sufficient here.

When the mass ratio is low, for example, when it equals to 3% and 4%, the improvement of ξ_{min} is not significant. Despite that, the performance of some harmonic indexes is better than others, for example when it equals to 3, 4, 5 or 6. These is consistent with the engine order of the modes with higher aeroelastic damping in the tuned case as shown in Figure 4b. Such an consistency is the same with previous studies concerning harmonic mistuning [14,15,33]. Under each level of mass ratio, the distribution of piezoelectric materials are optimized using the aforementioned approach. Therefore, a larger mass ratio means a greater gap between the OC and SC natural frequencies and mistuning amplitude A in Equation (18) is lager. Consequently, we can observe in both figures that the best ξ_{min} increase monotonously with the increase of mass ratio of piezoelectric materials. For example, in Figure 18, ξ_{min} equals to $-0.24‰$, $-0.23‰$, $-0.21‰$ and $-0.19‰$ when mass ratio R_{m} equals to 7%, 8%, 9% and 10%, respectively. In addition, the best choice of the harmonic index h does not vary significantly for different mass ratio, and the performance difference among harmonic indexes is less apparent when mass ratio is increasing. This trend is less significant for the bonded PZT case (Figure 17) because it provides weaker frequency gap than the embedded case under the same mass ratio.

Finally, the desired harmonic indexes are 6, 8, 13 and 16 for 10% mass ratio, when the piezoelectric material is embedded. Compared with the original results shown in Figure 4b, the amplitude of ξ_{min} has been reduced by half. When the piezoelectric material is bonded, the best harmonic index are 4 and 5. Detailed results are shown in Figures 20 and 21. We can see that these intentional mistuning can increase the minimum aerodynamic damping to higher levels compared with the tuned bladed disk. Moreover, the number of the unstable mode decreases after introducing the intentional mistuning. With h decided, the mistuning patterns can be implemented, as illustrated in Figure 22. Recall that the capacitance is computed by the blade frequency variation according to Figure 17. Please note that the required capacitance is varying from several to hundreds pF, and this can be purchased with easy.

We can decompose the mistuned modal shape corresponding to ξ_{min} by the tuned modal shapes, as shown in 23. We can see that a lot of modes are involved including those with positive aeroelastics damping ratios. With their contributions, the stability of mistuned modes are improved. This is also consistent with the existing literature.

4.3. Effects of Random Mechanical Mistuning

As mentioned, mechanical mistuning caused by the material disperse, wear and manufacturing tolerance is random and inevitable. Therefore, the performance of intentional mistuning should be further examined in the presence of random mechanical mistuning. We apply the optimum intentional mistuning to the bladed disk obtained in the last section and introduce random mechanical mistuning simultaneously. The former is realized by external capacitance and the latter is simulated by stiffness variance in the numerical analysis. The level of random mistuning is quantified by the standard deviation of the first modal frequency with cantilevered blade along sectors. There are many sources of mistuning in practice as mentioned. Quantifying each of them in real engineering scenario can be a difficult task, and it is out of the scope of this work. However, their influences can be eventually attributed to the non-periodic perturbation on the stiffness and inertial coefficients of the dynamic model. Therefore, we use stiffness mistuning as a representative case to study the influence of additional random mistuning caused by different sources. Such an idea can be seen in various published papers [37,38] particularly when the researchers aim to draw general conclusions concerning the influence of mistuning with respect to the dynamic characteristics.

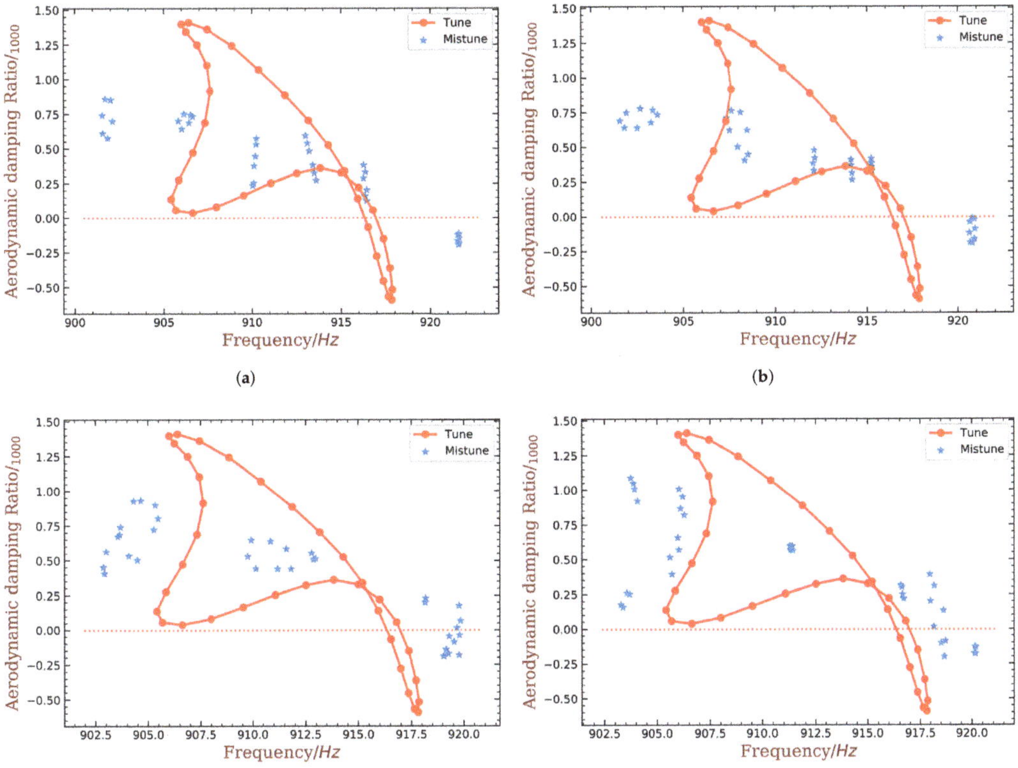

(**a**)

(**b**)

(**c**)

(**d**)

Figure 20. The distribution of aerodynamic damping ratio with different harmonic index. (mass ratio = 10%, embedded PZT). (**a**) $h = 6$, (**b**) $h = 8$, (**c**) $h = 13$, (**d**) $h = 16$.

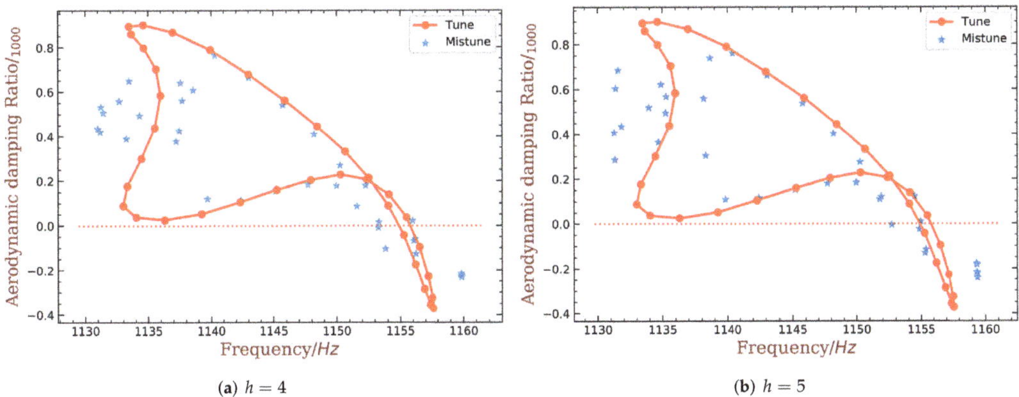

(**a**) $h = 4$

(**b**) $h = 5$

Figure 21. The distribution of aerodynamic damping ratio with different harmonic index. (mass ratio = 10%, embedded PZT). (**a**) $h = 4$, (**b**) $h = 5$.

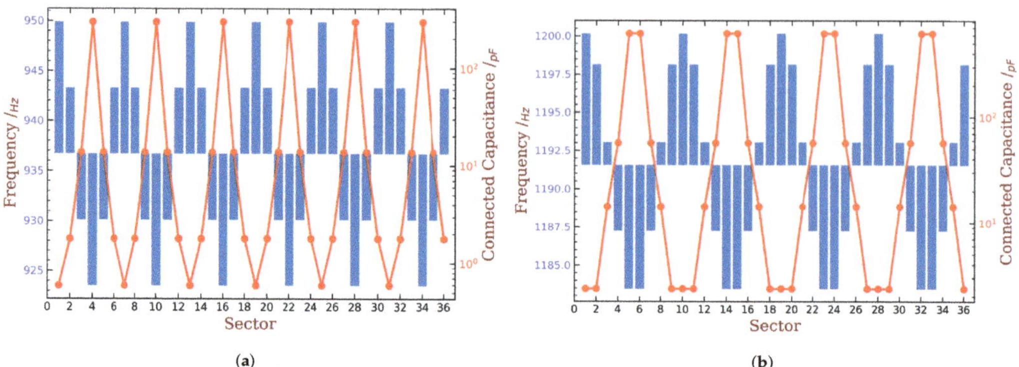

(a) (b)

Figure 22. Illustration of some intentional mistuning patterns (blue bars) and associated implementation by capacitance (red dots), mass ratio = 10%. (**a**) $h = 6$, embedded PZT, (**b**) $h = 4$, bonded PZT.

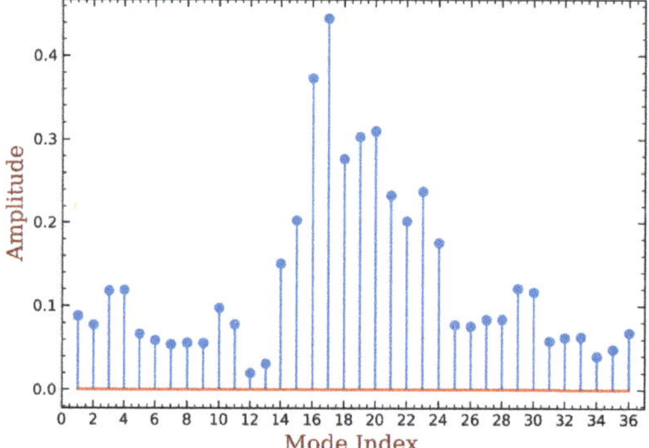

Figure 23. The contributions of tuned modes to the minimum aerodynamic damping (mass ratio = 10%, $h = 13$, bonded PZT).

At first, we only introduce mechanical mistuning into the tuned bladed disk. We take 100 groups of $\Delta\omega_j, j = 1, 2, \ldots, 36$ from the normal distribution with 0 mean expectation at each standard deviation changes from 0.1% to 0.5%. Note that when the standard deviation equals to 0.5%, most of the $\Delta\omega_j$ values will be located approximately in the same range as that can be expanded by the capacitance as shown in Figure 17. That is why we choose to analyze this range of standard deviation. The frequency graph is shown in Figure 24, where the original $\xi_{\min}$ value of the tuned bladed disk is also marked. The y-axis label 'times' in these figures refers to the count of the samples whose $\xi_{\min}$ is inside the corresponding small interval between two markers in the curves. Thus, the results shown in these figures can be interpolated as an approximation of the probability density function. We can see that mistuning is always beneficial to the improvement of aeroelastic stability in comparison with the tuned case, and this observation is consistent with the literature [1,2]. Specifically, the mean value of $\xi_{\min}$ in creases with respect to the mistuning level and the samples are distributed in a wider range. As mentioned, the variance range of blade frequencies with 0.5% random mistuning is equivalent to it shown in Figure 24. But 0.5% random mistuning can only have mean value of $\xi_{\min}$ around -0.44‰ and barely reaches -0.35‰. This is much weaker than the performance of intentional mistuning with harmonic form, where $\xi_{\min}$ can be increased to around -0.20‰ as shown in Figures 18 and 19. Such a comparison also indicate the advantage of intentional mistuning.

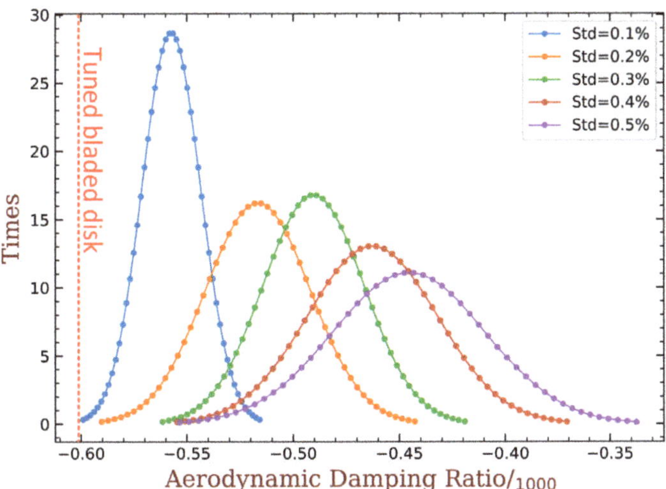

Figure 24. The frequency graph of ξ_{min} only with mechanical mistuning at different level.

Then we investigate the influence of mechanical mistuning when the intentional mistuning (implemented by capacitance) has already been imposed. The mass ratio of piezoelectric materials are both 10%, and we choose two typical cases from Figures 18 and 19 as the intentional mistuning: (1) $h = 6$ with embedded PZT, and (2) $h = 4$ with bonded PZT. The results are shown in Figure 25, where the ξ_{min} with only intentional mistuning are marked as reference. When the random mistuning is small, the influence is generally small. Especially in the case of embedded PZT, most of the samples are better than the reference. When the mistuning level increases, the results start to spread, and nearly half of samples have worse performance than the reference. Despite that the ξ_{min} decreases from -0.23‰ to around -0.30‰, and the worst case is still higher than the best case in Figure 25. This indicates that intentional mistuning can still significantly improve the aeroelastic stability with the presence of random mechanical mistuning.

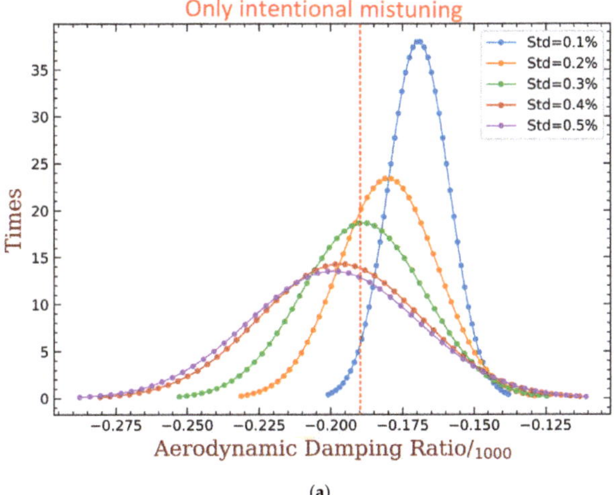

(**a**)

Figure 25. *Cont.*

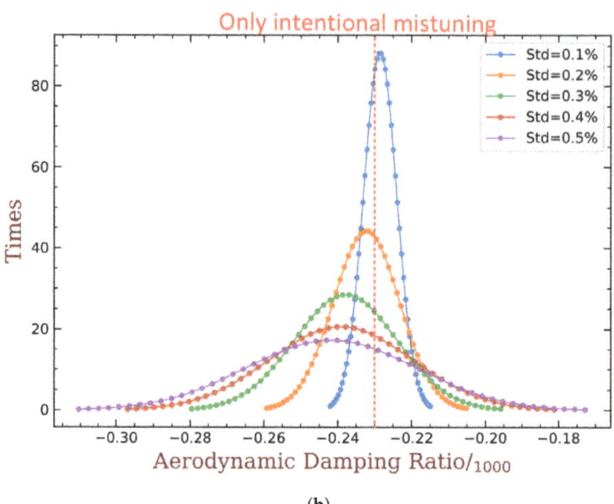

(**b**)

Figure 25. The frequency graph of $\xi_{\min}$ with best intentional mistuning (mass ratio = 10%) plus different level of mechanical mistuning. (**a**) $h = 6$, embedded PZT; (**b**) $h = 4$, bonded PZT.

5. Conclusions

1. Herein, an adaptive method based on the piezoelectric technique to improve the aeroelastic stability of the bladed disk is proposed. The basic idea is to bond or embed piezoelectric materials to each blade and use different shunt capacitance on each blade as the source of mistuning. We show that the required small difference of stiffness among blades is altered into a relatively larger difference of the shunt capacitance. This provides a more feasible and robust way to implement the intentional mistuning.
2. A method to determine the distribution of piezoelectric materials have been established. A linearly weight strain indicator is used as the optimization criterion, whose physical meaning is the absolute value of the electric field. The theoretical basis of this method is clarified. The method is adapted for FE model and has no assumptions on the geometrics of piezoelectric materials. It only needs a single modal analysis of the bladed disk and a given threshold of the mass ratio, and eventually yields the best distribution of piezoelectric materials for the targeting mode(s). The obtained shapes are described by the edges of the finite elements so they are non-smooth. One can choose to use the rounded or rectangular piezoelectric materials that can be purchased with ease, to cover the targeting area obtained by the proposed method. Alternatively, one can choose to customize the piezoelectric materials after smoothing the edges. In either way, the proposed method can provide a good starting point.
3. An empirical bladed disk with NASA-ROTOR37 profile is used as an example. For the first bending mode of the blade, only using piezoelectric materials with a mass of 10% of the blade mass can reduces the unstable margin from 0.6‰ to 0.2‰. The required capacitance is varying from several to hundreds pF, and this can be purchased with easy. These quantitative results demonstrate that the proposed method is very promising.
4. The proposed method shows good robustness with the presence of random mechanical mistuning. We show that additional (and inevitable) mistuning would somewhat weaken the performance of intentional mistuning imposed by capacitance variance. However, the stability of the system is still much better than the original case.
5. There are two main challenges to be resolved before the proposed method can be further applied to real engineering products. The first one is the installation of piezoelectric materials (PZT, MFC and so on). In the laboratory environment, they

are often glued to the surface of host structures, and additional protection is required when testing in rotation status [39]. Alternatively, the piezoelectric materials can be embedded into composites [40], and thus avoid negative affects to fluid field. In this way, the wire and electric components can also be packaged. The reliability of such installation strategies needs thorough validation before they can be used in real aero-engines. The second challenge is the strength of piezoelectric materials. We should locate them to the places where stress level is relatively high, to achieve a stronger ability of the electric field to tailor the mechanical properties. In this case, whether the piezoelectric materials and the glue can withstand the tress level for a satisfying duration should be investigated. We are addressing this issue with on-going experiments.

Author Contributions: Conceptualization, Y.F. and L.L.; methodology, X.L. and X.Y.; validation, X.L. and Y.F.; writing—original draft preparation, X.L.; writing—review and editing, Y.F.; supervision, L.L.; funding acquisition, Y.F. and L.L. All authors have read and agreed to the published version of the manuscript.

Funding: This work was funded by Aeronautical Science Foundation of China (2019ZB051002), and Advanced Jet Propulsion Creativity Center (Projects HKCX2020-02-013 and HKCX2020-02-016).

Institutional Review Board Statement: Not applicable.

Informed Consent Statement: Not applicable.

Data Availability Statement: The data presented in this study are available on request from the corresponding author..

Conflicts of Interest: The authors declare no conflicts of interest. The funders had no role in the design of the study; in the collection, analyses or interpretation of data; in the writing of the manuscript; or in the decision to publish the results.

Appendix A. Material Properties of PZT-5H

Mass density: $\rho = 7500\ \mathrm{kg/m^3}$.

Material stiffness matrix evaluated at constant electric field:

$$\mathbf{c}^{\mathrm{E}} = 10^{10} \times \begin{bmatrix} 12.6 & 7.95 & 8.41 & & & \\ 7.95 & 12.6 & 8.41 & & & \\ 8.41 & 8.41 & 11.7 & & & \\ & & & 2.3 & & \\ & & & & 2.3 & \\ & & & & & 2.35 \end{bmatrix} \mathrm{Pa}$$

Permittivity matrix evaluated at constant strain:

$$\varepsilon^{\mathrm{S}} = \varepsilon_0 \times \begin{bmatrix} 1700 & & \\ & 1700 & \\ & & 1470 \end{bmatrix}$$

where $\varepsilon_0 = 8.854 \times 10^{-12}\ \mathrm{C/(V{\cdot}m)}$.

Piezoelectric stress coupling matrix:

$$\mathbf{e} = \begin{bmatrix} & & -6.5 \\ & & -6.5 \\ & & 23.3 \\ & 0 & \\ & 17 & \\ 17 & & \end{bmatrix} \mathrm{N/(V \cdot m)}$$

References

1. Whitehead, D. Effect of mistuning on the vibration of turbo-machine blades induced by wakes. *J. Mech. Eng. Sci.* **1966**, *8*, 15–21. [CrossRef]
2. Bendiksen, O. Flutter of mistuned turbomachinery rotors. In Proceedings of the American Society of Mechanical Engineers, International Gas Turbine Conference and Exhibit, 28 th, Phoenix, AZ, USA, 27–31 March 1983.
3. Campobasso, M.; Giles, M. Analysis of the effect of mistuning on turbomachinery aeroelasticity. In Proceedings of the 9th International Symposium on Unsteady Aerodynamics, Aeroacoustics and Aeroelasticity of Turbomachines (ISUAAAT 2000), Lyon, France, 4–8 September 2000; pp. 4–8.
4. Kielb, R.E.; Hong, E.; Hall, K.C. Probabilistic Flutter Analysis of a Mistuned Bladed Disk. In Proceedings of the ASME Turbo Expo 2006: Power for Land, Sea, and Air, Barcelona, Spain, 8–11 May 2006; pp. 1–6.
5. Wang, P.Y.; Li, L. Parametric sensitivity for coupling vibration characteristics of bladed disk. *J. Aerosp. Power* **2014**, *29*, 81–90.
6. Deng, P.C.; Li, L.; Li, C. Study on vibration of mistuned bladed disk with bi-periodic piezoelectric network. *Proc. Inst. Mech. Eng. Part G J. Aerosp. Eng.* **2017**, *231*, 350–363. [CrossRef]
7. Nowinski, M.; Panovsky, J. Flutter Mechanisms in Low Pressure Turbine Blades. *J. Eng. Gas Turbines Power* **2000**, *122*, 82–88. [CrossRef]
8. Panovsky, J.; Kielb, R.E. A Design Method to Prevent Low Pressure Turbine Blade Flutter. *J. Eng. Gas Turbines Power* **1999**, *122*, 89–98. [CrossRef]
9. Srinivasan, A. *Influence of Mistuning on Blade Torsional Flutter*; NASA: Washington, DC, USA, 1980.
10. Crawley, E.; Hall, K. Optimization and mechanisms of mistuning in cascades. *J. Eng. Gas Turbines Power* **1985**, *107*, 418–426. [CrossRef]
11. Groth, P.; Mårtensson, H.; Andersson, C. Design and experimental verification of mistuning of a supersonic turbine blisk. *J. Turbomach.* **2010**, *132*, 011010. [CrossRef]
12. Biagiotti, S.; Pinelli, L.; Poli, F.; Vanti, F.; Pacciani, R. Numerical Study of Flutter Stabilization in Low Pressure Turbine Rotor with Intentional Mistuning. *Energy Procedia* **2018**, *148*, 98–105. [CrossRef]
13. Bleeg, J.M.; Yang, M.T.; Eley, J.A. Aeroelastic Analysis of Rotors With Flexible Disks and Alternate Blade Mistuning. *J. Turbomach.* **2008**, *131*, 011011. [CrossRef]
14. Martel, C.; Corral, R.; Llorens, J.M. Stability increase of aerodynamically unstable rotors using intentional mistuning. *J. Turbomach.* **2008**, *130*, 011006. [CrossRef]
15. Zhang, X.; Wang, Y. Mistuning Effects on Aero-elastic Stability of Contra-Rotating Turbine Blades. *Int. J. Aeronaut. Space Sci.* **2019**, *20*, 100–113. [CrossRef]
16. Fu, Z.; Wang, Y. Aeroelastic Analysis of a Transonic Fan Blade with Low Hub-to-Tip Ratio including Mistuning Effects. *J. Power Energy Eng.* **2015**, *3*, 362. [CrossRef]
17. Figaschewsky, F.; Kühhorn, A.; Beirow, B.; Nipkau, J.; Giersch, T. Design and Analysis of an Intentional Mistuning Experiment Reducing Flutter Susceptibility and Minimizing Forced Response of a Jet Engine Fan. In Proceedings of the ASME Turbo Expo 2017: Turbomachinery Technical Conference and Exposition, Charlotte, NC, USA, 26–60 June 2017; pp. 1–13.
18. Corral, R.; Khemiri, O.; Martel, C. Design of mistuning patterns to control the vibration amplitude of unstable rotor blades. *Aerosp. Sci. Technol.* **2018**, *80*, 20–28. [CrossRef]
19. Hagood, N.; von Flotow, A. Damping of structural vibrations with piezoelectric materials and passive electrical networks. *J. Sound Vib.* **1991**, *146*, 243 – 268. [CrossRef]
20. Padula, S.; Kincaid, R. *Optimization Strategies for Sensor and Actuator Placement*; NASA: Washington, DC, USA, 1999.
21. Frecker, M.I. Recent Advances in Optimization of Smart Structures and Actuators. *J. Intell. Mater. Syst. Struct.* **2003**, *14*, 207–216. [CrossRef]
22. Ducarne, J.; Thomas, O.; Deü, J.F. Placement and dimension optimization of shunted piezoelectric patches for vibration reduction. *J. Sound Vib.* **2012**, *331*, 3286–3303. [CrossRef]
23. Donoso, A.; Bellido, J.C. Tailoring distributed modal sensors for in-plane modal filtering. *Smart Mater. Struct.* **2009**, *18*, 037002. [CrossRef]
24. Kang, Z.; Wang, X. Topology optimization of bending actuators with multilayer piezoelectric material. *Smart Mater. Struct.* **2010**, *19*, 075018. [CrossRef]
25. Lee, S.; Youn, B.D. A design and experimental verification methodology for an energy harvester skin structure. *Smart Mater. Struct.* **2011**, *20*, 057001. [CrossRef]
26. Sun, H.; Yang, Z.; Li, K.; Li, B.; Xie, J.; Wu, D.; Zhang, L. Vibration suppression of a hard disk driver actuator arm using piezoelectric shunt damping with a topology-optimized PZT transducer. *Smart Mater. Struct.* **2009**, *18*, 065010. [CrossRef]
27. Belloli, A.; Ermanni, P. Optimum placement of piezoelectric ceramic modules for vibration suppression of highly constrained structures. *Smart Mater. Struct.* **2007**, *16*, 1662–1671. [CrossRef]
28. Aurélien, S.; Thomas, O.; Deü, J.F. Optimization of Shunted Piezoelectric Patches for Vibration Reduction of Complex Structures: Application to a Turbojet Fan Blade. In Proceedings of the 22nd International Conference on Design Theory and Methodology, Montréal, QC, Canada, 15–18 August 2010; Special Conference on Mechanical Vibration and Noise. ASMEDC, 2010; Volume 5, pp. 695–704. [CrossRef]

29. Pereira da Silva, L.; Larbi, W.; Deü, J.F. Topology optimization of shunted piezoelectric elements for structural vibration reduction. *J. Intell. Mater. Syst. Struct.* **2015**, *26*, 1219–1235. [CrossRef]
30. Li, L.; Yu, X.; Wang, P. Research on aerodynamic damping of bladed disk with random mistuning. In Proceedings of the ASME Turbo Expo 2017: Turbomachinery Technical Conference and Exposition, Charlotte, NC, USA, 26–30 June 2017; p. V07BT36A010. [CrossRef]
31. Hsu, K.; Hoyniak, D. A Fast Influence Coefficient Method for Linearized Flutter and Forced-Response Analysis. In Proceedings of the 49th AIAA Aerospace Sciences Meeting including the New Horizons Forum and Aerospace Exposition, Orlando, FL, USA, 4–7 January 2011; p. 229.
32. Hanamura, Y.; Tanaka, H.; Yamaguchi, K. A Simplified Method to Measure Unsteady Forces Acting on the Vibrating Blades in Cascade. *JSME Int. J.* **2008**, *23*, 880–887. [CrossRef]
33. Liu, X.; Li, L.; Fan, Y.; Deng, P. Improving the aero-elastic stability of bladed disks through parallel piezoelectric network. In Proceedings of the 2018 Joint Propulsion Conference, Cincinnati, OH, USA, 9–11 July 2018; p. 4738.
34. Trindade, M.; Benjeddou, A. Effective Electromechanical Coupling Coefficients of Piezoelectric Adaptive Structures: Critical Evaluation and Optimization. *Mech. Adv. Mater. Struct.* **2009**, *16*, 210–223. [CrossRef]
35. Ducarne, J.; Deü, J.F. Structural Vibration Reduction by Switch Shunting of Piezoelectric Elements: Modeling and Optimization. *J. Intell. Mater. Syst. Struct.* **2010**, *21*, 797–816. [CrossRef]
36. Thomas, O.; Ducarne, J.; Deü, J.-F. Performance of piezoelectric shunts for vibration reduction. *Smart Mater. Struct.* **2011**, *21*, 015008. [CrossRef]
37. Chan, Y.J.; Ewins, D. Management of the variability of vibration response levels in mistuned bladed discs using robust design concepts. Part 2. *Mech. Syst. Signal Process.* **2010**, *24*, 2792–2806. [CrossRef]
38. Chan, Y.J.; Ewins, D. Management of the variability of vibration response levels in mistuned bladed discs using robust design concepts. Part 1. *Mech. Syst. Signal Process.* **2010**, *24*, 2777–2791. [CrossRef]
39. Min, J.B.; Duffy, K.P.; Choi, B.B.; Provenza, A.J.; Kray, N. Numerical modeling methodology and experimental study for piezoelectric vibration damping control of rotating composite fan blades. *Comput. Struct.* **2013**, *128*, 230–242. [CrossRef]
40. Bachmann, F.; De Oliveira, R.; Sigg, A.; Schnyder, V.; Delpero, T.; Jaehne, R.; Bergamini, A.; Michaud, V.; Ermanni, P. Passive damping of composite blades using embedded piezoelectric modules or shape memory alloy wires: A comparative study. *Smart Mater. Struct.* **2012**, *21*, 075027. [CrossRef]

 materials

MDPI

Article

Damage Detection in Flat Panels by Guided Waves Based Artificial Neural Network Trained through Finite Element Method

Donato Perfetto, Alessandro De Luca *, Marco Perfetto, Giuseppe Lamanna and Francesco Caputo

Department of Engineering, University of Campania "L. Vanvitelli", Via Roma 29, 81031 Aversa, Italy; donato.perfetto@unicampania.it (D.P.); marco.perfetto1@studenti.unicampania.it (M.P.); giuseppe.lamanna@unicampania.it (G.L.); francesco.caputo@unicampania.it (F.C.)

* Correspondence: alessandro.deluca@unicampania.it

Abstract: Artificial Neural Networks (ANNs) have rapidly emerged as a promising tool to solve damage identification and localization problem, according to a Structural Health Monitoring approach. Finite Element (FE) Analysis can be extremely helpful, especially for reducing the laborious experimental campaign costs for the ANN development and training phases. The aim of the present work is to propose a guided wave-based ANN, developed through the use of the Finite Element Method, to determine the position of damages. The paper first addresses the development and assessment of the modeling technique. The FE model accuracy was proven through the comparison of the predicted results with experimental and analytical data. Then, the ANN was developed and trained on an aluminum plate and subsequently verified in a composite plate, as well as under different damage configurations. According to the results herein proposed, the ANN allowed to detect and localize damages with a high level of accuracy in all cases of study.

Keywords: Artificial Neural Network (ANN); guided waves; Structural Health Monitoring (SHM); Finite Element Analysis (FEA); damage detection; metals; composites

Citation: Perfetto, D.; De Luca, A.; Perfetto, M.; Lamanna, G.; Caputo, F. Damage Detection in Flat Panels by Guided Waves Based Artificial Neural Network Trained through Finite Element Method. *Materials* **2021**, *14*, 7602. https://doi.org/10.3390/ma14247602

Academic Editor: Bing Wang

Received: 19 November 2021
Accepted: 7 December 2021
Published: 10 December 2021

1. Introduction

Structural monitoring of primary and secondary structural elements is generally a complex, costly and time-consuming process. All structural components are inspected at regular intervals, using different non-destructive techniques (NDTs) that increase more and more the total duration of the downtime.

Such inspections for health-condition assessment are of the utmost importance for the safe and efficient operation of structures. A damage that cannot be reliably detected could not be repaired on time, leading to eventually catastrophic failures.

To overcome these problems, in view to ensure the safety of the structure, Structural Health Monitoring (SHM) systems able to monitor the actual structural integrity become necessary [1–3]. The use of a SHM system brings with it numerous benefits in terms of maintenance and repairing operations, including reduction in uncertainty in operators' decision-making process, quasi-real time control of deficiencies and, thus, greater safety through continuous monitoring. Moreover, SHM will provide a dataset useful for maintenance phases, improving in the meantime the engineering capabilities and the future design of primary components with relative reduction in both medium- and long-term costs.

In the execution of SHM strategies, the most commonly used approach is that based on guided (or Lamb) waves (GW) [4–6] because of their powerful capability of long-distance propagation with high speed and little loss of energy. Several useful pieces of information can be derived from implementing a Lamb-wave-based identification associated with a damage detection method, such as qualitative indication of the occurrence of damage, quantitative assessment of the position of damage, quantitative estimation of the severity of damage and prediction of structural safety like the residual service life [7,8].

Successful damage identification by using a GW-based SHM system can be performed with transducers in a sparse configuration and using well-calibrated signal-interpretation techniques. Some essential steps are the following: (i) activating the desired diagnostic Lamb wave signal, using an appropriate transmitter and capturing the damage-scattered wave signals using a sensor or a sensors network in accordance with either the pitch-catch or the pulse-echo configuration; (ii) extracting and evaluating the characteristics of the captured wave signals with appropriate signal post-processing tools; (iii) establishing quantitative and/or qualitative connections between the extracted signal characteristics and the damage parameters (presence, location, geometric identity, severity, etc.); and (iv) figuring out the damage parameters of interest in terms of captured signals, based on the quantitative connections established in Step (iii) [8].

However, the main challenge of guided waves is that they propagate with multiple modes and exhibit a strongly dispersive behavior when the excitation frequency increases [9]. Thus, the attention is mainly paid on the zero-order symmetric and antisymmetric modes, S_0 and A_0, respectively. Both the S_0 and A_0 modes are sensitive to structural damage, and both can be used for identifying damage, though the S_0 mode exhibits higher sensitivity to damage in the structural thickness, fatigue damage and delamination. Meanwhile, the interpretation of the signals can be very challenging, limiting the extraction of useful information on damages. Thus, a large number of experiments is required to properly identify GW characteristics. To reduce time and costs, chirp broadband signals can be used, together with the Finite Element Method (FEM), semi-analytical tools, etc.

With the latest advances in neurosciences and high-capability computing devices, machine learning (ML) algorithms based on Artificial Neural Networks (ANNs) have rapidly emerged as a promising tool to solve damage/defects identification and localization problem [8,10]. Various techniques, including mode shapes, natural frequencies, strain history and many others [11–15], have been used for damage identification during the years for different application fields. This was achieved by utilizing different supervised or unsupervised machine learning techniques for damage recognition [16–18].

Recent research has focused on damage quantitative estimation by using Lamb wave signals features in combination with ANN and FEM in order to ensure the accuracy of ANN (strictly related to the data used to train the network) [10,16]. A Lamb-wave-based damage evaluation method assisted by an ANN model was presented by Qian et al. [19], using Damage Indexes related to changes in amplitude and phase of recorded signals. They found a good agreement between experimental and predicted results in a carbon fiber-reinforced polymer composite plate. Sharif-Khodaei et al. [20,21] developed an Artificial Neural Network (ANN) that is able to locate impact of different energies in a complex structure, such as a composite-stiffened panel. Such an ANN was trained by means of a large number of impact simulations, based on FEM, covering a wide range of impact energies and locations. De Fenza et al. [10] used GW propagation and ANN to determine the location and the degree of damage in a metallic plate: it was divided into sectors, and for each one, the probability of occurrence of the damage was assessed.

Obviously, the performance of an ANN strictly depends on the size and complexity of the obtained dataset. A very large or high-dimensional dataset could affect the accuracy of the classification or clustering and make the model computationally expensive. Employing data-preprocessing methods when dealing with high-dimensional sensory data is also of great importance and impacts the ANN performance [18].

In this paper, a machine learning approach based on ANN is proposed in order to demonstrate and discuss the usability of artificial intelligence for the purpose of detecting and localizing a damage and its coordinates. A physical experiment and numerical simulations on an aluminum plate were carried out. The aluminum plate was equipped with four piezoelectric transducers to activate and receive the wave response in baseline (no damage) and actual (damaged) configurations of the structure.

As the first step, the development and assessment of the Finite Element modeling technique was addressed. The FE model accuracy was proven through the comparison of the predicted results with experimental and analytical data in terms of S_0 and A_0 group velocities' curves.

Once established with respect to GW propagation, the numerical data were used for the training and validation process of the ANN. In particular, starting from the S_0 wave packet extracted from the numerical dataset through a post-processing technique, Damage Indexes (DIs), under different damaged configurations (in terms of both position and dimension), were calculated. High value of DI means a damage very close to the corresponding actuator–receiver path [22]. This was possible through the development of an in-house Matlab® (The MathWorks Inc., Natick, MA, USA) code that automatically extracts the S_0 mode from the baseline and actual states of the structure and calculates the DIs. The DIs dataset was used as input for the ANN, while the targets are represented by the coordinates of the centers of the modeled damages. Achieved results prove that the ANN can predict the damage location precisely.

Then, the developed ANN was verified on a CFRP (carbon fiber-reinforced polymer) plate. According to the results herein proposed, the ANN allowed to detect and localize damages with a high level of accuracy in all cases of study.

The trained algorithm could be directly applied to experimental measurements without the need of retraining.

The layout of the paper is as follows. Details of the case study are outlined in Section 2. Section 3 deals with the numerical methodology, based on FE, for a proper modeling of the wave propagation mechanisms; dispersion curves (experimental, numerical and semi-analytical) of the zero-order modes are shown and compared. The description of the ANN-based damage detection procedure, using Damage Indexes, is detailed in Section 4. Finally, the performance of the ANN and achieved results are discussed in Section 5 for an aluminum and a composite panel under several damaged configurations. Section 6 concludes the paper.

2. Case Study

GW propagation mechanisms were investigated in a simple flat isotropic panel. For isotropic materials, Lamb waves travel with the same velocity omni-directionally, and the wavefronts form a circle. This case study, although simple, represents a starting point for the development of an ANN useful for damage detection.

The plate under investigation has a square shape with dimension L = 287 mm, and a thickness of t_a = 2 mm (Figure 1). The mechanical properties are listed in Table 1. A four Circular DuraAct (PI Ceramics) PIC255 piezoelectric transducers network was used for both the actuation and sensing of Lamb waves. The thickness and the radius of the PZT wafers are t_{PZT} = 0.5 mm and d_{PZT} = 10 mm, respectively. The PZTs, whose mechanical properties are listed in Table 1, were surface-mounted onto the specimen. In particular, they are located at a distance h = 55 mm from the edges (Figure 1). The adhesive EA 9466 from Loctite (Henkel AG & Co. KGaA, Düsseldorf, Germany) was used to bond the transducers.

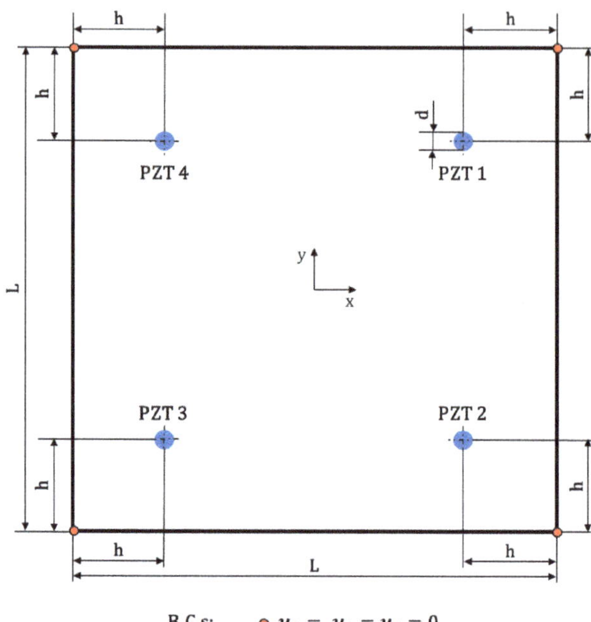

Figure 1. Schematic of the geometry and map of PZTs network.

Table 1. Material properties of Al 6061 plate and PIC255 sensors.

Material Properties	Symbol	Units	Al 6061	PIC255
Mass density	ρ	[kg m^{-3}]	2700	7850
Young's modulus	E	[GPa]	69	76
Shear modulus	G	[GPa]	26	29
Poisson's ratio	ν	–	0.33	0.32
Dielectric constant	K_3	–	–	1280
Piezoelectric charge constant	d_{31}	[10^{-9} mm V^{-1}]	–	−180

In order to reduce the number of experiments to carry out, a chirp signal was used. The chirp signal is given as follows:

$$V_{chirp}(t) = V_{in}\left[H(t) - H\left(t - t_{chirp}\right)\right] \sin\left(2\pi\left(f_0 t + \frac{f_1 - f_0}{t_{chirp}} t^2\right)\right), \quad (1)$$

where $t_{chirp} = 0.25$ ms is the duration of the chirp signal, f_0 = 50 kHz is the start frequency, f_1 = 500 kHz is the end frequency, V_{in} is the input amplitude and H is the Heaviside function. The chirp signal allows users to achieve in a single test all dispersion curves in the selected frequency band. The tone-burst response, preferred due to the dispersive nature of Lamb waves [23], was then extracted by using the reconstruction procedure described in Reference [24] to allow for the comparison for each frequency.

A 16 V peak-to-peak input amplitude was applied to the actuator PZT, using a TiePie waveform generator, and the TiePie Digital Oscilloscope was used to record the signals acquired at the sensor PZTs with a sampling frequency of 2 MHz. The total recording duration of the experimental signals is tot $= 2 \times 10^{-4}$ s, and each measurement is recorded 32 times and averaged to improve the signal to noise ratio. The acquired signals from all four channels have a resolution of 12 bit. Each measurement was 0.200 ms long.

3. Numerical Approach for Wave Propagation Modeling

In this section, at first, FE modeling techniques are compared in order to prove the efficiency of 2D-Shell elements [25,26] in simulating GW propagation in the isotropic plate under different frequencies against 3D ones. In fact, especially for complex structures, the modeling via 3D-Solid Finite Elements can be prohibitive in terms of computational costs. Thus, there is a need to adopt, as far as possible, shell elements.

Once established against experiment or analytical data, the FE model can be used to predict wave-propagation mechanisms in a damaged configuration of the plate, and to get a useful dataset for the training of the ANN.

In this work, numerical data were compared to the experimental ones, using sensor 1 as actuator and the others as receivers. Such a comparison, although simple and well-studied, offers the opportunity to validate the numerical models and evaluate their convergence by using an isotropic material.

The geometry of the panel under investigation, described in Section 2, was modeled in Abaqus® CAE environment (Dassault Systems Simulia Corp., Providence, RI, USA). Three different element types from the Abaqus® Finite Elements library were considered: (i) 2D-Shell (conventional shell), S4R element with 6 degrees of freedom per node; (ii) 3D-Shell (continuum shell), SC8R element with 3 degrees of freedom per node; and (iii) 3D-Solid (brick element), C3D8R element with 3 degrees of freedom per node.

In order to reduce the number of simulations, a chirp excitation signal was modeled in the frequency range (50 ÷ 500 kHz), as described in the previous section. Moreover, it is well-known from the literature that, to correctly characterize the scattering phenomena of Lamb waves across a modeled damaged area, a minimum number of 8–10 nodes per wavelength (NPW) should be set [8]. Thus, for the plate and sensor average element size evaluation (l_e^{plate} and l_e^{pzt}, respectively), two different values of NPW under 500 kHz carrier frequency (it is the actual end frequency of the chirp signal) were considered: 10 NPW, corresponding to $l_e^{plate} = 0.60$ mm, $l_e^{pzt} = 0.37$ mm and step size $t_{inc}(10\ \text{NPW}) = 1 \times 10^{-7}$ s; and 20 NPW, corresponding to $l_e^{plate} = 0.30$ mm, $l_e^{pzt} = 0.18$ mm and step size $t_{inc}(20\ \text{NPW}) = 5 \times 10^{-8}$ s.

This led to the development of different FE models: two 2D-Shell based, providing 10 and 20 NPW respectively; two 3D-Shell based, providing 10 and 20 NPW respectively; and only one 3D-Solid based, providing 10 NPW (to reduce the computational costs). In all the numerical analyses, the chirp actuation signal was imposed. This means that, for example, taking into account $l_e^{plate} = 0.60$ mm, the FE analysis discretises 100 NPW under 50 kHz reconstruction frequency. For the sake of simplicity, the FE models are just named 10 and 20 NPW, respectively.

The wave propagation is a dynamic phenomenon, so the explicit environment was chosen to simulate the actuation, propagation and sensing of GW. For this reason, to ensure the accuracy of the numerical solution, the time step must be less than the ratio of the minimum distance of any two adjacent nodes to the maximum wave velocity (often the group velocity of the S_0 mode), as follows:

$$t_{inc} = \frac{l_e^{min}}{C_g}, \tag{2}$$

where l_e^{min} is the size of the smallest element, and C_g is the group velocity [27].

The FE model is reported in Figure 2. For the actuation and receiving signal, the sensors network was numerically modeled as in the experiment in terms of location and material/geometry configurations. Each sensor was modeled by means of reduced integration solid elements (C3D8R), while 3 elements were modeled through the thickness. To ensure the contact between sensors and plate, a node-to-surface contact formulation was employed at the "tied" interfaces to simulate the adhesive layer between sensors and plate (not here modeled) [25].

Finally, the translational degrees of freedom of the 4 corners of the plate were constrained as in the experiment, while, relative to the GW propagation, radial displacements equivalent to the input voltage of Equation (1) were calculated through Equation (3):

$$d_r = Q_V V, \tag{3}$$

where Q_V is a conversion constant for the actuation, depending on the plate and sensor geometry and material properties, as widely described in Reference [8].

This effective displacement was applied on the upper actuator edge (see also Figure 2b) after having defined a proper polar coordinate system at the center of the actuator, as effectively reported in Reference [8].

The implicit scheme could be used to handle either the wave propagation or mechanical–piezoelectric problems. In fact, there is the option of using electric coupling PZT Finite Elements. That means that the voltage can be directly applied to the terminals of the transducers, and the corresponding strain response in the sensors is acquired through the electro-mechanical coupling of the elements (Equation (3)). Such elements, however, are not available for the explicit procedure used herein to perform the analyses. A comprehensive overview about the modeling of PZT can be found in References [8,23,28–31].

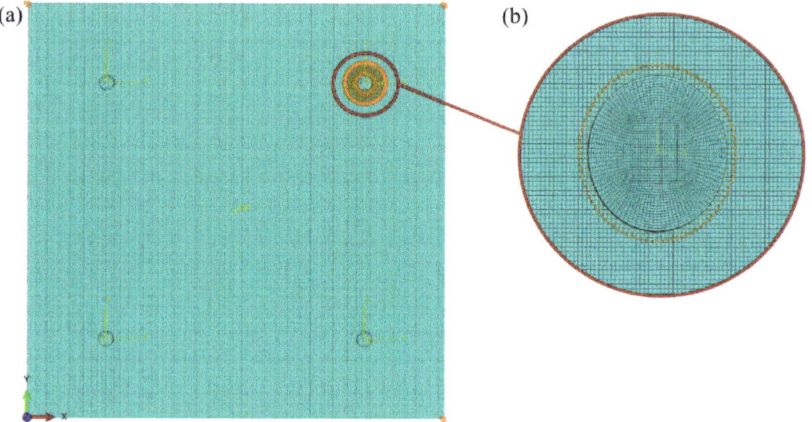

Figure 2. (**a**) FE model of the aluminum plate under study in Abaqus® CAE. (**b**) Focus on sensor mesh and GW input signal.

Dispersion Curves

The wave propagation was simulated in the Abaqus® explicit environment. The test case consists of a multi-frequency analysis with a chirp actuation signal, activated by 'PZT 1', due to isotropy of the plate, according to Equation (1). To avoid mode superimposition, low-frequency Lamb waves should usually be chosen for damage detection. In this study, the excited signal is a chirp one limited to the frequency range 50 ÷ 500 kHz. Then, in order to concentrate the majority of the wave energy on a specific central frequency, recorded data were reconstructed by means of a n-cycles sinusoidal tone-burst Hanning windowed signal with a step of 25 kHz (50:25:500 kHz), allowing reducing spectrum leakage.

Numerical results in the selected frequency range were compared to the experimental ones and also to those obtained by means of the Dispersion Calculator (Center of Lightweight Production Technology, German Aerospace Center (DLR), Augsburg, Germany) [32]. It is a Matlab®-based general-purpose tool that interactively allows users to create dispersion curves that simply define the material model and properties. It computes the phase and group velocity dispersion, as well as internal stress and displacement fields (mode shape) of Lamb and shear horizontal waves in isotropic and multilayered composites.

Displacements field contour plots in Figure 3 show the wave propagation under 300 kHz carrier test for the three investigated FE models. As expected for an isotropic material, the propagation speed is equal along all material directions and the guided waves propagate in a circular pattern. Differences between 2D-Shell, 3D-Shell and 3D-Solid can be attributed to the different representation of the panel boundaries, as also noted in References [33,34].

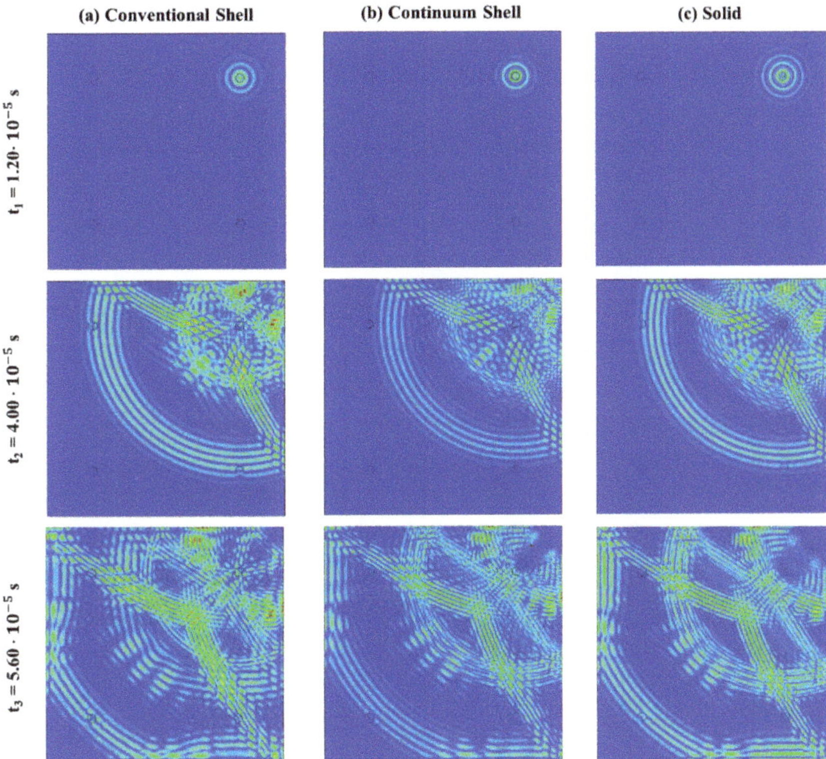

Figure 3. Guided wave propagation at 3 time instants in the aluminum plate modeled by the three different element types—300 kHz carrier.

Predicted signals, recorded at sensor locations, were calculated as the average of the in-plane strains, $\overline{\varepsilon}$, reads by all nodes defining each sensor. Similar to Equation (3), accordingly to piezoelectric relations, the voltage in PZTs can be calculated through the strain measurements as follows:

$$V = Q_s\,\overline{\varepsilon}, \tag{4}$$

where Q_s is a conversion constant for the sensing (further mathematical details can be found in Reference [8]). Then, converted signals were processed by means of the developed code and compared in terms of ToF and amplitudes.

To validate the Finite Element models, the responses of the transducers were compared for the numerical and experimental results. In particular, for the sake of brevity, in Figure 4, only signals reconstructed under the 300 kHz carrier were compared to the respective experimental ones. The same level of accuracy was achieved for all frequencies. Due to the material symmetry, signals recorded at PZT 2 and PZT 4 are the same. In all three cases, the numerical predictions are able to approximate the arrival of the first wave packet with adequate accuracy for both 10 (Figure 4a) and 20 (Figure 4b) NPW. According to Figure 4, each signal was normalized with respect to the maximum amplitude of its own S_0 wave packet. This allows highlighting the differences between predicted and experimental data.

It is noted that attenuation phenomena were not considered in this study. Additionally, the adhesive layer between the PZT wafer and the panel was modeled as a rigid link. These effects can influence the amplitude of the received waves and justify the slight mismatch existing between measured and predicted data [33,34].

Once the distances between actuator/receivers and the ToF on all paths are known, it is possible to calculate the velocity of Lamb waves packet (c_g) [35]. Considering that for an isotropic material, c_g does not depend on propagation's direction, to compute the group velocity, only the path actuator 1-receiver 2 was considered. As highlighted in Table 2, all three element types can make group velocity estimations with small relative error compared to the experimental measurement. As expected, the 3D-Solid modeling procedure permit to achieve more accurate results, but the error associated with the 2D-Shell elements is relatively small and can be considered widely acceptable.

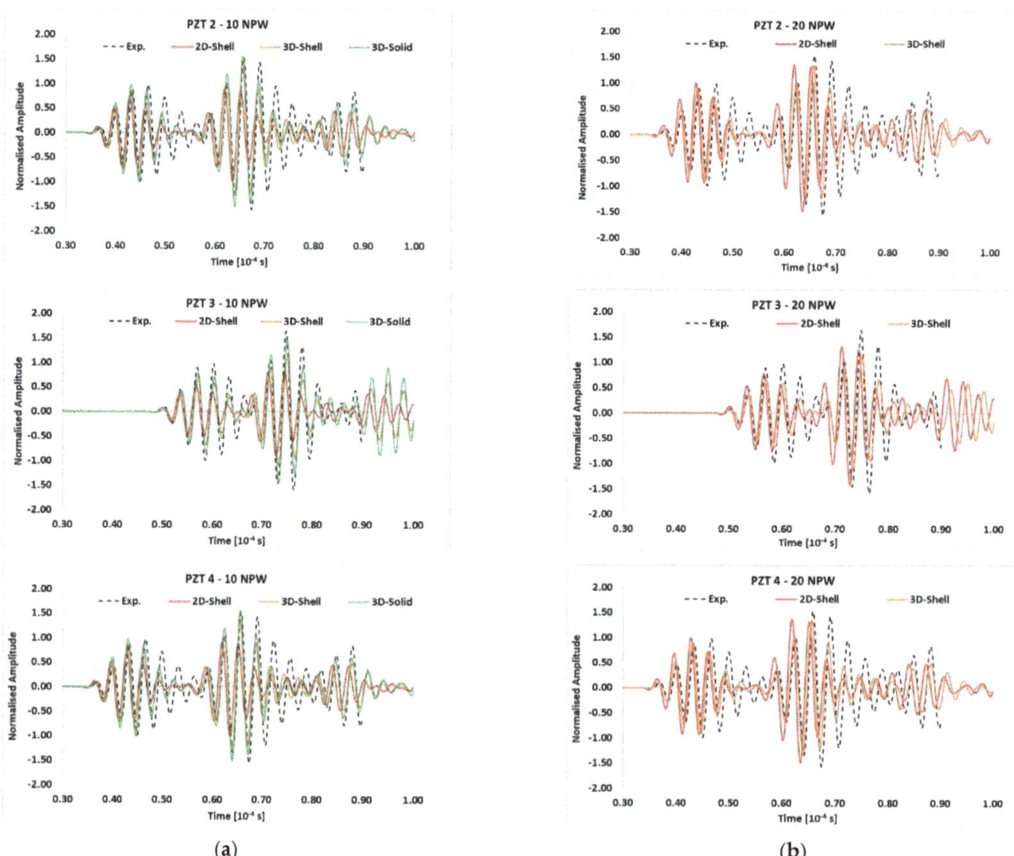

Figure 4. Comparison of the received signals at each PZT location for the conventional shell, continuum shell and solid FE models with the experimental data adopting (**a**) 10 NPW and (**b**) 20 NPW—300 kHz carrier.

Table 2. Comparison of the relative error for the estimation of the S_0 mode group velocity for each element type with respect to the experiment—300 kHz carrier.

Data	c_g (m/s)	Error (%)
Experimental	5201	–
Semi-Analytical	5268	−1.28
2D-Shell (20 NPW)	4996	3.94
3D-Shell (20 NPW)	5098.9	1.96
3D-Solid (10 NPW)	5117.4	1.60

To assess the proposed numerical procedure, the extracted dispersion curves were validated against the experimental one and those provided by Dispersion Calculator Software. Effectively, curves extracted from Dispersion Calculator were defined through a semi-analytical method.

Guided wave dispersion curves for the S_0 and A_0 modes are reported in Figure 5, in which dots represent the extracted values. As visible from Figure 5—10 NPW—the conventional shell FE model (red dots) is not capable of accurately predicting the group velocity of the A_0 mode. Moving to 20 NPW, a good agreement between experimental, semi-analytical and numerical data was found for all of the developed FE models, demonstrating the good modeling of the wave-propagation phenomenon.

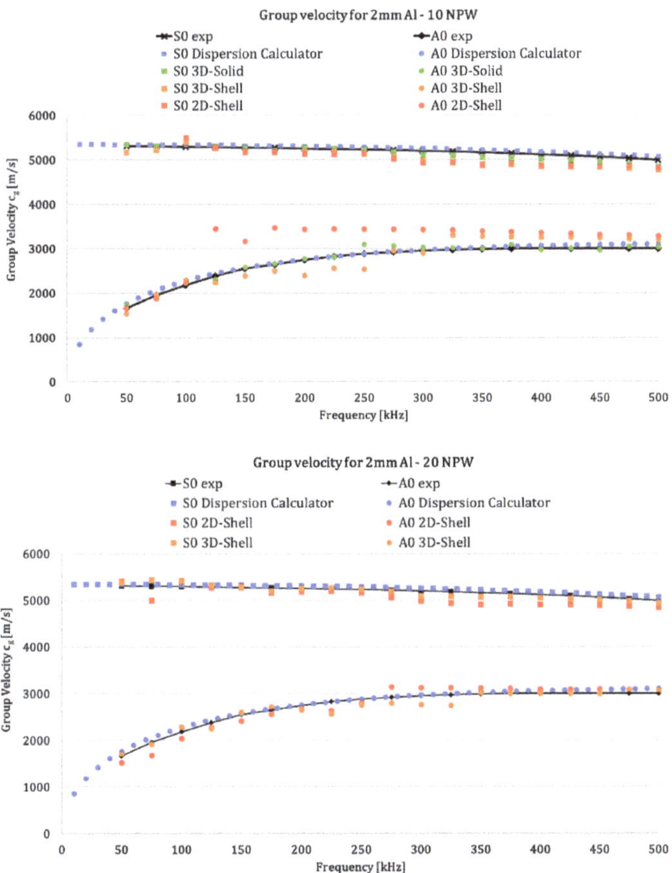

Figure 5. S_0 and A_0 modes' dispersion curves in the flat aluminum panel for 10 and 20 NPW.

Because of the good level of agreement, in order to reduce the computational costs with respect to the 3D modeling technique, the 20 NPW 2D-Shell FE model can be adopted to easily and accurately extract the dispersion curves for both the S_0 and the A_0 modes in an isotropic panel, as is visible from Figure 5.

Focusing on the 20 NPW 2D-Shell FE model and its comparison to the experimental and semi-analytical data, it can be observed that, by interpolating these points with an 2nd-order polynomial curve, the trend of the dispersion curve can be obtained. As it can be seen from Figure 6, the numerical analyses provide very accurate results: the trend of the polynomial curves (solid lines) interpolating numerical data (dots of Figure 5) matches very well the 2nd-order polynomial related to experiments and semi-analytical values, and this is true for both S_0 and A_0 modes. Even if the A_0 mode is slower than S_0, due to reflections being more difficult to detect, the numerical trend is very similar to the actual one, confirming the capability of the algorithm.

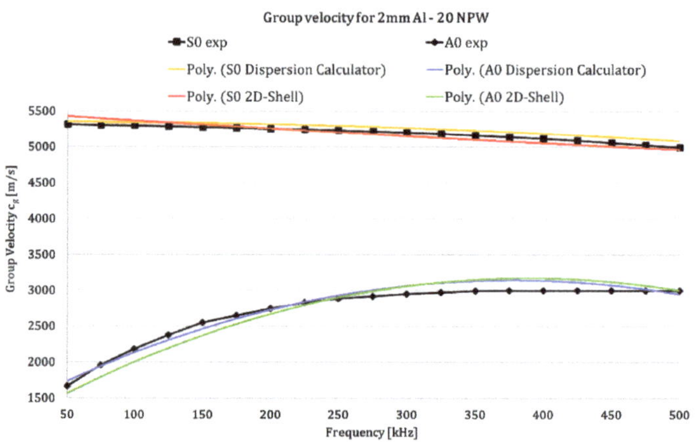

Figure 6. Dispersion curves for the S_0 and A_0 modes—2nd-order polynomial fitting curves—20 NPW.

Lastly, considering the 2nd-order polynomial interpolations, the percentage differences with respect to the experimental and semi-analytical values were computed (Tables 3 and 4, respectively). For the S_0 mode, the absolute percentage difference does not exceed the 3%, while, for the A_0 mode, it is limited to 12%, demonstrating, again, the efficiency of the numerical modeling and the post-processing procedure.

Table 3. Percentage deviation between 20 NPW 2D-Shell FE and experimental group velocity values extracted from the 2nd-order polynomial interpolation.

Frequency (kHz)	c_g[m/s]S_0Mode			c_g[m/s]A_0Mode		
	Experimental	20 NPW 2D-Shell	Difference (%)	Experimental	20 NPW 2D-Shell	Difference (%)
50	5315	5427	−2.10	1665	1561	6.25
100	5296	5369	−1.36	2184	2127	2.61
150	5277	5314	−0.69	2526	2550	−0.95
200	5256	5258	−0.02	2748	2888	−5.08
250	5231	5205	0.51	2883	3141	−8.94
300	5201	5154	0.92	2950	3310	−12.2
350	5164	5103	1.19	2990	3110	−4.01
400	5118	5054	1.26	2995	3118	−4.11
450	5061	5006	1.11	2996	3054	−1.94
500	4993	4960	0.67	2958	2920	2.60

Table 4. Percentage deviation between 20 NPW 2D-Shell FE and semi-analytical group velocity values extracted from the 2nd-order polynomial interpolation.

Frequency (kHz)	c_g[m/s]S_0Mode			c_g[m/s]A_0Mode		
	Semi-Analytical	20 NPW 2D-Shell	Difference (%)	Semi-Analytical	20 NPW 2D-Shell	Difference (%)
50	5352	5427	−1.40	1735	1561	10.03
100	5348	5369	−0.39	2133	2127	0.28
150	5337	5314	0.43	2465	2550	−3.45
200	5320	5258	1.17	2732	2888	−5.71
250	5296	5205	1.72	2932	3141	−7.13
300	5266	5154	2.13	3068	3310	−7.89
350	5229	5103	2.41	3138	3110	0.89
400	5186	5054	2.55	3142	3118	0.76
450	5136	5006	2.53	3081	3054	0.88
500	5080	4960	2.36	2955	2920	1.18

4. ANN-Based Damage Detection Procedure

4.1. Damage Indexes

Once we established the capability of the shell elements in simulating GW propagation behavior, square-shaped damages were introduced by degrading the elastic material properties of the corresponding Finite Elements (softening technique) [36] of about 70% (Figure 7). This technique allowed achieving a good agreement with reference to the experiments, as well as reducing the modeling efforts with reference to the deleting technique [36].

A wide numerical campaign was set up in order to get a useful dataset for the training of the ANN. In particular, the dataset consists of the signals reconstructed under a central frequency 300 kHz. The chosen frequency enables the localization of damage with a diameter greater than 7 mm more accurately (in recognition of the fact that the half wavelength of a selected wave mode must be shorter than or equal to the damage size to allow the wave to interact with the damage) [37,38]. On the other hand, as the frequency decreases, the wavelength increases, making the minimum size of localized damage higher and higher.

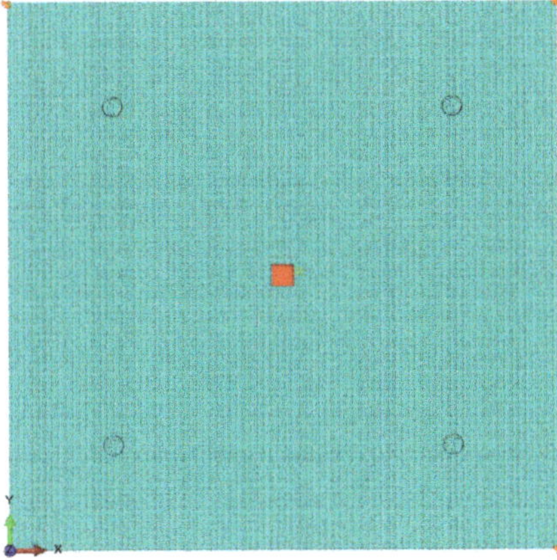

Figure 7. Example of FE damage modeling: degradation of the elastic properties (−70%) of the elements set highlighted in red.

Thus, square-shaped damages with a size of 5 and 10 mm were modeled at different points of the plate, as better specified in Table 5. The coordinates are expressed by considering the bottom left-hand corner of the plate as the origin of the Cartesian coordinates system, Figure 7. Figure 8 shows the positions of the damages (Table 5) on the plate (those common to both damage dimensions). As visible, some damages external to the area covered by the sensors network were also considered for the definition of the training set in order to improve ANN efficiency.

Table 5. Analyzed damaged configurations.

Configuration #	Damage Size (mm)	Center Coordinates (mm)
d_1	5/10	[143.5, 259.5]
d_2	5/10	[259.5, 204.5]
d_3	5/10	[232, 143.5]
d_4	5/10	[204.5, 55]
d_5	5/10	[143.5, 82.5]
d_6	5/10	[70.5, 70.5]
d_7	5/10	[40, 143.5]
d_8	5/10	[82.5, 204.5]
d_9	5/10	[143.5, 143.5]
d_{10}	10	[60, 120]
d_{11}	10	[55, 165]
d_{12}	10	[143, 212]
d_{13}	10	[168, 222]
d_{14}	10	[216, 216]
d_{15}	10	[215, 100]
d_{16}	10	[190, 180]
d_{17}	10	[110, 110]
d_{18}	10	[175, 115]
d_{19}	10	[120, 168]

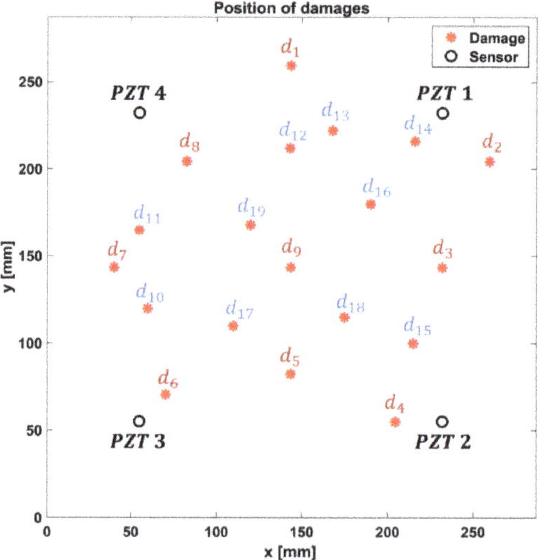

Figure 8. Sensors (black circles) and damages (red stars) map.

The signals were collected in a round-robin manner, where one transducer acts as actuator and the others act as sensors until the signals from all the receivers are collected. The presence of a damage alters the wave propagation, causing clear changes in the damaged wave packets with respect to the healthy ones. In particular, the change in amplitude and time of flight can be used to quantitatively indicate the most affected path [22].

Thus, the first step in the damage-detection phase is to extract the S_0 mode in the pristine configuration and then in the "faulty" configuration. This was achieved by implementing an in-house code that allowed us to extract the S_0 mode automatically without the need for visual graphical support. For the sake of brevity, an example is reported in Figure 9. The blue lines represent the recorded signals, the red one is the envelope and the yellow-colored part of the signal is the extracted S_0 mode.

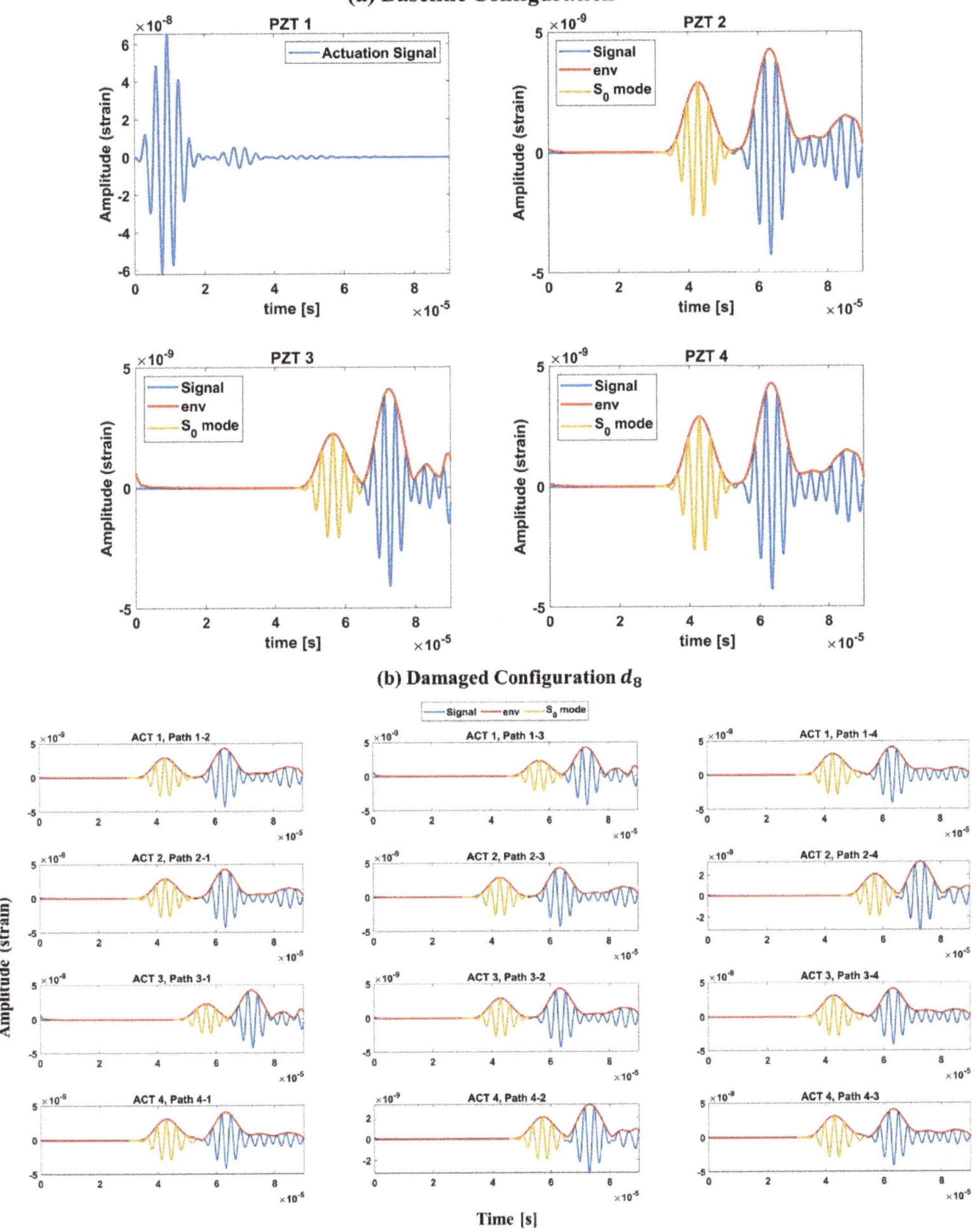

Figure 9. S_0 mode extraction from baseline and actual configurations. Example for damage in position d_8.

Figure 10 shows a comparison between the S_0 mode extracted from the two configurations, pristine and damaged, in which it is possible to notice how the amplitude of the signal recorded in the actual configuration changes. Such a comparison is quantified through the evaluation of a Damage Index, DI. Specifically, in this work, the DI given by Equation (5) was used [22]:

$$DI = \sqrt{\sum_i \frac{\left(C_p^2 - C_d^2\right)}{C_p^2}}, \quad (5)$$

where C_p and C_d are respectively the amplitude of the signal, at same time, in the configuration "pristine" and "damaged". The high value of the Damage Index means that the damage is placed along or close to the corresponding actuator–sensor path. Conversely, low Damage Index means that the damage is far from the actuator–sensor path.

The developed code also allows automatically calculating the Damage Index for each actuator–sensor path. The Damage Indexes obtained for each damage condition were normalized with respect to the maximum value among all paths for the different actuators. An example is reported in Figure 11 for the damage in position d_8. As visible from the figure, the DI enables the identification of the most affected paths. Moreover, by knowing the damage position, the DIs suggest that the damage is localized closer to the PZT 4, since, when this sensor is used as an actuator, the values of DI are much higher.

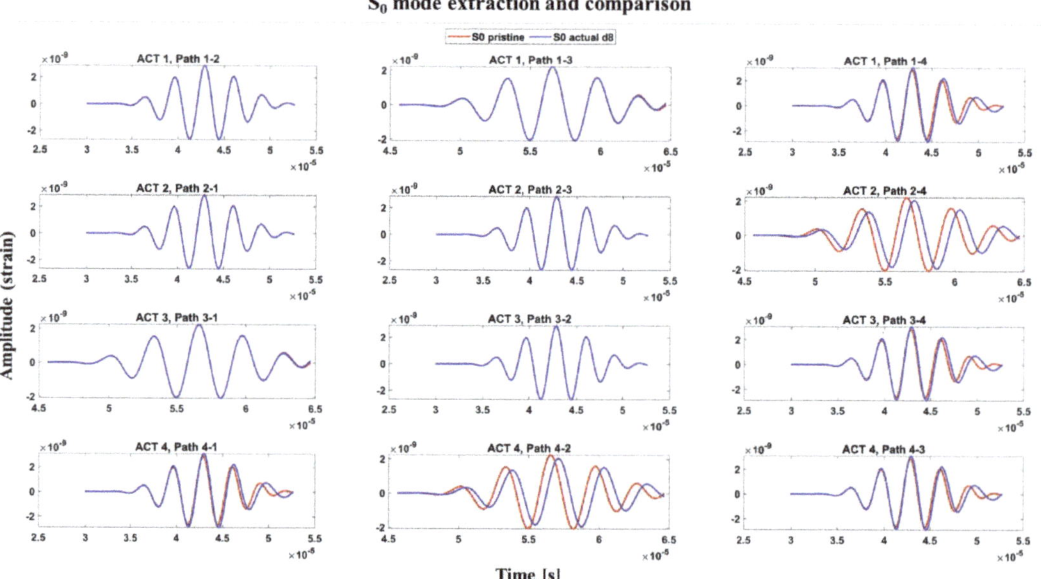

Figure 10. Comparison between the S_0 modes for the baseline and the actual configurations. Example for damage in position d_8.

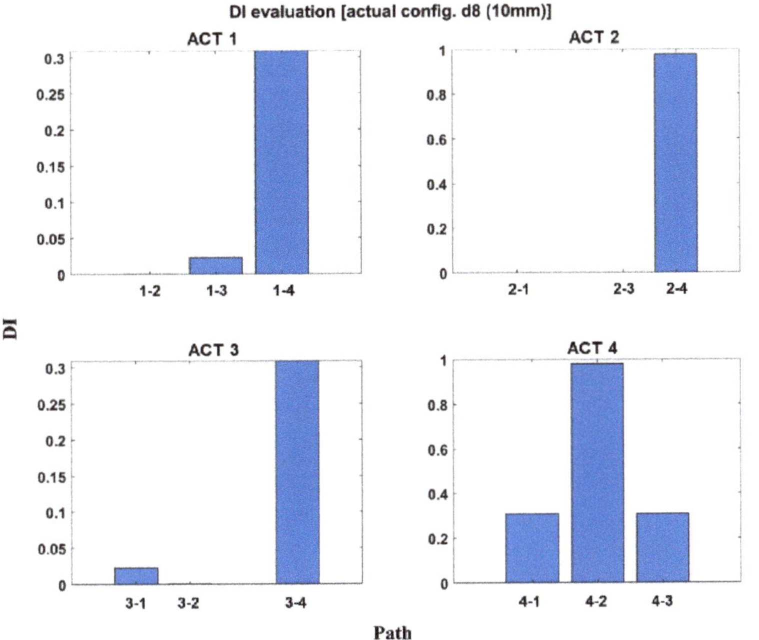

Figure 11. DI evaluation (damage of 10 mm). Example for damage in position d_8.

4.2. ANN Modeling

An Artificial Neural Network requires datasets for building up the input-output relation. In particular, as previously described, through an in-house code, it was possible to extract signal features for the input vector x_j. The ANN approach used in this work is based on the Damage Index of Equation (5) [22] that was described in the previous section. Therefore, the N-dimensional input vector consists of the DIs related to each sensing (actuator–receiver) path. The output vector instead includes the (x, y) coordinates of the localized damage. After generating the input and output vectors, the architecture of the ANN can be defined. It is clearly stated that the efficiency of the ANN increases when the dimension of the dataset increases. However, at the same time, this leads to an increase of the computational cost.

As shown in Figure 12, the feed-forward neural network herein developed contains one input layer of 12 features (4 sensors and 3 paths, for a total of 12 inputs for each damaged configuration) and one output layer of two features, namely the damage center coordinates (x, y). Two hidden layers with multiple (10) neurons were defined to connect the input and output data, resulting in a fully connected neural network, according to the literature [39]. Each k-th neuron of the hidden layer is connected with all neurons of the previous layer, and its output z_k can be described as follows:

$$z_k = \varphi\left(\sum_{j=0}^{N} W_{kj} x_j + b_k\right), \tag{6}$$

where x_j is the value of the input vector at the discrete position, with $j = 1, 2, \ldots N$; W_{kj} is the weight connected to each neuron k at the discrete position, with $j = 1, 2, \ldots N$; φ is the transfer (activation) function; and b_k is the bias value.

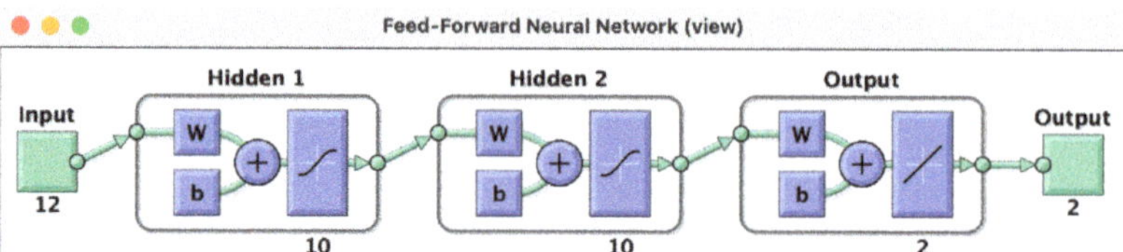

Figure 12. Architecture of the ANN includes one input layer, two hidden layers and one output layer; w, weights vector; b, bias vector (the image was produced by the MATLAB's ANN toolbox).

5. ANN Results and Discussion

5.1. Training of the ANN

In the current work, the Levenberg–Marquardt algorithm was used to train the ANN, as it appears to be the fastest method for training moderate-sized feedforward neural networks [40]. This algorithm is one of the variations of the back propagation algorithm, which is a gradient descent method in which the network weights are moved along the negative of the gradient of the performance function. In the current case, the chosen performance function is the mean squared error (MSE). The learning problem is considered to be solved when the combination of weights is able to minimize the error function.

The results of this neural network (Figure 13a) show that the gradient is close to zero, as expected, as well as the performance parameter. The u (Mu) parameter is the algorithm damping factor: it decreases after each successful step (reduction in the performance function), while it increases when a tentative step would result in a performance function increment. In this case, it starts from 0.001 as initial default value, and it decreases with a factor of 0.1, ending with a value of 0.0001.

The neural network was trained with fifteen of the nineteen 10 mm damage conditions, while the remaining four were used for validation and testing purposes. Figure 13 shows, as well, the performance obtained during the training, considering that the training was carried out in several steps on different datasets (always 15 of 19) in order to avoid the overspecialization of the network.

The ANN validation step consisted of the evaluation of the regression plot, which simply shows the relationship between the outputs of the network and the defined targets [41]. In detail, if the training were perfect, the network outputs and the targets would be exactly equal, but the relationship is rarely perfect in practice, especially when a limited dataset is adopted. Three regression plots for training, validation and testing data, together with the one for the overall dataset, are reported in Figure 13b. The dashed line in each plot represents the perfect result, i.e., outputs = targets. The solid line represents the best fit linear regression line between outputs and targets. The R value is an indication of the relationship between the outputs and targets. $R = 1$ indicates that there is an exact linear relationship between outputs and targets, while if R is close to zero, then there is no linear relationship between outputs and targets. For the network herein proposed, the data indicate a good fit: the network is sufficiently accurate even though a limited dataset was used for the training phase.

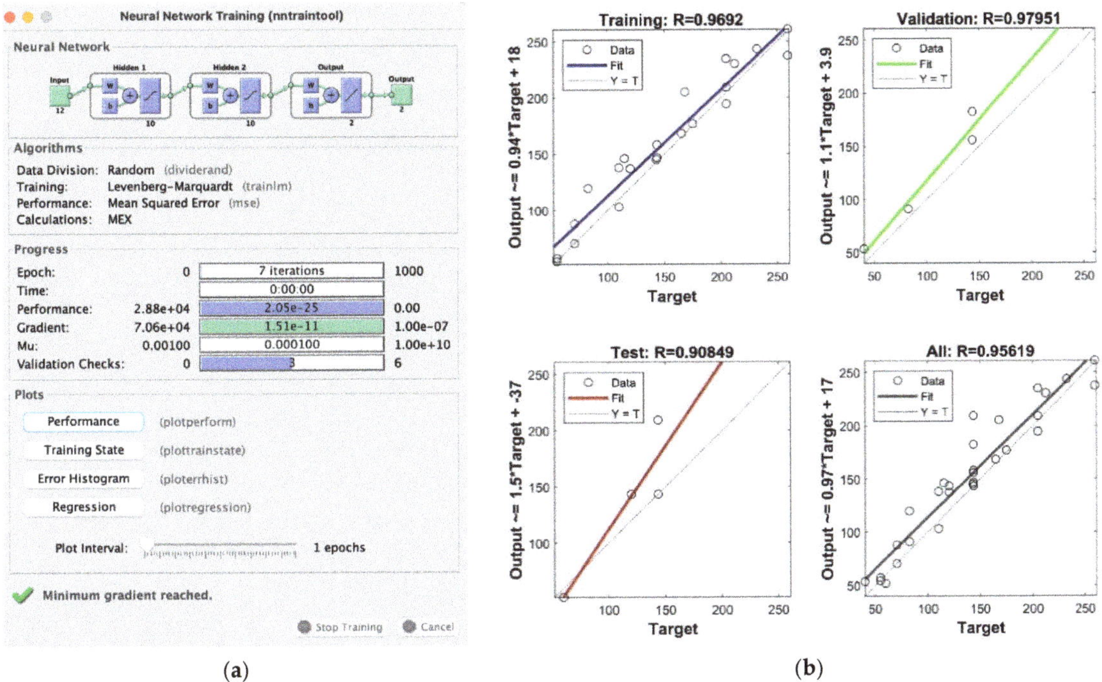

(**a**) (**b**)

Figure 13. (**a**) ANN performance and (**b**) regression plot.

5.2. ANN Validation and Tests

In this section, the results provided by ANN are shown. Specifically, the fully trained ANN was validated by means of the remaining four 10 mm damaged test cases. Figure 14 shows a good agreement between the predicted damage locations and the modeled ones. When the inputs do not belong to the training set or to a neighborhood of it, the networks provide results that are different from the expected ones but comparable to them (Figure 14). Thus, the 5 mm damage dataset, not adopted for the training procedure, was used to test the ANN. It is herein underlined that, under 300 kHz carrier frequency, the minimum detectable damage should be 7 mm. Thus, the capability of the ANN to identify and successfully localize damages smaller than the defined threshold must be verified. According to Figure 15, the ANN is also capable of predicting smaller damages; however, it does so with a reduced but acceptable accuracy.

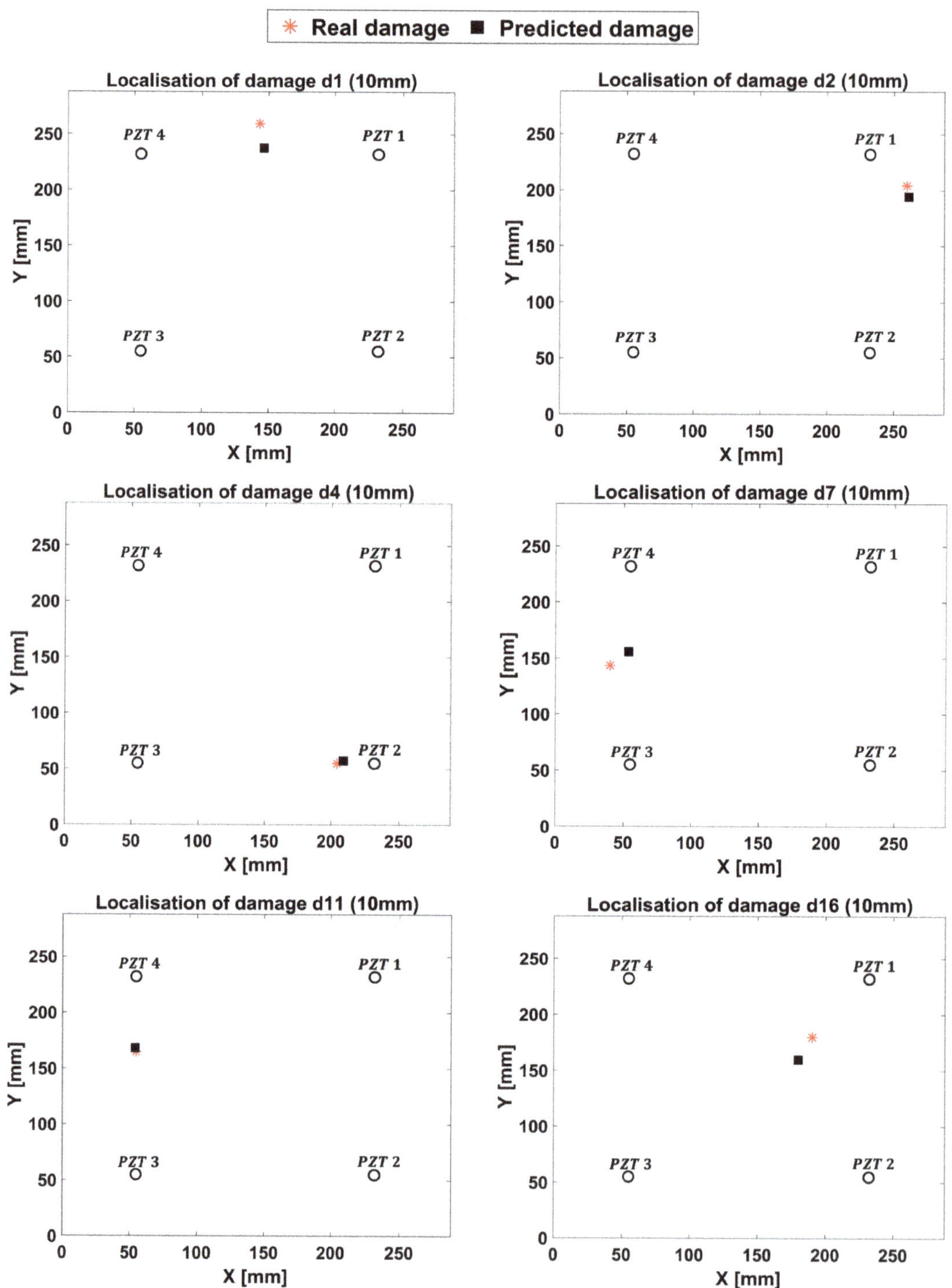

Figure 14. Results of ANN (10 mm damage).

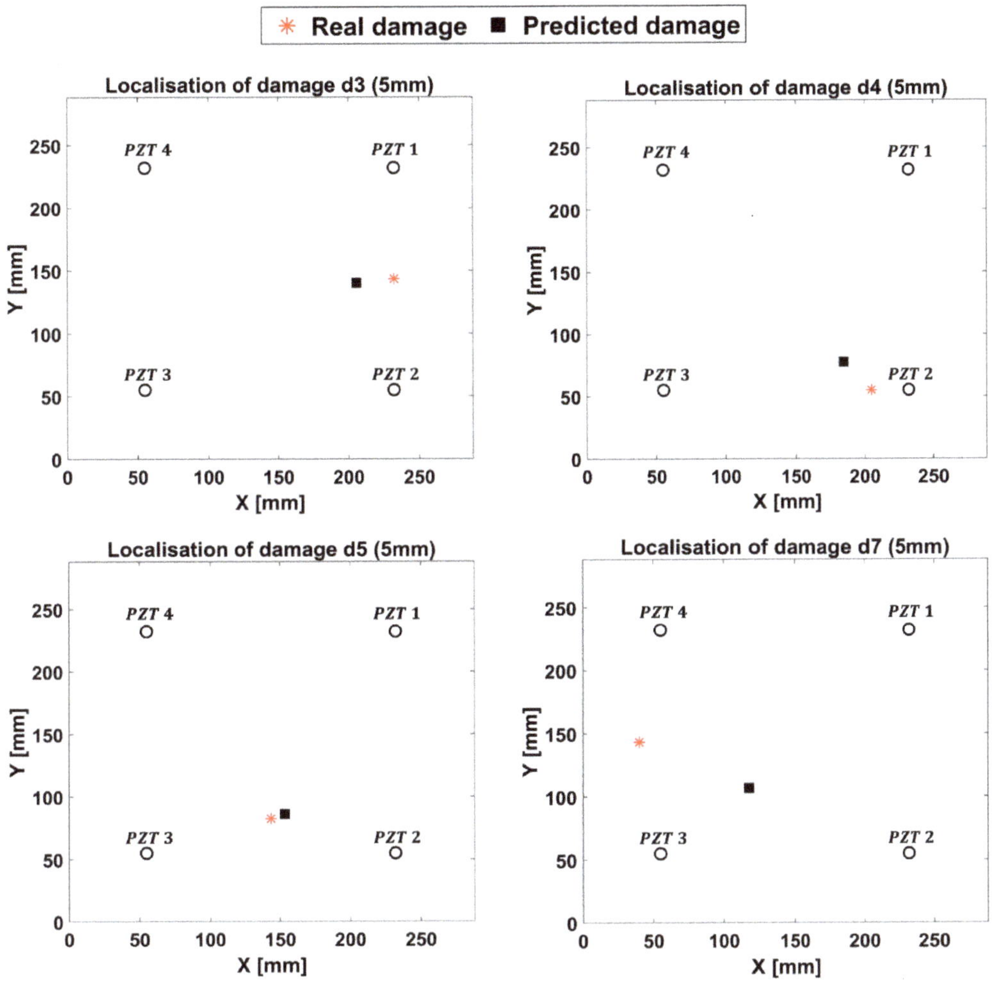

Figure 15. Results of ANN for the 5 mm damaged configurations.

5.3. ANN for a Damaged Composite Panel

The ANN developed, tested and used for damage detection for the aluminum panel was herein used also to predict the damage location in a CFRP (carbon fiber-reinforced polymer) composite panel.

The CFRP composite panel is made of eight layers with stacking sequence $[0/90/+45/-45]_S$. The thickness of the plate is t = 1.5 mm, while the horizontal and vertical dimensions are the same of the aluminum panel of Figure 1. The 0° fiber direction is oriented along the x-axis. Each layer of the prepreg is assumed to behave as an orthotropic material, and the mechanical properties of the lamina are listed in Table 6.

Table 6. Lamina mechanical properties of CFRP composite.

E_{11} (GPa)	E_{22} (GPa)	E_{33} (GPa)	G_{12} (GPa)	G_{13} (GPa)	G_{23} (GPa)	ν_{12}	ν_{13}	ν_{23}	ρ (kg m^{-3})
105	7.7	7.7	3.6	3.6	2.7	0.36	0.36	0.4	1540

In detail, the 20 NPW 2D-Shell FE modeling approach was chosen for the plate, while the 10 mm damage was modeled again by degrading the elastic material properties of the corresponding Finite Elements (softening technique) of about 70%. In this case, four damaged configurations, as shown in Figure 16, were investigated separately. The ANN was applied to all of these new cases without any further training step with respect to the ANN applied to the aluminum plate.

The main idea is to verify if the ANN developed and trained for an isotropic plate is capable of predicting damage also for a composite one. According to the results shown in Figure 16, the ANN produced accurate results. Moreover, the distances between the real damage coordinates and the predicted ones are reported in Table 7. As visible, the distance (error) is lower than 73 mm, and that can be considered to be acceptable, taking into account the previous considerations about the ANN development and the fact that the ANN was trained by using only the aluminum dataset. The error is expected to decrease with an improved training phase. It must also be noticed that the error equal to 73 mm represents a singularity. In fact, according to Table 7, the other errors range between 7.8 and 29.3 mm.

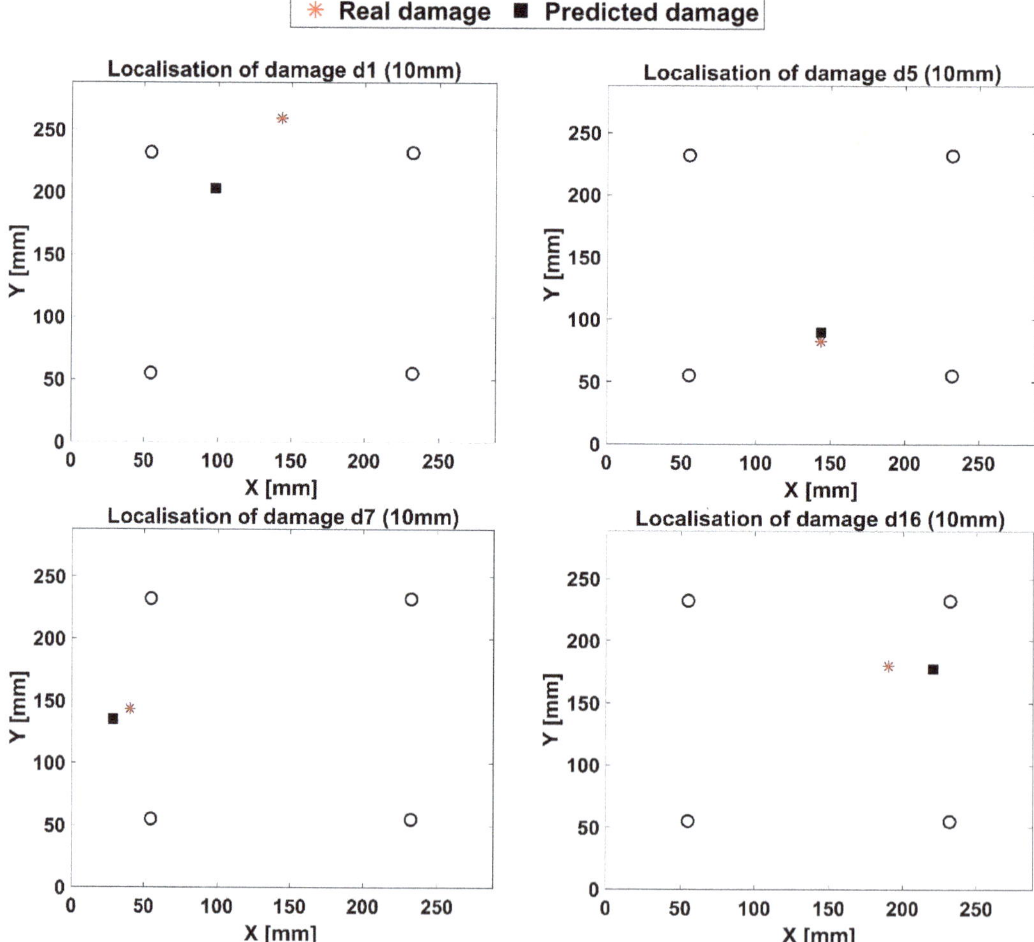

Figure 16. Results of ANN for the composite 10 mm damaged configurations.

Table 7. ANN accuracy for damage prediction in the composite panel.

Configuration #	Real Damage Coordinates (mm)	ANN Predicted Damage Coordinates (mm)	Distance (Error) between Points (mm)
d_1	[143.5, 259.5]	[96.4, 203.5]	73.1
d_5	[143.5, 82.5]	[141.9, 90.2]	7.8
d_7	[40, 143.5]	[27.7, 135.3]	14.8
d_{16}	[190, 180]	[219.2, 177.7]	29.3

6. Conclusions

In this paper, a machine learning approach based on ANN for damage detection and localization was proposed. Specifically, this paper dealt with the development and the assessment of an ANN for the damage detection based on the guided wave propagation method. The ANN was developed and assessed through the Finite Element Method and then used to simulate guided wave propagation in an aluminum plate under different damage configurations. A first step of the research activities herein proposed was addressed to the development of a reliable FE model aimed to calculate the S_0 and A_0 dispersion curves. In particular, three FE modeling techniques were investigated: the former based on 2D-Shell elements, the second based on 3D-Shell elements and the latter based on 3D-Solid elements. The reliability of the modeling techniques was assessed against experimental data, as well as semi-analytic ones provided by Dispersion Calculator software. According to the numerical–experimental–analytical comparison of the results, 2D-Shell FE model was chosen for the development of the dataset useful for the ANN, allowing achieving the highest level of accuracy to the computational time ratio. The possibility to use an accurate and faster method to train an ANN is of relevant importance, enabling the training with respect to a wider dataset.

Concerning the development of the ANN, it was trained with respect to an aluminum plate under different damage configurations. Square-shaped damages with a size of 5 and 10 mm were modeled at different points of the plate. The Damage Index dataset was used as the input vector for the ANN, while the coordinates of the modeled damages were the targets. The procedure is based on an in-house code that automatically extracts the S_0 mode from the baseline and actual states of the structure and calculates the DIs.

The trained ANN showed a good performance in terms of regression (even if a small dataset was used for the training). The ANN was found to be able to detect all damages with an acceptable level of accuracy. Because of the proven ANN reliability, its detection capability was also assessed on a different plate made of CFRP laminate, modeled through FEM. It is important to highlight that data achieved by the simulations involving the CFRP plate were not used for the ANN training phase. Nevertheless, the ANN showed its reliability in damage detection yet again.

Author Contributions: Conceptualization, A.D.L. and D.P.; methodology, A.D.L., D.P. and M.P.; software D.P. and M.P.; validation, F.C. and G.L.; investigation, A.D.L. and D.P.; writing—original draft preparation, D.P.; writing—review and editing, A.D.L., F.C. and G.L.; supervision, F.C. and G.L.; project administration, A.D.L.; funding acquisition, A.D.L. All authors have read and agreed to the published version of the manuscript.

Funding: This research was funded by the University of Campania "Luigi Vanvitelli" in the framework of "SAFES—Smart pAtch For active Shm" funded research project, as part of V:ALERE 2020 program.

Institutional Review Board Statement: Not applicable.

Informed Consent Statement: Not applicable.

Data Availability Statement: Not applicable.

Conflicts of Interest: The authors declare no conflict of interest.

References

1. Diamanti, K.; Soutis, C. Structural Health Monitoring Techniques for Aircraft Composite Structures. *Prog. Aerosp. Sci.* **2010**, *46*, 342–352. [CrossRef]
2. Doebling, S.W.; Farrar, C.R.; Prime, M.B.; Shevitz, D.W. *Damage Identification and Health Monitoring of Structural and Mechanical Systems from Changes in Their Vibration Characteristics: A Literature Review*; Los Alamos National Lab. (LANL): Los Alamos, NM, USA, 1996.
3. Wang, B.; Zhong, S.; Lee, T.-L.; Fancey, K.S.; Mi, J. Non-Destructive Testing and Evaluation of Composite Materials/Structures: A State-of-the-Art Review. *Adv. Mech. Eng.* **2020**, *12*, 1687814020913761. [CrossRef]
4. Raghavan, A.; Cesnik, C.E.S. Review of Guided-Wave Structural Health Monitoring. *Shock. Vib. Dig.* **2007**, *39*, 91–114. [CrossRef]
5. Rocha, H.; Semprimoschnig, C.; Nunes, J.P. Sensors for Process and Structural Health Monitoring of Aerospace Composites: A Review. *Eng. Struct.* **2021**, *237*, 112231. [CrossRef]
6. Gorgin, R.; Luo, Y.; Wu, Z. Environmental and Operational Conditions Effects on Lamb Wave Based Structural Health Monitoring Systems: A Review. *Ultrasonics* **2020**, *105*, 106114. [CrossRef]
7. Rose, J.L. A Baseline and Vision of Ultrasonic Guided Wave Inspection Potential. *J. Press. Vessel. Technol.* **2002**, *124*, 273–282. [CrossRef]
8. Su, Z.; Ye, L. *Identification of Damage Using Lamb Waves: From Fundamentals to Applications*; Pfeiffer, F., Wriggers, P., Eds.; Lecture Notes in Applied and Computational Mechanics; Springer: London, UK, 2009; Volume 48.
9. Wilcox, P.D. A Rapid Signal Processing Technique to Remove the Effect of Dispersion from Guided Wave Signals. *IEEE Trans. Ultrason. Ferroelectr. Freq. Control.* **2003**, *50*, 419–427. [CrossRef]
10. De Fenza, A.; Sorrentino, A.; Vitiello, P. Application of Artificial Neural Networks and Probability Ellipse Methods for Damage Detection Using Lamb Waves. *Compos. Struct.* **2015**, *133*, 390–403. [CrossRef]
11. Califano, A.; Chandarana, N.; Grassia, L.; D'Amore, A.; Soutis, C. Damage Detection in Composites by Artificial Neural Networks Trained by Using in Situ Distributed Strains. *Appl. Compos. Mater.* **2020**, *27*, 657–671. [CrossRef]
12. Fotouhi, S.; Pashmforoush, F.; Bodaghi, M.; Fotouhi, M. Autonomous Damage Recognition in Visual Inspection of Laminated Composite Structures Using Deep Learning. *Compos. Struct.* **2021**, *268*, 113960. [CrossRef]
13. Movsessian, A.; García Cava, D.; Tcherniak, D. An Artificial Neural Network Methodology for Damage Detection: Demonstration on an Operating Wind Turbine Blade. *Mech. Syst. Signal. Process.* **2021**, *159*, 107766. [CrossRef]
14. Padil, K.H.; Bakhary, N.; Abdulkareem, M.; Li, J.; Hao, H. Non-Probabilistic Method to Consider Uncertainties in Frequency Response Function for Vibration-Based Damage Detection Using Artificial Neural Network. *J. Sound Vib.* **2020**, *467*, 115069. [CrossRef]
15. Mishra, M. Machine Learning Techniques for Structural Health Monitoring of Heritage Buildings: A State-of-the-Art Review and Case Studies. *J. Cult. Herit.* **2021**, *47*, 227–245. [CrossRef]
16. Su, Z.; Ye, L. Lamb Wave-Based Quantitative Identification of Delamination in CF/EP Composite Structures Using Artificial Neural Algorithm. *Compos. Struct.* **2004**, *66*, 627–637. [CrossRef]
17. De Oliveira, R.; Marques, A.T. Health Monitoring of FRP Using Acoustic Emission and Artificial Neural Networks. *Comput. Struct.* **2008**, *86*, 367–373. [CrossRef]
18. Andrej, K.; Bešter, J.; Kos, A. Introduction to the Artificial Neural Networks. In *Artificial Neural Networks: Methodological Advances and Biomedical Applications*; Suzuki, K., Ed.; InTech: Shanghai, China, 2011; pp. 1–18.
19. Qian, C.; Ran, Y.; He, J.; Ren, Y.; Sun, B.; Zhang, W.; Wang, R. Application of Artificial Neural Networks for Quantitative Damage Detection in Unidirectional Composite Structures Based on Lamb Waves. *Adv. Mech. Eng.* **2020**, *12*, 1687814020914732. [CrossRef]
20. Ghajari, M.; Khodaei, Z.S.; Aliabadi, M.H. Impact Detection Using Artificial Neural Networks. *Key Eng. Mater.* **2012**, *488–489*, 767–770. [CrossRef]
21. Sharif-Khodaei, Z.; Ghajari, M.; Aliabadi, M.H. Determination of Impact Location on Composite Stiffened Panels. *Smart Mater. Struct.* **2012**, *21*, 105026. [CrossRef]
22. De Luca, A.; Perfetto, D.; De Fenza, A.; Petrone, G.; Caputo, F. Guided Wave SHM System for Damage Detection in Complex Composite Structure. *Theor. Appl. Fract. Mech.* **2020**, *105*, 102408. [CrossRef]
23. Giurgiutiu, V. *Structural Health Monitoring with Piezoelectric Wafer Active Sensors*, 2nd ed.; Academic Press, Elsevier Inc.: Oxford, UK, 2014.
24. Michaels, J.E.; Lee, S.J.; Croxford, A.J.; Wilcox, P.D. Chirp Excitation of Ultrasonic Guided Waves. *Ultrasonics* **2013**, *53*, 265–270. [CrossRef]
25. Abaqus 6.14. *Abaqus 6.14 Analysis User's Guide*; Dassault Systems Simulia Corp.: Providence, RI, USA, 2014.
26. Reddy, J.N. *An Introduction to the Finite Element Method*, 3rd ed.; McGraw-Hill: Bengaluru, India, 2006.
27. Duczek, S.; Joulaian, M.; Düster, A.; Gabbert, U. Numerical Analysis of Lamb Waves Using the Finite and Spectral Cell Methods. *Int. J. Numer. Methods Eng.* **2014**, *99*, 26–53. [CrossRef]
28. Yang, C.; Ye, L.; Su, Z.; Bannister, M. Some Aspects of Numerical Simulation for Lamb Wave Propagation in Composite Laminates. *Compos. Struct.* **2006**, *75*, 267–275. [CrossRef]
29. Giurgiutiu, V. *Structural Health Monitoring of Aerospace Composites*; Academic Press, Elsevier Inc.: Oxford, UK, 2015.
30. Paget, C.A. Active Health Monitoring of Aerospace Composite Structures by Embedded Piezoceramic Transducers. Ph.D. Thesis, Royal Institute of Technology, Stockholm, Sweden, 2001.

31. Slatte, F.; Jovanov, K. Structural Health Monitoring for Aerospace Composite Structures—An Investigation of the Potential Using the Finite Element Method. Master's Thesis, Chalmers University of Technology, Gothenburg, Sweden, 2012.
32. German Aerospace Center (DLR)-Institute of Structures and Design-Center for Lightweight Technology the Dispersion Calculator. Available online: https://www.dlr.de/zlp/en/desktopdefault.aspx/tabid-14332/24874_read-61142/#/gallery/33485 (accessed on 5 April 2021).
33. De Luca, A.; Perfetto, D.; De Fenza, A.; Petrone, G.; Caputo, F. Guided Waves in a Composite Winglet Structure: Numerical and Experimental Investigations. *Compos. Struct.* **2019**, *210*, 96–108. [CrossRef]
34. Sharif-Khodaei, Z.; Aliabadi, M.H. Assessment of Delay-and-Sum Algorithms for Damage Detection in Aluminium and Composite Plates. *Smart Mater. Struct.* **2014**, *23*, 075007. [CrossRef]
35. Perfetto, D.; De Luca, A.; Lamanna, G.; Caputo, F. On the Prediction of Guided Wave Dispersion Curves in Plates for Health Monitoring Applications. *Macromol. Symp.* **2021**, *396*, 2000333. [CrossRef]
36. De Luca, A.; Perfetto, D.; De Fenza, A.; Petrone, G.; Caputo, F. A Sensitivity Analysis on the Damage Detection Capability of a Lamb Waves Based SHM System for a Composite Winglet. *Procedia Struct. Integr.* **2018**, *12*, 578–588. [CrossRef]
37. Lee, B.C.; Staszewski, W.J. Modelling of Lamb Waves for Damage Detection in Metallic Structures: Part I. Wave Propagation. *Smart Mater. Struct.* **2003**, *12*, 804–814. [CrossRef]
38. Lee, B.C.; Staszewski, W.J. Modelling of Lamb Waves for Damage Detection in Metallic Structures: Part II. Wave Interactions with Damage. *Smart Mater. Struct.* **2003**, *12*, 815–824. [CrossRef]
39. Pham, D.T.; Pham, P.T.N. Artificial Intelligence in Engineering. *Int. J. Mach. Tools Manuf.* **1999**, *39*, 937–949. [CrossRef]
40. Lourakis, M.I.A. A Brief Description of the Levenberg-Marquardt Algorithm Implemened by Levmar. *Matrix* **2005**, *3*, 2.
41. MathWorks. Matlab User's Guide (R2020a). Available online: https://it.mathworks.com/help/matlab/ (accessed on 30 November 2021).

 materials

Review

Textile-Based Mechanical Sensors: A Review

Zaiwei Zhou [1,†], Nuo Chen [2,†], Hongchuan Zhong [1], Wanli Zhang [1], Yue Zhang [1,3,*], Xiangyu Yin [3,4,*] and Bingwei He [1,3,*]

1 College of Mechanical Engineering and Automation, Fuzhou University, Fuzhou 350108, China; zzw2400@163.com (Z.Z.); N190220082@fzu.edu.cn (H.Z.); 13215041192@163.com (W.Z.)
2 Department of Mechanical and Energy Engineering, Southern University of Science and Technology, Shenzhen 518055, China; 11810222@mail.sustech.edu.cn
3 Fujian Engineering Research Center of Joint Intelligent Medical Engineering, Fuzhou 350108, China
4 College of Chemical Engineering, Fuzhou University, Fuzhou 350108, China
* Correspondence: yuezhang@fzu.edu.cn (Y.Z.); xyin65@fzu.edu.cn (X.Y.); bw_he@aliyun.com (B.H.)
† These authors contributed equally to this work.

Abstract: Innovations related to textiles-based sensors have drawn great interest due to their outstanding merits of flexibility, comfort, low cost, and wearability. Textile-based sensors are often tied to certain parts of the human body to collect mechanical, physical, and chemical stimuli to identify and record human health and exercise. Until now, much research and review work has been carried out to summarize and promote the development of textile-based sensors. As a feature, we focus on textile-based mechanical sensors (TMSs), especially on their advantages and the way they achieve performance optimizations in this review. We first adopt a novel approach to introduce different kinds of TMSs by combining sensing mechanisms, textile structure, and novel fabricating strategies for implementing TMSs and focusing on critical performance criteria such as sensitivity, response range, response time, and stability. Next, we summarize their great advantages over other flexible sensors, and their potential applications in health monitoring, motion recognition, and human-machine interaction. Finally, we present the challenges and prospects to provide meaningful guidelines and directions for future research. The TMSs play an important role in promoting the development of the emerging Internet of Things, which can make health monitoring and everyday objects connect more smartly, conveniently, and comfortably efficiently in a wearable way in the coming years.

Keywords: textile-based mechanical sensors; mechanism; preparation; advantages; applications

Citation: Zhou, Z.; Chen, N.; Zhong, H.; Zhang, W.; Zhang, Y.; Yin, X.; He, B. Textile-Based Mechanical Sensors: A Review. *Materials* **2021**, *14*, 6073. https://doi.org/10.3390/ma14206073

Academic Editor: Daniela Iannazzo

Received: 8 September 2021
Accepted: 6 October 2021
Published: 14 October 2021

1. Introduction

As measuring devices, mechanical sensors can obtain mechanical stimulus information and convert it into electrical signals to achieve functions such as stimulus acquisition, transmission, and storage. When mechanical stimuli, such as pressure, strain, motion, etc., is applied, the mechanical sensor can convert it into electrical signals such as resistance, voltage, and capacitance, and based on the mechanical stimulus-electrical signal regression curve, we can infer the magnitude of the mechanical stimulus. In accordance with different mechanical stimuli, mechanical sensors can be substantially divided into strain sensors, pressure sensors, position sensors, velocity sensors, tactile sensors, etc. Traditional mechanical sensors made of metal, semiconductor, or ceramic have stable performance and mature preparation technology, which have played a major role in traditional industries. However, their inherent shortcomings, such as rigidity, large size and weight, and small deformability, indicate that they are not applicable to the fields of electronic skins, human physiological signals, intelligent robots, etc. [1]. Flexible materials usually feature softness, easy deformation, and light weight [2,3]. Some novel flexible materials exhibit unique properties such as self-healing, hydrophobicity, biocompatibility, and biodegradability [4–6], which make them more competitive when compared with rigid mechanical sensors [7–10].

Textile materials are considered as promising new versions of silicon wafers in wearable electronics, not only because they have the properties of most flexible materials, but also because they hold the advantages of low cost, good conformality, comfort, and wearability [11]. Therefore, textile materials have shown great application potential in wearable electronics, human-machine interaction, smart fabrics, etc. [12–14].

Some applications of textile-based mechanical sensors (TMSs), such as monitoring heartbeat, pulse, and other health signals [15–17], constructing robot human-like tactile perception function [18–20], and fabricating human-movement monitor smart clothing [21–23], are already research hotspots for wearable electronics. In these applications, epidermic sensors are also strong competitors. Recent studies suggest that ultrathin multimodal devices based on piezoelectric or triboelectric polymers may be attractive candidates for the development of imperceptible mechanical sensors in direct contact with the human body [24]. Unlike TMSs, the imperceptible epidermic sensors fit directly on the skin due to their thinness, small size, and even transparency, whereas TMSs are woven into clothing to be hidden. Similarly, tattoo sensors, which can be sub-micron thick and monitor the human body information without affecting the senses and aesthetics, are attracting the interest of researchers [25]. Compared to other types of mechanical sensors, TMSs have natural advantages in monitoring large areas, large deformation mechanical signals, and biocompatibility due to their weaving method and textile substrates.

However, the research on TMSs is still in the laboratory stage, and there are many scientific and technical problems to be solved, such as the inability to balance good electrical conductivity and air permeability, poor durability, and improved integration of active materials and fabrics, etc. Although some reviews have been reported on the principles, materials, preparation methods, and application areas of textile-based sensors [26–29], there are few reviews that systematically introduce TMSs. Here, we adopt a novel approach to introduce different kinds of TMSs by combining sensing mechanisms, textile structure, and novel fabricating strategies. In particular, we highlight the unique pros and cons of the TMSs, and summarize methods to improve existing deficiencies. To summarize the recent progress, we focus on the latest results and applications of TMSs, and for these, we present the prospects and challenges of TMSs.

In this review, we systematically introduce features and latest achievements in TMSs. Four types of sensing mechanisms of resistance, capacitance, piezoelectricity, and triboelectricity of TMSs as well as the configuration features of each type of device and their corresponding preparation and integration methods are introduced. The commonly used substrate materials, sensor materials, and preparation methods of conductive fiber, yarn, or fabric are also summarized. Next, the advantages and sensing performances of TMSs are analyzed, and the designs related to improving the performance and advantages of TMSs are discussed. The advanced applications based on TMSs are then described in detail. Finally, the challenges and development directions of the device are further discussed.

2. Sensing Mechanisms, Materials, and Preparations

TMSs can introduce electrical conductivity in three structural forms: fiber, yarn, and fabric [25]. Among them, conductive fibers can be inherently conductive metal wires of silver, copper, and stainless steel, or they can be made by blending conductive materials, such as polyaniline (PANI), polypyrrole (PPy), and graphite, carbon nanotubes (CNTs) into natural (e.g., cotton, silk, and bamboo) and synthetic (e.g., nylon, vinylon, and polypropylene) fibers, or by directly coating or plating these conductive materials onto polymer (polypropylene (PP), polyethylene terephthalate (PET), polyimide (PI), etc.) fibers [30]. The obtained conductive fibers can be twisted to make conductive yarns, and by subsequent weaving and knotting can be made into conductive fabrics. In addition, conductive fabrics can also be made by directly dip-coating, spin coating, or printing conducting inks onto normal fabrics. TMSs in three structural forms of fiber, yarn, and fabric can be integrated into textiles or woven into clothing. Fibers or yarns are the first choice for TMSs as they can be easily converted into fabrics or combined with conventional fabrics by textile techniques (e.g., weaving or knitting). TMSs are mainly categorized as resistive, capacitive, piezoelectric, and triboelectric [31] according to their sensing mechanism. The different sensing mechanisms and their implementation methods are described in detail below.

2.1. Resistive Sensor

Resistive TMSs convert mechanical stimuli, such as displacement or force to a resistance change, using piezoresistive materials. As the resistance of a conductive material is defined as $R = \rho L/S$, when a mechanical stimulus causes changes of the piezoresistive materials in resistivity (ρ), length (L), and/or cross-sectional area (S), it will bring out a resistance change. The sensing response of resistive TMSs depends on the interaction of these main factors: (1) intrinsic changes in the resistance of sensing elements in response to mechanical stimuli; (2) geometric variation of assembled devices; and (3) changes in the conduction network. Based on a regression curve, the mechanical stimuli and their degrees can be determined. Because resistive TMSs often have the advantages of high sensitivity, wide detection range, high precision, and simple measurement circuits, they have received the most extensive attention and study [32]. However, large signal drift, poor durability, and obvious hysteresis are the key issues that restrict their practical application [33].

Resistive TMSs are typically composed of a soft substrate and a sensing material [34]. The soft substrate needs have properties such as a certain elasticity, good flexibility, and long-term stability. These properties can provide a carrier for sensing materials, and endow the sensing materials and subsequent textiles with piezoresistive properties. They can also reduce the stress concentration of TMSs when subjected to mechanical stimuli. Commonly used substrate materials for TMSs include silk, cotton, polydimethylsiloxane (PDMS), polyurethane (PU), etc. Correspondingly, commonly used sensing materials have carbon materials, metal materials, conductive polymers, etc. The sensing materials need to be both conductive and mechanically robust, which are typically prepared by coating, depositing, winding, or electroplating functional conductive layer on fibers, yarns, or fabrics, and they can also be prepared by wet spinning or 3D printing processes. Functional fibers and yarns can be attached to complex surfaces or woven into fabrics, which makes the sensors adaptable to different application scenarios by changing their shape and can be prepared in a sustainable and large-scale way. Neves et al. [35] produced conductive fibers by coating graphene onto polymer fibers that could be bent, stretched, compressed, twisted, and deformed into complex shapes while still maintaining good performance and reliability (Figure 1a). These graphene-based conductive fibers can be utilized as a platform for constructing integrated electronics directly in textiles. The stretch property of the conductive coating layer is often significantly lower than that of the textile substrate, which often leads to the formation of cracks in large deformations, resulting in reduced sensing stability. Additionally, as the textile surface is coated with a relatively rigid layer of conductive material, the feel and comfort of the fabric will be different. Composite conductive fiber—fiber adding conductive materials—can maintain the characteristics of the main fiber, such as the feeling and wear resistance. The structure of composite fibers is more integrated than that of coated textiles, effectively improving the problem of micro-cracks on the surface of coated fibers. Li et al. [36] used a syringe to extrude a mixture of conductive multiwalled carbon nanotubes (MWCNTs) and PDMS through a mesh with micron-sized holes to fabricate functional fibers (Figure 1b). The fibers as part of a wearable sensor were then integrated into a smart glove to recognize finger dexterity, gestures, and temperature signals. This preparation method is simple and convenient, but it is not suitable for large scale preparation of functional fibers. Wang et al. [37] used a concentric nozzle to rapidly and precisely print nanofibers with a bilayer structure (Figure 1c). The inner layer of the nanofiber acts as a sensing layer composed of metal or conductive polymer materials, whereas the outer layer acts as a protective and supportive layer that consists of long-chain polymer materials. The smart mask made from the nanofibers by a one-step progress can be used to detect whether the mask is worn properly and whether breathing is abnormal. Additionally, traditional textile manufacturing technologies of wet spinning and electrostatic spinning are also applicable to fabricate functional fibers. Sheng et al. [38] prepared porous fiber-based strain sensors by the wet-spinning method, wherein thermoplastic polyurethane was used as elastomer and carbon nanotubes (CNTs) and graphene as conductive fillers (Figure 1d). Before wet-spinning, dispersants and binders were introduced to improve the interaction between the elastomer and the conductive fillers to achieve the purpose of effectively withstanding external forces. Qi et al. [39] used a simple electrostatic spinning technique to prepare nanofiber sensing yarn, which was composed of a fibrous core electrode wrapped and wound by piezoresistive elastic nanofibers (Figure 1e). The yarn showed a fine layered structure, and could be woven into fabrics to achieve multi-mode sensing of various mechanical stimuli.

Fabric-based mechanical sensors can also be designed and prepared by the methods of coating, deposition, inkjet printing, screen printing, etc. Among them, directly coating sensing material onto common fabrics is the simplest and easiest method to achieve large-scale TMSs preparation. However, this method will bring about a poor bond between the sensing material and the flexible fabrics, so that the stability and durability of the prepared TMSs cannot be guaranteed. Thus, how to improve the adhesion between the two materials has become the first problem to be solved in the preparation of high-performance TMSs, wherein functionalized molecular grafting sensing materials is one of the preferred methods. Liu et al. [40] coated fluorinated MXene nanosheets onto 15 different fabrics, because the surface of the MXene is rich in a large amount of functional groups that interact with the fabric surface to improve adhesion between the two. It has been experimentally proven that the MXene formed a strongest bond with pure cotton and will not come off even after washing and ultrasonic processing. In addition, the adhesion can be enhanced by improving the preparation process of the TMSs. A multifunctional mechanical-sensitive fabric is prepared via ultrasonically spraying reduced graphene oxide (rGO) and silver nanowires (AgNWs) onto synthetic and 100% natural cotton fabrics (Figure 1f). The obtained fabrics show a good durability and can be washed repeatedly without performance degradation [41]. Luo et al. [42] used simple and efficient screen printing to transfer high-performance AgNW inks onto stretchable fabrics (Figure 1g), which presented excellent tensile properties and sensing performance. Conveniently, sensing materials with different patterns can be printed by simply changing the screen with different shapes, and fabric-based mechanical sensors prepared by printing processes can be designed into desired patterns to improve the sensing range, sensitivity, and other properties of resistive sensors. In summary, the coating and screen-printing method is easy to implement TMSs with superior sensitivity and a relatively large sensing range. However, the low linear correlation and cyclic stability of the sensing layer greatly limit its practical application.

2.2. Capacitive Sensor

Capacitive TMSs are realized based on the capacitance changes of the sensing devices induced by external mechanical stimuli. The capacitance of a sensing device is defined as $C = \varepsilon_r \varepsilon_o A/d$, where ε_r is the relative permittivity, ε_o is the vacuum permittivity, A is the effective area of the electrode, and d is the pole-plate spacing [43]. Therefore, the change of one or more parameters of the permittivity, spacing, or effective area will cause a change in capacitance of the device, and then the magnitude of the mechanical stimulus that causes the parameter change can be quantified. Capacitive TMSs feature properties of high response repeatability, small signal drift, long term cycle stability, and low energy consumption, but they are susceptible to external field interference, relatively low sensitivity, and limited sensing range [44].

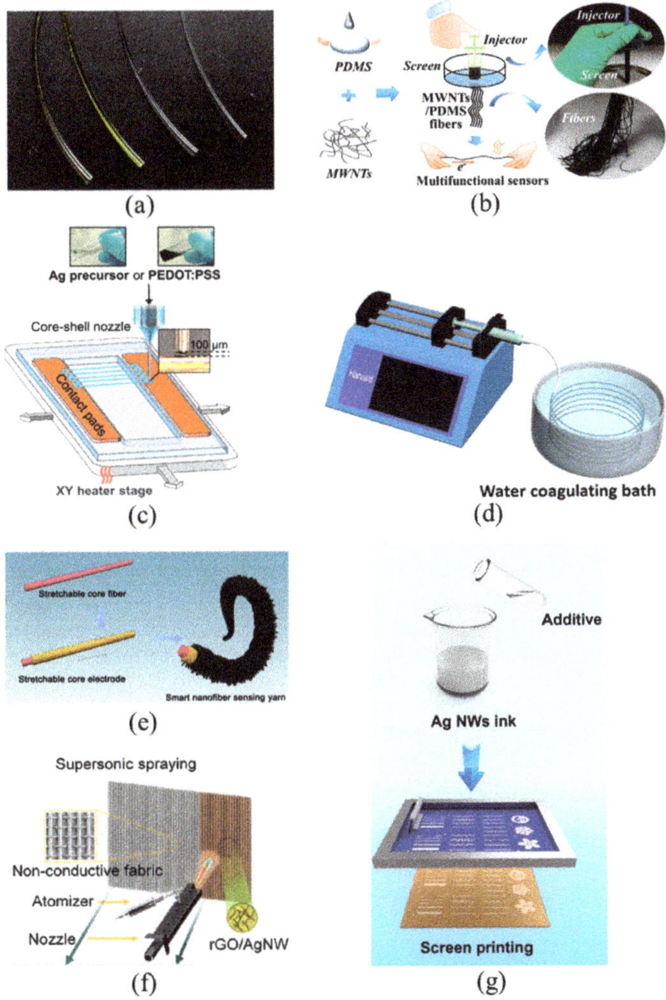

Figure 1. Illustration of the manufacturing process of various resistive textile-based mechanical sensors (TMSs). PDMS means polydimethylsiloxane, MWNTs means multi-walled carbon nanotubes, PEDOT:PSS means poly(3,4-ethylenedioxythiophene)/poly(styrenesulfonate), rGO means reduced graphene oxide and AgNWs means silver nanowires. Fiber-based sensors prepared by (**a**) coating, reproduced with permission from [35]; (**b**) extrusion, reproduced with permission from [36]; (**c**) printing, reproduced with permission from [37]; and (**d**) wet spinning, reproduced with permission from [38]; (**e**) Yarn-based sensors prepared by electrostatic spinning, reproduced with permission from [39]; Fabric-based sensors prepared by (**f**) spraying, reproduced with permission from [41]; and (**g**) screen printing, reproduced with permission from ref. [42].

In contrast, the fabrication of capacitive TMSs is more challenging than for resistive ones, because capacitive TMSs usually consist of two electrode layers and a dielectric layer [45], wherein the electrodes require good electrical conductivity; commonly used electrode materials include conductive fabrics, metal wires, carbon materials, etc. Although each electrode needs to be conductive, the capacitive response is irrelevant to the change in resistance of the electrode during exposure to mechanical stimulation. At the same time, the materials used as dielectric layers usually possess a large dielectric constant to reduce leakage current. Commonly used dielectric materials are elastic polymers, fabric gaskets, ionic gels, etc.

The preparation method of the capacitive TMSs is similar to that of the resistive ones, but the configurations of the two devices are different. The device configuration of a capacitive TMS can be roughly divided into two types. The first configuration is constructed as a sandwich structure, which consists of two flat electrodes composed of conductive fabrics and a dielectric layer composed of common fiber membranes or ionic gel membranes sandwiched between the electrodes. This type of structure is the most common one, which usually endows the device with a large sensing range. Keum et al. [46] prepared sandwich-structure ion-electron pressure TMSs using silver-plated compound silk fibers as electrodes and high-permittivity ion gel membranes as the dielectric material. The composition and membrane thickness of the ionic gel were designed to maximize the change of the contact area between the conductive fabrics and the ionic gel under external forces, which in turn optimized the sensing performance of the devices. Fu et al. [47] prepared a flexible pressure TMS by using fabrics sprayed with AgNWs as the flexible electrodes and a ceramic nanofiber film fabricated via an electrostatic spinning process as the dielectric layer (Figure 2a). The obtained pressure sensor can be used to detect human health conditions and motion, such as pulse, vocal cord vibration, and body movement, etc. The second type of configuration of capacitive TMSs is equipped with a core-sheath structure fabricated by coating and coaxial spinning. Compared with the capacitive TMSs with a sandwich structure, the TMSs with a core-sheath structure are smaller and easier to embed into clothes. Capacitive TMSs can be obtained by fiber crossing. Guan et al. [48] prepared silver nanowire-bacterial cellulose fibers with porous structures using a wet-spinning process and then coaxially coated the fibers with PDMS to produce functional fibers with a core-sheath structure (Figure 2b). A capacitive multifunctional sensor was fabricated by arranging the functional fibers crosswise to form an interpenetrating network, in which the AgNWs-bacterial cellulose fibers served as electrodes and the PDMS coating acted as the dielectric layer. By detecting changes in capacitance, the sensor could detect both the pressure and the position of objects. Additionally, Zhang et al. [49] prepared a high-performance capacitive strain sensor by twisting two core-spun yarns into a fine double-ply yarn (Figure 2c). The core-spun yarns were fabricated by wrapping silver-coated nylon fibers with cotton fibers, and then they were fixed with polyurethane. The sensor exhibits good linearity and tensile properties and can be blended into wearable fabrics to monitor athletes and patients without compromising lifestyle or comfort.

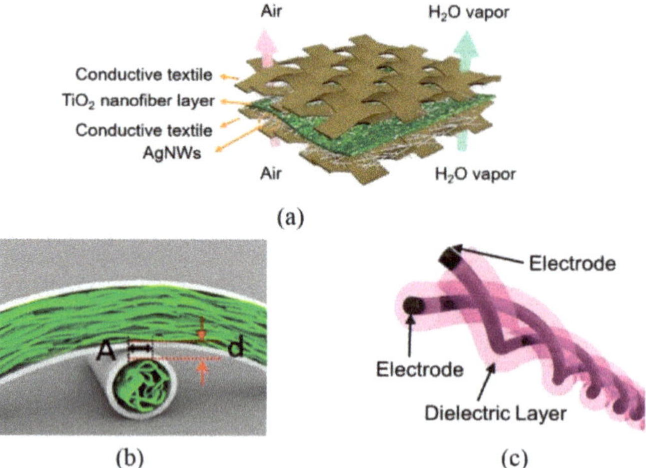

Figure 2. *Cont.*

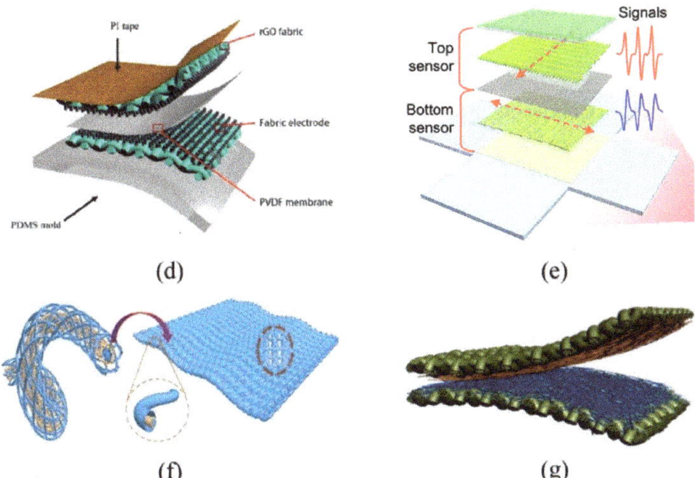

(d) (e)

(f) (g)

Figure 2. Schematic diagrams of TMSs configurations based on capacitive, piezoelectric, and triboelectric sensing mechanisms. PDMS means polydimethylsiloxane, PI means polyimide, rGO means reduced graphene oxide and PVDF means polyvinylidene fluoride. (**a**) sandwich structure fabric capacitive sensors, reproduced with permission from [47]; (**b**) crossed fiber capacitive sensors, reproduced with permission from [48]; (**c**) helix fiber capacitive sensors, reproduced with permission from [49]; (**d**) sandwich structure piezoelectric sensors, reproduced with permission from [50]; (**e**) double-layer piezoelectric sensors with vertical arrangement, reproduced with permission from [51]; (**f**) coaxial fiber triboelectric sensors, reproduced with permission from [52]; and (**g**) double-layer triboelectric sensors, reproduced with permission from ref. [53].

2.3. *Piezoelectric Sensor*

Piezoelectric TMSs are produced from flexible materials with piezoelectric effects, which work by converting mechanical stimuli into voltage signals [54]. The piezoelectric constant of the piezoelectric material determines the performance of a piezoelectric sensor in converting mechanical energy into electrical energy. Commonly used piezoelectric materials include composites, polymers, ceramics, single crystals, etc. [55].

Piezoelectric TMSs can generate internal voltage when subjected to external pressure, which makes them self-powered while achieving pressure sensing. In addition, such TMSs often present the advantages of fast response time and high sensitivity [56], giving them great prospects in wearable devices. Tan et al. [50] prepared piezoelectric TMSs using the piezoelectric effect of the single-crystalline ZnO nanorods grown on conductive rGO-PET fabric (Figure 2d). The piezoelectric TMS is constructed with three layers consisting of polyvinylidene fluoride (PVDF) membrane, the top and bottom electrode layers of conductive rGO-PET fabrics with self-orientation ZnO nanorods. When subjected to an external force, the piezoelectric configuration deformed, leading to a potential difference between the two electrode layers so the magnitude of external force can be obtained by detecting the voltage change. Hong et al. [51] provided a new solution of designing anisotropic kirigami structures and manufacturing a functional piezoceramic network to monitor joint motions and distinguish between different motion modes, in which the piezoelectric composite is the core sensory element for the sensors, formed by a lead zirconate titanate (PZT) ceramic network with nylon textile with kirigami-structured honeycomb grids and a PDMS matrix (Figure 2e). Piezoelectric sensors show obvious advantages in measurement range, piezoelectric anisotropy, multifunctional measurement, and long-term monitoring, which greatly enhance their practical application range.

2.4. Triboelectric Sensor

The frictional initiation effect is a normal phenomenon in daily life. It occurs when a material is subjected to normal contact, sliding, or twisting. The combining of the electrification effect/triboelectric effect and electrostatic induction is the principle of triboelectric TMSs that occurs among a broad range of materials, including synthetic polymers and natural silk [57], wool [58], and cotton [59]. Similar to piezoelectric sensors, triboelectric TMSs can convert mechanical motion into electrical signals, and in turn, by analyzing the obtained signals, dynamic mechanical motions can be interpreted. By correlating the mechanical input with the corresponding parameters, a series of triboelectric TMSs have been fabricated, including pressure sensors, strain sensors, and vibration sensors. Triboelectric TMSs are generally composed of two electrodes with different tribo-polarities; the greater the difference in tribo-polarities between the two electrode materials, the better the electrical performance of the sensors [60]. Commonly used positively charged materials include nylon, cotton, silver, and copper, whereas commonly used negatively charged materials include PDMS, PVDF, polytetrafluoroethylene (PTFE), etc. [61]. The advantages of triboelectric TMSs, such as low cost, simple preparation process, high output voltage, and self-power, will strongly promote the construction of the Internet of Things [62].

Triboelectric TMSs can be divided into two kinds. One kind is a single yarn device with two frictional electrical sequences that are woven into fabrics or textiles. Zhang et al. [52] prepared a coaxial triboelectric yarn by sequentially wrapping PTFE and Ag yarns around axial metallized silver yarn via a winding machine (Figure 2f). The fabricated triboelectric yarn was then woven into wearable, multifunctional textile by needles to harvest the mechanical energy from human body motions. Since the tribo-polarities between PTFE and Ag materials differ greatly, charge transfer was easily achieved in repeated contact–separation processes. The second kind of triboelectric TMSs is obtained by directly weaving two types of yarns or fibers with inherently different tribo-polarities. Fan et al. [63] wove terylene wrapped stainless steel conductive yarns and nylon yarns into an all-textile triboelectric sensing array. Guo et al. [53] fabricated a textile based wearable hybrid triboelectric-piezoelectric TMS composed of silk fibroin nanofibers and PVDF nanofibers that were electrospun onto conductive fabrics as the triboelectric pair. Before fabricating a cloth-shape smart device, the two triboelectric fabrics were attached to separate substrates to realize effective contact and separation (Figure 2g). The hybrid TMS is capable of generating both triboelectricity and piezoelectricity at the same time to realize high power generation that enables its use as a sensor to identify various types of body motion without another power supply.

Each of the different types of TMSs has its own pros and cons. Resistor TMSs are by far the most widely studied and applied due to the simple principle and read-out circuit. Compared with other types, capacitor TMSs have a longer lifetime due to lower heat generation. In addition, they also show applications in non-contact measurements because of the sensing mechanism. Piezoelectric and triboelectric TMSs are difficult to detect static forces due to charge loss. Despite the fact that piezoelectric and triboelectric TMSs are limited by the inability to detect static forces and their applications are restricted, they are still a current research hotspot owing to their huge advantages of self-powering. How to overcome the shortcomings of piezoelectric and triboelectric TMSs that cannot detect static forces is a major challenge. The resistive and capacitive TMSs have wider applications in most scenarios. However, in some harsh environments, they need to replace the power supply in time, which necessitates more labor and material resources. Whereas the piezoelectric and triboelectric TMSs are self-powered, and the signal can be detected in the field for a long time, which greatly saves cost.

3. Advantages and Performance

3.1. Advantages

To identify and quantify mechanical stimuli in space, the method of arranging sensing units in membranes or blocks on a flexible substrate is usually adopted. This method is indeed very simple and practical in the application situation where the precision of mechanical stimuli is not high. However, the large area of each sensing unit, as well as the messy electrode port (generally, n sensing units need n + 1 electrode ports) and complex circuit layout result in the sensing units not being sufficiently close, hence the integrated devices often show a low spatial resolution in detecting mechanical stimuli. In contrast, TMSs offer great advantages over other flexible mechanical sensors in terms of large area manufacturing. The TMSs, using fine fibers, yarns, or fabrics as a distributed sensing network, greatly reduce the number of electrode ports (generally, m × n sensing points can be arranged with m + n electrodes) and achieve a high-resolution spatial recognition of mechanical stimuli. Sundaram et al. [64] assembled a sensor array on a knitted glove to detect tactile information (Figure 3a). The sensing array consists of a piezoresistive membrane connected by a network of conductive wire electrodes. This approach allows the detection of mechanical stimuli with very high spatial resolution. In addition to the use of piezoresistive membranes, sensing arrays can also be prepared directly using piezoresistive fibers. Luo et al. [65] proposed a strategy to prepare fibers by coating conductive stainless steel wires with piezoresistive nanocomposites using an automated coating technique (Figure 3b). Taking advantage of this strategy, the fibers can be fabricated into a sensing unit by a simple vertical overlapping progress, and can be subsequently woven into large-scale sensing textiles with arbitrary 3D shapes for spatially accurate detection of mechanical stimuli. In addition to preparing an array sensor, fibers are also able to detect mechanical stimuli in space by using an electrical time-domain reflectometer to send pulses to a separate transmission line, and the amplitude of the step indicates the magnitude of the pressure and the occurrence time of the step indicates the distance. For example, Leber et al. [66] prepared an elastic fiber that integrated dozens of liquid metal conductors with a uniform, complex cross-sectional structure. The fiber was arranged in a snake shape onto stretchable fabric, which was connected to an electrical time-domain reflectometer via a single contact (Figure 3c). An electrical time-domain reflectometer could detect the location and magnitude of mechanical stimuli by transmitting high-frequency pulses to a transmission line and then reflecting them at discontinuities. This approach makes the structure of the sensing fabric simpler, with only one electrode port. However, it is difficult to integrate the sensing fabric into a wearable device due to the need of an electrical time-domain reflectometer.

During practical applications, mechanical sensors are subject to various mechanical stimuli such as stretching, squeezing, bending, etc., which require a good deformability of the electrodes and sensing elements in mechanical sensors, a stable sensing function in a large deformation state, and a structural and electrical performance reversibility after the release of deformation [67]. TMSs have unique advantages in terms of deformability, which is generally obtained by optimal structural and material design. Gao et al. [68] uniformly sprayed CNTs dispersion onto PU nanofiber substrate and then obtained CNTs/PU fibers with helical structure by twisting to an over-twisting state. The CNTs/PU fibers could reach 900% deformation range (Figure 3d), because the helical structure has the advantage in tensile strain and strength compared to fiber film and fiber bundle structures, which underwent unscrewing during the stretching process. Importantly, the spiral fibers could recover their original structure after stress release due to the PU elastic substrate, and the cracks in the conductive network of CNTs were repaired. Therefore, sensors are highly reversible in terms of structural and electrical properties. In addition to improving the deformation performance by designing the fiber structure, the deformation capacity of the fabric can also be improved by studying the weaving process and weaving structure. Rezaei et al. [69] twisted cotton threads with eight copper wires and then coated them with reactive triboelectric materials poly vinyl chloride (PVC) and polyamide 6 (PA 6), respectively. The two types of fibers were twisted and prepared into a double twisted thread as a triboelectric sensor. By studying the weaving process, the ribbed weave was chosen to endow the fabric with excellent tensile properties and flexibility (Figure 3e). Another key to improving the deformability of TMSs is material optimization through the development of materials with inherent deformation characteristics or the integration of deformable materials. Commonly used materials are hydrogels, PDMS, etc. Ye et al. [70] prepared a conductive hydrogel by integrating polymer, silk protein, and carbon materials, which had a strain range of 600% and showed a rapid recovery ability during large deformations.

For wearable applications, TMSs are required to have the feel of textiles and be comfortable to wear. When sensors are worn or fitted to the body, they should not cause any discomfort to the body, and therefore require good biocompatibility, skin friendliness, and comfort. Mechanical sensors that use textile materials as substrates usually possess these characteristics [71]. TMSs can be sewn on the garment manually by attaching them to an elastic backing or by using a specially designed support frame. Further, TMSs are woven directly into textiles that can be worn directly without support frames, backing, or other clothing items, allowing wearable TMSs to take a step forward in comfort and appeal [65]. In addition, breathability is also a major focus of the sensors. Fabrics woven with fiber-based TMSs usually present good air breathability due to the voids between the fibers, whereas TMSs with a coated sensing layer usually need to be designed as a porous or mesh structure to provide breathability. For instance, Ma et al. [72] fabricated a breathable liquid metal-fiber layer by coating liquid metal on an elastic fiber layer (Figure 3f). During the fabricating process, pre-stretching was applied to make the liquid metal suspended between the elastic fibers self-organize into a transverse mesh and a vertically curved structure, thus providing the liquid metal-fiber layer with good breathability. The combination of hydrophobicity and washability, which are not available in other types of flexible mechanical sensors, makes TMSs more attractive. The combined characteristics can be possessed by material design without affecting breathability. Hu et al. [73] obtained graphene coated e-textiles through dipping, water treatment, and a subsequent drying step (Figure 3g). After dipping, the fabric is immersed in a water bath where the rapid separation of solvent and water causes the coating to shrink, allowing the fabric structure to remain breathable. At the same time methyltrichlorosilan reacts with water to form a methyltrihydroxysilane precursor, which undergoes condensation to form a highly hydrophobic and sticky polymethylsiloxane structure in a high temperature drying step. Even when textiles are immersed in detergent under ultrasonic washing conditions, the coating remains stable and does not peel off, demonstrating excellent washability. The microstructure design also allows the fabric to be hydrophobic. Inspired by the "papillae structure" on the surface of lotus leaf, Song et al. [74] successfully prepared a waterproof multimode sensor by constructing two-dimensional MXene nanosheets and zero-dimensional silicon nanoparticles on a cotton fiber substrate. The conductivity can be maintained even under wet and corrosive conditions.

Compared with epidermic sensors and tattoo sensors, which are also hotspots in wearable device research, TMSs have the inherent advantage of monitoring large-area, large-deformation mechanical signals and biocompatibility because of their weaving method and textile substrate. Even though TMSs may not be as thin and transparent as other sensors, TMSs can be woven into clothing to hide themselves quite well. TMSs integrated in clothing have more comfort to monitor mechanical signals; additionally, although the skin-conformality is not as good as epidermic sensors and tattoo sensors, TMSs can also be attached to the skin directly for more accurate signals by traditional adhesives. The weaker adhesion between the TMSs and the body causes loss of contact of TMSs from the human skin, resulting in larger noise and inaccurate mechanical signals. To solve this problem, it is necessary for new technologies to replace traditional adhesives to ensure conformal contact between TMSs and human skin.

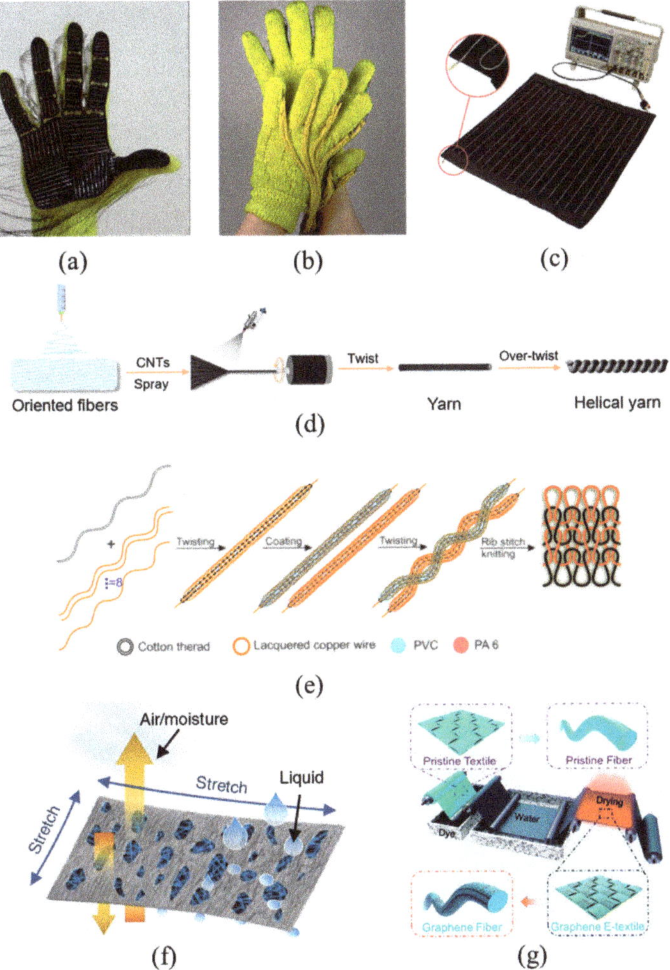

Figure 3. Advantages. CNTs means carbon nanotubes, PVC means poly vinyl chloride and PA6 means Polyamide 6 (**a**) large area array sensors based on piezoresistive film, reproduced with permission from [64]; (**b**) large area array sensors based on fibers, reproduced with permission from [65]; (**c**) large area detection mechanical stimulation using electrical time domain reflectometer, reproduced with permission from [66]; (**d**) large deformation sensors based on spiral structure, reproduced with permission from [68]; (**e**) large deformation sensors based on ribbed weave, reproduced with permission from [69]; (**f**) fabric sensors breathability based on pre-stretching, reproduced with permission from [72]; and (**g**) fabric sensors breathability based on coating shrinkage, reproduced with permission from ref. [73].

3.2. Performance

So far, a lot of work has been carried out on the developing and research of TMSs to improve the sensing performance of the devices, such as sensitivity, response range, response time, stability, etc., because these properties determine the practical application capabilities of the sensors [75]. Sensing performance can be improved by introducing special geometric structures, such as microarrays [76,77], microcracks [78], micropatterns [79,80], pleated structures [81,82], porous structures [83,84], spiral structures [85,86], etc. The innovation of materials is also a major key point to improve the sensing performance. For example, new materials such as graphene [87,88] and MXene [89,90] have good electrical conductivity and mechanical properties and are widely used in sensors.

Sensitivity is a key factor in evaluating the performance of various sensors. Therefore, many scholars focus on pursuing high sensitivity. The formula for sensitivity is $S = (\Delta X/X_0)/Y$, where S represents the sensitivity, X_0 represents the initial value of the electrical signal, resistance, capacitance, voltage, etc., ΔX represents the amount of change in the electrical signal, and Y represents the mechanical stimulus applied to the sensor, such as pressure, strain, etc. The usual methods to improve sensitivity include the introduction of microstructures [78], the use of new sensing materials [80], and the employment of multilayer structures [91]. Li et al. [92] prepared a folded core-sheath structure fiber strain sensor by pre-stretching and releasing ultra-light MWCNTs/themoplasticelastomer (TPE) composite film wrapped TPE fiber core. After releasing the pre-stretched TPE fiber core, a periodic bending structure was formed along the fiber axial direction (Figure 4a), which combined with the super-elasticity of the TPE core and provided this fiber high strain sensitivity.

The sensing range refers to the detection range of sensors working reliably, which should cover the full range of the desired application. The sensing range of a sensor determines its application field. Take gripping as an example: the finger joints bend by up to 30% [34]. Therefore, textile strain sensors should be fully functional in the strain range of 0 to 30% to monitor these relative finger movements. Maintaining a high sensitivity over a wide sensing range is one of the properties pursued by scientists. Using multi-layer structures and increasing the contact area are the main methods to improve detection range at present. Pyo et al. [91] developed a resistive tactile sensor by alternately stacking CNTs and Ni fabrics (Figure 4b). The graded structure of the fabric provides a large surface area and microscopic roughness, thereby significantly increasing contact area in response to pressure. The design of multi-layer structures can further increase the contact area and effectively distribute stress to each layer, consequently dramatically increasing the pressure detection range as well as the device's sensitivity. The sensor shows a sensitivity of 26.13 kPa^{-1} over a wide pressure range of 0.2–982 kPa. Liu et al. [93] successfully fabricated a microcracked nonwoven strain sensor with a wide operating range and a high sensitivity (Figure 4c). Taking advantage of the difference in modulus between the electrically conductive cellulose nanocrystal/graphene coating and the nonwoven fabric, microcrack structures were constructed by a simple dip-coating and pre-strain technique, which determined the sensor's detecting range and sensitivity, making it possible to prepare sensors with designed sensing capability by adjusting the density of the microcrack structure.

Response time of a mechanical sensor is defined as the time to achieve a steady-state response upon mechanical stimulation, and it is usually defined as 90% time to reach stability. The response time of a sensor is especially important in applications that require real-time data processing: the shorter the response time, the better the real-time performance of the results. The response time of TMSs is typically in the millisecond range level, which can of course be effectively reduced by structural design and material optimization. Yu et al. [94] prepared an ultrathin all-fabric capacitive sensor with two AgNWs electrodes and a breathable micropatterned nanofiber dielectric layer sandwiched between them (Figure 4d). Due to the unique structure of the micropatterned nanofiber dielectric layer, the sensor shows a response time of 27.3 ms. Xu et al. [95] prepared fabric sensors using laser engraved silver-plated fabric as electrodes and graphite flake modified nonwoven fabric as the sensing material. Due to the unique structure of the electrodes and the random rough surface of the sensing material, the sensor has a fast response time of 4 ms.

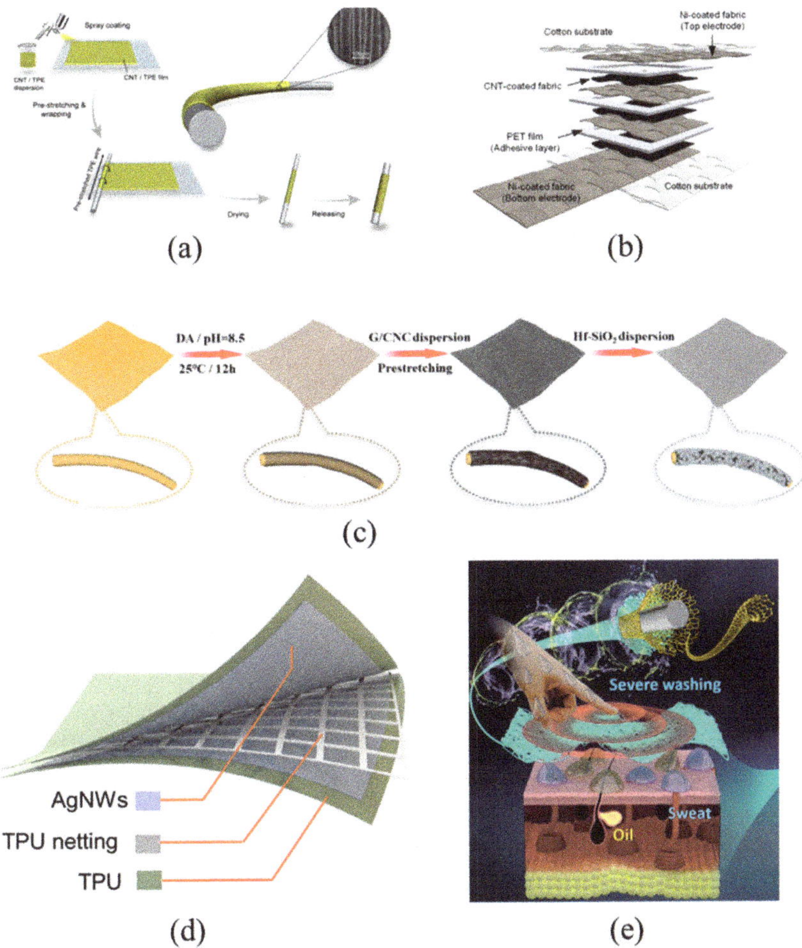

Figure 4. Performance. TPE means thermoplastic elastomer, CNT means carbon nanotube. PET means polyethylene terephthalate, G/CNC means conductive cellulose nanocrystal/graphene, AgNWs means silver nanowires and TPU means thermoplastic polyurethanes. (**a**) high sensitivity based on folded structure, reproduced with permission from [92]; (**b**) large response range based on hierarchical and multilayer structure, reproduced with permission from [91]; (**c**) weighing sensitivity and response range based on microcrack density, reproduced with permission from [93]; (**d**) short response time based on micropatterned dielectric.layer, reproduced with permission from [94]; (**e**) excellent stability based on layered protection structure, reproduced with permission from ref. [96].

Stability is a key factor to ensure that a sensor can effectively acquire mechanical stimuli and respond accordingly. The stability of TMSs includes mechanical stability, chemical stability, and cyclic stability, etc., which can also be improved by optimizing structure and materials. Mechanical and chemical stability refers to the ability of the sensor to resist wear, corrosion, and other external environmental factors. To address the vulnerability of textile-based sensors to external mechanical, chemical, and environmental interference, especially human sweat, grease, and wear and tear, Zhang et al. [96] combined carbon nanotube networks, polymer layers, and textile substrates to form self-protected and reproducible e-textiles with a layered structure (Figure 4e). The sensor exhibits excellent characteristics of superhydrophobicity, wear resistance, and mechanical and chemical stability, and its response resistance does not change significantly during 3000 compression cycles. Cycling stability is the ability of the sensing response to return to its original value after unloading. Resistive TMSs usually rely on large deformation to change the resistance to play the role of sensing. Due to irreversible deformation, they are less cyclically stable compared to other types of sensors. Thus, improved cycling stability can also be done by changing the sensing mechanism. Wu et al. [97] transferred water-soluble poly(vinyl alcohol) template-assisted silver nanofibers onto a fabric surface to serve as sensor electrodes, and used a highly elastic three-dimensional penetration fabric as a dielectric layer. The integrated capacitive TMS were prepared with good dimensional stability and excellent cycling stability ($\geq$20,000).

In addition to some of the properties discussed in detail above, the reply ability [98], crosstalk problems [99], linearity [100], and anti-interference [101] of TMSs are also key factors that determine the overall performance and practical applicability of the sensor. For example, it is highly desirable that the sensing response (resistance, etc.) is linearly related to the strain, as this will allow easy prediction of the strain from the sensing response. Sometimes, the required sensing performance is not consistent for different application scenarios, which requires us to design and prepare TMSs that meet the requirements according to the actual situation. However, high-performance TMSs are still a research hotspot. Therefore, the design of innovative textile materials and structures is necessary to achieve TMSs with high sensitivity and other high performance. From a production point of view, the manufacture of TMSs should be easily scalable and economically feasible, which are important factors that limit the large-scale adoption of TMSs.

4. Applications

TMSs have great advantages of easy large-scale preparation, flexibility, and biocompatibility. Combined with integrated circuits, they can obtain real-time data for machine learning and artificial intelligence [102], and thus are receiving increasing attention from academia and industry. The assembly of TMSs with units such as power supplies, signal processing components, communications units, and data management software is also a major difficulty. Although TMSs have not yet been marketed on a large scale like most other flexible electronic devices, a great number of studies have shown that TMSs have promising application prospects in wearable electronics, smart fabrics, robotics, and other fields. It is foreseeable that such devices will change people's lives and improve people's quality of life in the near future.

4.1. Health Monitoring

Health issues are the most important concern for human beings. Traditional medical services usually require professional physicians and large testing and rehabilitation equipment, which requires patients to go to hospitals for testing and treatment on each occasion. Additionally, it increases the burden of patients and physicians and reduces the timeliness of disease detection and treatment. TMSs are very suitable for being applied in the field of health monitoring, because doctors get health data even without patients leaving home combined with Internet of Things (IOT) technology and TMSs [103]. Utilizing the inherent flexibility and comfort of fabric-based sensors, TMSs can not only be used to detect the wearer's blood pressure, heart rate, and certain diseases, but also can be used to monitor the wearer's disease progression or motor symptoms for a long time, will provide an alert of possible health threats to patients or an objective basis for doctors to guide the wearer in medication or rehabilitation training, and will even greatly influence the reform of the medical industry and home care industry. In personal health care monitoring, it is important to identify the location and number of sensors. Common body monitoring areas include the soles of the feet, hands, legs, chest, neck, pulse, etc.

Wicaksono et al. [104] reported a large-scale e-textile-based smart clothing (Figure 5a), which is able to perform multi-modal physiological (temperature, heart rate, and respiration) monitoring. Moreover, customized clothing with various forms, sizes, and functions could be made using standard, adjustable, and high-yield textile manufacturing processes and garment patterning techniques. Similar to corsets, the soft and stretchable features of customized smart clothing allow for close contact between the electronic device and human skin, providing physical comfort and improving the anti-interference of sensor. These make it suitable not only in hospitals and laboratories, but also in home care for mobile, comfortable, and continuous physical activity monitoring, which has huge potential in healthcare, rehabilitation, and scientific training.

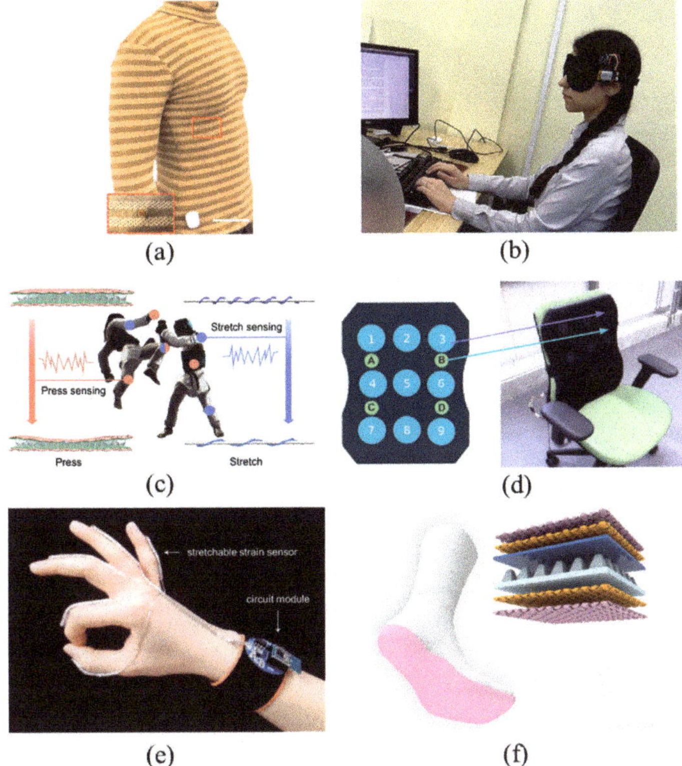

Figure 5. Applications: (**a**) smart clothing applied to health monitoring, reproduced with permission from [104]; (**b**) smart eye mask applied to health monitoring, reproduced with permission from [105]; (**c**) smart clothing applied to motion analysis, reproduced with permission from [106]; (**d**) smart cushion applied to sitting analysis, reproduced with permission from [107]; (**e**) smart glove applied to sign language recognition, reproduced with permission from [108]; and (**f**) smart socks applied to human-machine interaction, reproduced with permission from ref. [109].

Most users are very sensitive to the monitoring tools placed on the face or head, and those tools can cause psychological fluctuations that affect the measurement results. The combination of a fabric sensor and clothing can be designed into a wearable device that is not easy to detect or even unnoticeable, so that when it is used as a health detection device, it increases the comfort of the wearer and the accuracy of the test data. For example, when a reusable hydrogel wet electrode and a full-fabric ionic pressure sensor were integrated into an eye mask, the eye mask could track and monitor eye movements and intraocular pressure in daily use [105] (Figure 5b). The technology is virtually non-irritating and non-invasive to the wearer, coupled with its aesthetics and washing stability. When combined with artificial intelligence algorithms, it can be used to accurately monitor the ophthalmology and heart signals required for sleep quality and mental health research.

4.2. Motion Recognition for Analysis

Tracking physical activity and habitual movements and analyzing them to extract useful information can be used to scientifically correct training modes and intensity, monitor sitting posture, guide ergonomic design, recognize sign language, etc. Those have become major applications of TMSs. The realization of motion analysis through TMSs first lies in high-precision detection and quantification of various motion variables, such as position, angle, and pressure, and then converting them into visualized electrical signals in real time. Subsequently comprehensively analyzing and establishing the relationship between these electrical signals and human body posture is the key to realizing the high-precision recognition of human spatial motion characteristics.

Smart clothing is often used to monitor human health or exercise. While conventional rigid sensors typically monitor only small strains, TMSs can withstand larger deformations. To detect athletes' movement in high-intensity physical exercises similar to taekwondo, the smart sportswear is required to withstand large-scale deformation (strain > 50%) and heavy blows (>100 kPa) while maintaining good performance stability. Based on this, Ma et al. [106] used a composite fabric of core-sheath yarn and spacer fabric to prepare an "all-in-one" electronic fabric with a dual tactile and tension stimulus response (Figure 5c). In general, retractable strain sensors with a core-sheath yarn structure can easily measure large movements, yet with a "double solution phenomenon", that is, a non-monotonic electrical signal response to tensile deformation. To solve this problem, an insulating PU layer was pre-coated on the core-sheath yarns before twisting. This allows the TMS to accurately monitor the exercise action and intensity, thereby implementing its potential application in taekwondo and high-intensity physical exercise analysis.

Low back pain is the most common work-related physical injury among people and is directly related to working in a twisted and bent position for long periods of time. In addition, long-term incorrect sitting posture is not good for people, especially teenagers, because it can cause bone deformation, hunching, curvature of the spine, myopia, and other health problems. To address this issue, Ishac et al. [107] designed a smart cushion, which is flexible, adaptable, and portable (Figure 5d). The cushion consists primarily of a new conductive-fabric pressure sensing array designed and arranged based on human biomechanics, which was used to sense and classify sitting posture (98.1% in accuracy) and adjust upright posture of the user through vibrotactile feedback.

By collecting and analyzing human movement information, TMSs can not only be used to monitor human movement status, assist in training, and guide ergonomic design, but also assist in communication between people. Sign language is a basic communication method for a considerable portion of the population, but it can only be used by trained individuals. To overcome this limitation, Han et al. [108] developed a wearable system that integrates yarn-based stretchable sensors on five fingers (Figure 5e). With this wearable system, sign language gestures can be converted into analog electrical signals and transmitted wirelessly to a portable electronic device. Real-time speech translation can then be achieved using machine learning algorithms and a graphical user interface with recognition rates >98% and recognition times <1 s. An automated sign language recognition system can obviously strengthen the communication function of sign language, making it easier, smoother, and more effective for deaf-mute people to communicate with the outside world.

4.3. Human-Machine Interaction

Human-machine interaction is the study of the interaction between a system and its users. In the past, the realization of human-machine interaction usually required bulky, large, and hard devices. Due to the characteristics of E-textile itself, TMSs, as the core of electronic fabrics or wearable devices, provide a more convenient and comfortable channel than other flexible electronics for human-machine interaction. It is possible to control machines or virtual reality (VR) games using TMSs without affecting human physical activity, which has huge impact on industries such as entertainment and leisure and robotics. More and more work is carried out to study and improve the TMSs performance for highly effective human-machine interaction.

TMSs have great potential for human-machine interaction scenarios such as smart furniture, VR games, and robotic control, etc. Gesture recognition occupies an important position in the field of human-computer interaction technology because of its intuitive and convenient advantages. Shuai et al. [110] prepared stretchable, conductive, and self-healing hydrogel sensing fibers by continuous spinning. When the fiber sensors were separately fixed on five fingers, it could monitor and distinguish the movement of each finger by monitoring the electrical signal changes caused by fingers bring bent, and then judge the gestures such as "OK", "Victory", and "claw". The strain sensing capability of this fiber sensor shows its great potential in human-computer interaction. Gait analysis is also commonly used in human-computer interaction and personal healthcare. As we know, the sole of the human foot has different pressure zones; TMSs can generate a map of the pressure distribution on the bottom of the foot. Zhang et al. [109] simulate a virtual reality (VR) fitness game by mapping gait information collected by smart socks to virtual space (Figure 5f). The smart socks were developed by embedding textile-based triboelectric pressure sensors into commercial socks; combined with deep learning, five different gaits were recognized with an accuracy of 96.67%, ensuring the feasibility of the gait control interfaces and promoting the practical application of flexible electronics in the field of the smart home. TMSs are also finding applications in the entertainment field, such as portable textile keyboards. Chen et al. [111] obtained flexible hierarchical helical yarns via coiling the conductive electrodes around a highly stretchable PU (the first helix layer) and super flexible silicon rubber (the second helix layer), respectively. The yarn can work in a strain range of up to 120% and obtain bioenergy to convert into electrical signals that are used to control household devices without amplifying the signal. Interactive electronic pianos and robotic arms can be precisely controlled using the sensors with little noticeable delay, which makes them promising as portable self-powered wearable electronics.

The application of TMSs boils down to acquiring mechanical signals from the human body or robot, and through the processing of data, status information and control signals can be obtained. As artificial intelligence continues to develop, it may even replace humans in making decisions through the mechanical signals acquired.

5. Challenges and Prospects

Ingenious structures, perfect processes, and outstanding materials endow TMSs with excellent performance and more prominent advantages over other flexible sensors. The great advantages of TMSs in terms of cost, large-area array, deformation, conformability, wearability, and comfort make TMSs attract extensive research in academia. However, like most flexible electronics, TMSs suffer from a short lifetime compared to rigid sensors, and many performance issues still need to be addressed. Additionally, their roll-to-roll production technology is immature and most TMSs are fabricated in the laboratory or research stages. Many of the reported TMSs are manufactured using manual techniques rather than mechanical engineering. The large-scale manufacturability and reproducibility of e-textiles are uncertain. The future of TMSs depends on the ability to mass produce seamless embedded electronics that meet everyday needs. There are many challenges in moving TMSs to market commercialization: (1) The complex manufacturing methods and expensive materials of some sensors greatly increase cost and reduced practicality; (2) It is difficult to prepare sensors with uniform performance on a large scale; (3) They lack excellent repeatability and robustness, and are often accompanied by poor reliability and life span; (4) The lack of standards makes it difficult for them to be accepted by the general public and the development of production standards will help reduce manufacturing time and final manufacturing costs; (5) The development of flexible integrated circuits and flexible batteries has limited the practical application of TMSs. Once all these challenges are met, mass production of e-textiles will become a reality, which will be a major milestone for wearable devices. New wearable e-textiles provide an opportunity for the market as well as a challenge for interacting with traditional electronic devices. In addition, most of the current TMSs are still used to detect a single or dual mode of mechanical stimuli, mostly pressure or strain, but the mechanical stimuli associated with the human body are much more complex than these. Therefore, how to collect more mechanical stimulus information, how to decouple different types of signals, and how to quantify and spatially resolve multiple mechanical stimuli are the current difficulties faced by TMSs. In the future, textile-based mechanical sensors will move in the direction of high integration, which not only refers to the high integration of sensor fabrication technology and functions, but also refers to the integration of flexible electronic technology with other technologies such as flexible circuits, machine learning, and drive control. What is more, the smart fabric has the functions of responding to environmental stimuli, sensing and driving, and can even adapt to the environment. To address these challenges, with the joint efforts of scientists and engineers from many different disciplines, we believe that TMSs have a bright future and will contribute to the next generation of health monitoring, motion recognition, and human-computer interaction.

6. Conclusions

In this work, we have given a comprehensive review of TMSs. Various types of TMSs including resistive, capacitive, piezoelectric, and triboelectric have been introduced combined with materials, structures, and processes. In particular, the advantages and performance of TMSs and their improvement methods are described in detail. We then summarized the latest applications of TMSs in different fields. Furthermore, challenges and perspectives in TMSs were also presented for the current state of TMSs.

Author Contributions: Conceptualization, Y.Z., X.Y. and B.H.; methodology, Z.Z.; writing—original draft, Z.Z., N.C., H.Z. and W.Z.; writing—review and editing, Z.Z., H.Z., N.C. and W.Z.; supervision, Y.Z., B.H. and X.Y. All authors have read and agreed to the published version of the manuscript.

Funding: This research was funded by "Study on One-Step Construction of Capacitive Ionic Skins via 3D Printing and Its Key Technology", grant number 52103025.

Institutional Review Board Statement: Not applicable.

Informed Consent Statement: Not applicable.

Data Availability Statement: Data sharing not applicable.

Acknowledgments: This publication is the result of the implementation of the following project: NSFC NO.: 52103025 "Study on one-step construction of capacitive ionic skins via 3D printing and its key technology".

Conflicts of Interest: The authors declare no conflict of interest.

References

1. Ilami, M.; Bagheri, H.; Ahmed, R.; Skowronek, E.O.; Marvi, H. Materials, actuators, and sensors for soft bioinspired robots. *Adv. Mater.* **2021**, *33*, 2003139. [CrossRef]
2. Cheng, M.; Zhu, G.; Zhang, F.; Tang, W.-L.; Jianping, S.; Yang, J.-Q.; Zhu, L.-Y. A review of flexible force sensors for human health monitoring. *J. Adv. Res.* **2020**, *26*, 53–68. [CrossRef]
3. Wolterink, G.; Sanders, R.; van Beijnum, B.-J.; Veltink, P.; Krijnen, G. A 3D-Printed Soft Fingertip Sensor for Providing Information about Normal and Shear Components of Interaction Forces. *Sensors* **2021**, *21*, 4271. [CrossRef]
4. Huang, W.; Cheng, S.; Wang, X.; Zhang, Y.; Chen, L.; Zhang, L. Noncompressible Hemostasis and Bone Regeneration Induced by an Absorbable Bioadhesive Self-Healing Hydrogel. *Adv. Funct. Mater.* **2021**, *31*, 2009189. [CrossRef]
5. Song, M.; Yu, H.; Zhu, J.; Ouyang, Z.; Abdalkarim, S.Y.H.; Tam, K.C.; Li, Y. Constructing stimuli-free self-healing, robust and ultrasensitive biocompatible hydrogel sensors with conductive cellulose nanocrystals. *Chem. Eng. J.* **2020**, *398*, 125547. [CrossRef]
6. Li, X.; Sun, H.; Li, H.; Hu, C.; Luo, Y.; Shi, X.; Pich, A. Multi-Responsive Biodegradable Cationic Nanogels for Highly Efficient Treatment of Tumors. *Adv. Funct. Mater.* **2021**, *31*, 2100227. [CrossRef]
7. Lou, Z.; Wang, L.; Jiang, K.; Wei, Z.; Shen, G. Reviews of wearable healthcare systems: Materials, devices and system integration. *Mat. Sci. Eng. R* **2020**, *140*, 100523. [CrossRef]

8. Chen, H.; Bao, S.; Lu, C.; Wang, L.; Ma, J.; Wang, P.; Lu, H.; Shu, F.; Oetomo, S.B.; Chen, W. Design of an integrated wearable multi-sensor platform based on flexible materials for neonatal monitoring. *IEEE Access* **2020**, *8*, 23732–23747. [CrossRef]
9. Lim, H.R.; Kim, H.S.; Qazi, R.; Kwon, Y.T.; Jeong, J.W.; Yeo, W.H. Advanced soft materials, sensor integrations, and applications of wearable flexible hybrid electronics in healthcare, energy, and environment. *Adv. Mater.* **2020**, *32*, 1901924. [CrossRef] [PubMed]
10. Xu, H.; Xie, Y.; Zhu, E.; Liu, Y.; Shi, Z.; Xiong, C.; Yang, Q. Supertough and ultrasensitive flexible electronic skin based on nanocellulose/sulfonated carbon nanotube hydrogel films. *J. Mater. Chem. A* **2020**, *8*, 6311–6318. [CrossRef]
11. Heo, J.S.; Hossain, M.F.; Kim, I. Challenges in design and fabrication of flexible/stretchable carbon-and textile-based wearable sensors for health monitoring: A critical review. *Sensors* **2020**, *20*, 3927. [CrossRef] [PubMed]
12. Islam, G.N.; Ali, A.; Collie, S. Textile sensors for wearable applications: A comprehensive review. *Cellulose* **2020**, *27*, 6103–6131. [CrossRef]
13. Possanzini, L.; Tessarolo, M.; Mazzocchetti, L.; Campari, E.G.; Fraboni, B. Impact of fabric properties on textile pressure sensors performance. *Sensors* **2019**, *19*, 4686. [CrossRef] [PubMed]
14. Atalay, O. Textile-based, interdigital, capacitive, soft-strain sensor for wearable applications. *Materials* **2018**, *11*, 768. [CrossRef]
15. Koyama, Y.; Nishiyama, M.; Watanabe, K. Smart textile using hetero-core optical fiber for heartbeat and respiration monitoring. *IEEE Sens. J.* **2018**, *18*, 6175–6180. [CrossRef]
16. Yamada, Y. Textile-integrated polymer optical fibers for healthcare and medical applications. *Biomed. Phys. Eng. Express* **2020**, *6*, 062001. [CrossRef]
17. Patiño, A.G.; Menon, C. Inductive textile sensor design and validation for a wearable monitoring device. *Sensors* **2021**, *21*, 225. [CrossRef]
18. Xiong, J.; Chen, J.; Lee, P.S. Functional fibers and fabrics for soft robotics, wearables, and human-robot interface. *Adv. Mater.* **2021**, *33*, 2002640. [CrossRef]
19. Farrow, N.; McIntire, L.; Correll, N. Functionalized textiles for interactive soft robotics. In Proceedings of the 2017 IEEE International Conference on Robotics and Automation (ICRA), Singapore, 29 May–3 June 2017; pp. 5525–5531.
20. Zhou, B.; Altamirano, C.A.V.; Zurian, H.C.; Atefi, S.R.; Billing, E.; Martinez, F.S.; Lukowicz, P. Textile pressure mapping sensor for emotional touch detection in human-robot interaction. *Sensors* **2017**, *17*, 2585. [CrossRef]
21. Lu, W.; Yu, P.; Jian, M.; Wang, H.; Wang, H.; Liang, X.; Zhang, Y. Molybdenum disulfide nanosheets aligned vertically on carbonized silk fabric as smart textile for wearable pressure-sensing and energy devices. *ACS Appl. Mater. Interfaces* **2020**, *12*, 11825–11832. [CrossRef]
22. Zhao, J.; Fu, Y.; Xiao, Y.; Dong, Y.; Wang, X.; Lin, L. A naturally integrated smart textile for wearable electronics applications. *Adv. Mater. Technol.* **2020**, *5*, 1900781. [CrossRef]
23. Alkhader, A.S.; Saikia, M.J.; Driscoll, B.; Mankodiya, K. Design and characterization of a helmet-based smart textile pressure sensor for concussion. *Preprints* **2020**, 2020070629. [CrossRef]
24. Mariello, M.; Fachechi, L.; Guido, F.; de Vittorio, M. Conformal, ultra-thin skin-contact-actuated hybrid piezo/triboelectric wearable sensor based on AlN and parylene-encapsulated elastomeric blend. *Adv. Funct. Mater.* **2021**, *31*, 2101047. [CrossRef]
25. Gogurla, N.; Kim, S. Self-powered and imperceptible electronic tattoos based on silk protein nanofiber and carbon nanotubes for human-machine interfaces. *Adv. Energy Mater.* **2021**, *11*, 2100801. [CrossRef]
26. Zhang, J.-W.; Zhang, Y.; Li, Y.-Y.; Wang, P. Textile-based flexible pressure sensors: A review. *Polym. Rev.* **2021**, 1–31. [CrossRef]
27. Gonçalves, C.; da Silva, A.F.; Gomes, J.; Simoes, R. Wearable e-textile technologies: A review on sensors, actuators and control elements. *Inventions* **2018**, *3*, 14. [CrossRef]
28. Heo, J.S.; Eom, J.; Kim, Y.H.; Park, S.K. Recent progress of textile-based wearable electronics: A comprehensive review of materials, devices, and applications. *Small* **2018**, *14*, 1703034. [CrossRef]
29. Wilson, S.; Laing, R. Fabrics and garments as sensors: A research update. *Sensors* **2019**, *19*, 3570. [CrossRef]
30. Kim, B.; Koncar, V.; Devaux, E.; Dufour, C.; Viallier, P. Electrical and morphological properties of PP and PET conductive polymer fibers. *Synthetic Met.* **2004**, *146*, 167–174. [CrossRef]
31. Nag, A.; Mukhopadhyay, S.C.; Kosel, J. Wearable flexible sensors: A review. *IEEE Sens. J.* **2017**, *17*, 3949–3960. [CrossRef]
32. Pizarro, F.; Villavicencio, P.; Yunge, D.; Rodríguez, M.; Hermosilla, G.; Leiva, A. Easy-to-build textile pressure sensor. *Sensors* **2018**, *18*, 1190. [CrossRef]
33. Gong, Z.; Xiang, Z.; OuYang, X.; Zhang, J.; Lau, N.; Zhou, J.; Chan, C.C. Wearable fiber optic technology based on smart textile: A review. *Materials* **2019**, *12*, 3311. [CrossRef]
34. El Gharbi, M.; Fernández-García, R.; Ahyoud, S.; Gil, I. A review of flexible wearable antenna sensors: Design, fabrication methods, and applications. *Materials* **2020**, *13*, 3781. [CrossRef] [PubMed]
35. Neves, A.I.; Rodrigues, D.P.; de Sanctis, A.; Alonso, E.T.; Pereira, M.S.; Amaral, V.S.; Melo, L.V.; Russo, S.; de Schrijver, I.; Alves, H. Towards conductive textiles: Coating polymeric fibres with graphene. *Sci. Rep.* **2017**, *7*, 4250. [CrossRef] [PubMed]
36. Li, Y.; Zheng, C.; Liu, S.; Huang, L.; Fang, T.; Li, J.X.; Xu, F.; Li, F. Smart glove integrated with tunable MWNTs/PDMS fibers made of a one-step extrusion method for finger dexterity, gesture, and temperature recognition. *ACS Appl. Mater. Interfaces* **2020**, *12*, 23764–23773. [CrossRef] [PubMed]
37. Wang, W.; Ouaras, K.; Rutz, A.L.; Li, X.; Gerigk, M.; Naegele, T.E.; Malliaras, G.G.; Huang, Y.Y.S. Inflight fiber printing toward array and 3D optoelectronic and sensing architectures. *Sci. Adv.* **2020**, *6*, eaba0931. [CrossRef]

38. Sheng, N.; Ji, P.; Zhang, M.; Wu, Z.; Liang, Q.; Chen, S.; Wang, H. High sensitivity polyurethane-based fiber strain sensor with porous structure via incorporation of bacterial cellulose nanofibers. *Adv. Electron. Mater.* **2021**, *7*, 2001235. [CrossRef]
39. Qi, K.; Zhou, Y.; Ou, K.; Dai, Y.; You, X.; Wang, H.; He, J.; Qin, X.; Wang, R. Weavable and stretchable piezoresistive carbon nanotubes-embedded nanofiber sensing yarns for highly sensitive and multimodal wearable textile sensor. *Carbon* **2020**, *170*, 464–476. [CrossRef]
40. Liu, R.; Li, J.; Li, M.; Zhang, Q.; Shi, G.; Li, Y.; Hou, C.; Wang, H. MXene-coated air-permeable pressure-sensing fabric for smart wear. *ACS Appl. Mater. Interfaces* **2020**, *12*, 46446–46454. [CrossRef]
41. Kim, T.; Park, C.; Samuel, E.P.; An, S.; Aldalbahi, A.; Alotaibi, F.; Yarin, A.L.; Yoon, S.S. Supersonically sprayed washable, wearable, stretchable, hydrophobic, and antibacterial rGO/AgNW fabric for multifunctional sensors and supercapacitors. *ACS Appl. Mater. Interfaces* **2021**, *13*, 10013–10025. [CrossRef] [PubMed]
42. Luo, C.; Tian, B.; Liu, Q.; Feng, Y.; Wu, W. One-step-printed, highly sensitive, textile-based, tunable performance strain sensors for human motion detection. *Adv. Mater. Technol.* **2020**, *5*, 1900925. [CrossRef]
43. Duan, L.; D'hooge, D.R.; Cardon, L. Recent progress on flexible and stretchable piezoresistive strain sensors: From design to application. *Prog. Mater. Sci.* **2020**, *114*, 100617. [CrossRef]
44. Chen, L.; Lu, M.; Yang, H.; Salas Avila, J.R.; Shi, B.; Ren, L.; Wei, G.; Liu, X.; Yin, W. Textile-based capacitive sensor for physical rehabilitation via surface topological modification. *ACS Nano* **2020**, *14*, 8191–8201. [CrossRef]
45. Ferri, J.; Llinares Llopis, R.; Moreno, J.; Ibañez Civera, J.; Garcia-Breijo, E. A wearable textile 3D gesture recognition sensor based on screen-printing technology. *Sensors* **2019**, *19*, 5068. [CrossRef]
46. Keum, K.; Eom, J.; Lee, J.H.; Heo, J.S.; Park, S.K.; Kim, Y.-H. Fully-integrated wearable pressure sensor array enabled by highly sensitive textile-based capacitive ionotronic devices. *Nano Energy* **2021**, *79*, 105479. [CrossRef]
47. Fu, M.; Zhang, J.; Jin, Y.; Zhao, Y.; Huang, S.; Guo, C.F. A highly sensitive, reliable, and high-temperature-resistant flexible pressure sensor based on ceramic nanofibers. *Adv. Sci.* **2020**, *7*, 2000258. [CrossRef] [PubMed]
48. Guan, F.; Xie, Y.; Wu, H.; Meng, Y.; Shi, Y.; Gao, M.; Zhang, Z.; Chen, S.; Chen, Y.; Wang, H. Silver nanowire-bacterial cellulose composite fiber-based sensor for highly sensitive detection of pressure and proximity. *ACS Nano* **2020**, *14*, 15428–15439. [CrossRef]
49. Zhang, Q.; Wang, Y.L.; Xia, Y.; Zhang, P.F.; Kirk, T.V.; Chen, X.D. Textile-only capacitive sensors for facile fabric integration without compromise of wearability. *Adv. Mater. Technol.* **2019**, *4*, 1900485. [CrossRef]
50. Tan, Y.; Yang, K.; Wang, B.; Li, H.; Wang, L.; Wang, C. High-performance textile piezoelectric pressure sensor with novel structural hierarchy based on ZnO nanorods array for wearable application. *Nano Res.* **2021**, 1–8. [CrossRef]
51. Hong, Y.; Wang, B.; Lin, W.; Jin, L.; Liu, S.; Luo, X.; Pan, J.; Wang, W.; Yang, Z. Highly anisotropic and flexible piezoceramic kirigami for preventing joint disorders. *Sci. Adv.* **2021**, *7*, eabf0795. [CrossRef] [PubMed]
52. Zhang, X.; Wang, J.; Xing, Y.; Li, C. Woven wearable electronic textiles as self-powered intelligent tribo-sensors for activity monitoring. *Glob. Chall.* **2019**, *3*, 1900070. [CrossRef]
53. Guo, Y.; Zhang, X.-S.; Wang, Y.; Gong, W.; Zhang, Q.; Wang, H.; Brugger, J. All-fiber hybrid piezoelectric-enhanced triboelectric nanogenerator for wearable gesture monitoring. *Nano Energy* **2018**, *48*, 152–160. [CrossRef]
54. Li, J.; Fang, L.; Sun, B.; Li, X.; Kang, S.H. Recent progress in flexible and stretchable piezoresistive sensors and their applications. *J. Electrochem. Soc.* **2020**, *167*, 037561. [CrossRef]
55. Zhang, C.; Fan, W.; Wang, S.; Wang, Q.; Zhang, Y.; Dong, K. Recent progress of wearable piezoelectric nanogenerators. *ACS Appl. Electron. Mater.* **2021**, *3*, 2449–2467. [CrossRef]
56. Park, C.; Kim, H.; Cha, Y. Fiber-based piezoelectric sensors in woven structure. In Proceedings of the 2020 17th International Conference on Ubiquitous Robots (UR), Kyoto, Japan, 22–26 June 2020; pp. 351–354.
57. Wen, D.-L.; Liu, X.; Deng, H.-T.; Sun, D.-H.; Qian, H.-Y.; Brugger, J.; Zhang, X.-S. Printed silk-fibroin-based triboelectric nanogenerators for multi-functional wearable sensing. *Nano Energy* **2019**, *66*, 104123. [CrossRef]
58. Jeon, S.-B.; Kim, W.-G.; Park, S.-J.; Tcho, I.-W.; Jin, I.-K.; Han, J.-K.; Kim, D.; Choi, Y.-K. Self-powered wearable touchpad composed of all commercial fabrics utilizing a crossline array of triboelectric generators. *Nano Energy* **2019**, *65*, 103994. [CrossRef]
59. Zhu, M.; Shi, Q.; He, T.; Yi, Z.; Ma, Y.; Yang, B.; Chen, T.; Lee, C. Self-powered and self-functional cotton sock using piezoelectric and triboelectric hybrid mechanism for healthcare and sports monitoring. *ACS Nano* **2019**, *13*, 1940–1952. [CrossRef]
60. Liu, J.; Gu, L.; Cui, N.; Xu, Q.; Qin, Y.; Yang, R. Fabric-based triboelectric nanogenerators. *Research* **2019**, *2019*, 1091632. [CrossRef] [PubMed]
61. Liu, L.; Shi, Q.; Sun, Z.; Lee, C. Magnetic-interaction assisted hybridized triboelectric-electromagnetic nanogenerator for advanced human-machine interfaces. *Nano Energy* **2021**, *86*, 106154. [CrossRef]
62. Chen, G.; Au, C.; Chen, J. Textile triboelectric nanogenerators for wearable pulse wave monitoring. *Trends Biotechnol.* **2021**, *39*, 1078–1092. [CrossRef] [PubMed]
63. Fan, W.; He, Q.; Meng, K.; Tan, X.; Zhou, Z.; Zhang, G.; Yang, J.; Wang, Z.L. Machine-knitted washable sensor array textile for precise epidermal physiological signal monitoring. *Sci. Adv.* **2020**, *6*, eaay2840. [CrossRef]
64. Sundaram, S.; Kellnhofer, P.; Li, Y.; Zhu, J.-Y.; Torralba, A.; Matusik, W. Learning the signatures of the human grasp using a scalable tactile glove. *Nature* **2019**, *569*, 698–702. [CrossRef] [PubMed]
65. Luo, Y.; Li, Y.; Sharma, P.; Shou, W.; Wu, K.; Foshey, M.; Li, B.; Palacios, T.; Torralba, A.; Matusik, W. Learning human-environment interactions using conformal tactile textiles. *Nat. Electron.* **2021**, *4*, 193–201. [CrossRef]

66. Leber, A.; Dong, C.; Chandran, R.; Gupta, T.D.; Bartolomei, N.; Sorin, F. Soft and stretchable liquid metal transmission lines as distributed probes of multimodal deformations. *Nat. Electron.* **2020**, *3*, 316–326. [CrossRef]
67. Lin, W.; Wang, B.; Peng, G.; Shan, Y.; Hu, H.; Yang, Z. Skin-inspired piezoelectric tactile sensor array with crosstalk-free row+column electrodes for spatiotemporally distinguishing diverse stimuli. *Adv. Sci.* **2021**, *8*, 2002817. [CrossRef] [PubMed]
68. Gao, Y.; Guo, F.; Cao, P.; Liu, J.; Li, D.; Wu, J.; Wang, N.; Su, Y.; Zhao, Y. Winding-locked carbon nanotubes/polymer nanofibers helical yarn for ultrastretchable conductor and strain sensor. *ACS Nano* **2020**, *14*, 3442–3450. [CrossRef] [PubMed]
69. Rezaei, J.; Nikfarjam, A. Rib stitch knitted extremely stretchable and washable textile triboelectric nanogenerator. *Adv. Mater. Technol.* **2021**, *6*, 2000983. [CrossRef]
70. He, F.; You, X.; Gong, H.; Yang, Y.; Bai, T.; Wang, W.; Guo, W.; Liu, X.; Ye, M.; Stretchable, B. Multifunctional silk fibroin-based hydrogels toward wearable strain/pressure sensors and triboelectric nanogenerators. *ACS Appl. Mater. Interfaces* **2020**, *12*, 6442–6450. [CrossRef]
71. Tseghai, G.B.; Malengier, B.; Fante, K.A.; Nigusse, A.B.; van Langenhove, L. Integration of conductive materials with textile structures, an overview. *Sensors* **2020**, *20*, 6910. [CrossRef]
72. Ma, Z.; Huang, Q.; Xu, Q.; Zhuang, Q.; Zhao, X.; Yang, Y.; Qiu, H.; Yang, Z.; Wang, C.; Chai, Y. Permeable superelastic liquid-metal fibre mat enables biocompatible and monolithic stretchable electronics. *Nat. Mater.* **2021**, *20*, 859–868. [CrossRef]
73. Hu, X.; Huang, T.; Liu, Z.; Wang, G.; Chen, D.; Guo, Q.; Yang, S.; Jin, Z.; Lee, J.-M.; Ding, G. Conductive graphene-based E-textile for highly sensitive, breathable, and water-resistant multimodal gesture-distinguishable sensors. *J. Mater. Chem. A* **2020**, *8*, 14778–14787. [CrossRef]
74. Wang, S.; Du, X.; Luo, Y.; Lin, S.; Zhou, M.; Du, Z.; Cheng, X.; Wang, H. Hierarchical design of waterproof, highly sensitive, and wearable sensing electronics based on MXene-reinforced durable cotton fabrics. *Chem. Eng. J.* **2021**, *408*, 127363. [CrossRef]
75. Zhou, Z.; Li, Y.; Cheng, J.; Chen, S.; Hu, R.; Yan, X.; Liao, X.; Xu, C.; Yu, J.; Li, L. Supersensitive all-fabric pressure sensors using printed textile electrode arrays for human motion monitoring and human-machine interaction. *J. Mater. Chem. C* **2018**, *6*, 13120–13127. [CrossRef]
76. Li, W.; Jin, X.; Han, X.; Li, Y.; Wang, W.; Lin, T.; Zhu, Z. Synergy of porous structure and microstructure in piezoresistive material for high-performance and flexible pressure sensors. *ACS Appl. Mater. Interfaces* **2021**, *13*, 19211–19220. [CrossRef]
77. Wang, S.; Li, D.; Zhou, Y.; Jiang, L. Hierarchical $Ti_3C_2T_x$ MXene/Ni Chain/ZnO array hybrid nanostructures on cotton fabric for durable self-cleaning and enhanced microwave absorption. *ACS Nano* **2020**, *14*, 8634–8645. [CrossRef]
78. Zhang, H.; Liu, D.; Lee, J.-H.; Chen, H.; Kim, E.; Shen, X.; Zheng, Q.; Yang, J.; Kim, J.-K. Anisotropic, wrinkled, and crack-bridging structure for ultrasensitive, highly selective multidirectional strain sensors. *Nano Micro Lett.* **2021**, *13*, 1–15. [CrossRef]
79. Ford, M.J.; Patel, D.K.; Pan, C.; Bergbreiter, S.; Majidi, C. Controlled assembly of liquid metal inclusions as a general approach for multifunctional composites. *Adv. Mater.* **2020**, *32*, 2002929. [CrossRef]
80. Zulqarnain, M.; Stanzione, S.; Rathinavel, G.; Smout, S.; Willegems, M.; Myny, K.; Cantatore, E. A flexible ECG patch compatible with NFC RF communication. *NPG Flex. Electron.* **2020**, *4*, 1–8.
81. Chu, Z.; Jiao, W.; Huang, Y.; Zheng, Y.; Wang, R.; He, X. Superhydrophobic gradient wrinkle strain sensor with ultra-high sensitivity and broad strain range for motion monitoring. *J. Mater. Chem. A* **2021**, *9*, 9634–9643. [CrossRef]
82. Xu, L.; Yang, L.; Yang, S.; Xu, Z.; Lin, G.; Shi, J.; Zhang, R.; Yu, J.; Ge, D.; Guo, Y. Earthworm-inspired ultradurable superhydrophobic fabrics from adaptive wrinkled skin. *ACS Appl. Mater. Interfaces* **2021**, *13*, 6758–6766. [CrossRef]
83. Sun, Z.; Feng, L.; Wen, X.; Wang, L.; Qin, X.; Yu, J. Nanofiber fabric based ion-gradient-enhanced moist-electric generator with a sustained voltage output of 1.1 volts. *Mater. Horiz.* **2021**, *8*, 2303–2309. [CrossRef]
84. He, W.; Wang, C.; Wang, H.; Jian, M.; Lu, W.; Liang, X.; Zhang, X.; Yang, F.; Zhang, Y. Integrated textile sensor patch for real-time and multiplex sweat analysis. *Sci. Adv.* **2019**, *5*, eaax0649. [CrossRef] [PubMed]
85. Yang, Z.; Zhai, Z.; Song, Z.; Wu, Y.; Liang, J.; Shan, Y.; Zheng, J.; Liang, H.; Jiang, H. Conductive and elastic 3d helical fibers for use in washable and wearable electronics. *Adv. Mater.* **2020**, *32*, 1907495. [CrossRef] [PubMed]
86. Zhang, D.; Yang, W.; Gong, W.; Ma, W.; Hou, C.; Li, Y.; Zhang, Q.; Wang, H. Abrasion resistant/waterproof stretchable triboelectric yarns based on fermat spirals. *Adv. Mater.* **2021**, *33*, 2100782. [CrossRef] [PubMed]
87. Zheng, Q.; Lee, J.-H.; Shen, X.; Chen, X.; Kim, J.-K. Graphene-based wearable piezoresistive physical sensors. *Mater. Today* **2020**, *36*, 158–179. [CrossRef]
88. Yu, R.; Zhu, C.; Wan, J.; Li, Y.; Hong, X. Review of graphene-based textile strain sensors, with emphasis on structure activity relationship. *Polymers* **2021**, *13*, 151. [CrossRef]
89. Ma, C.; Ma, M.G.; Si, C.; Ji, X.X.; Wan, P. Flexible MXene-based composites for wearable devices. *Adv. Funct. Mater.* **2021**, *31*, 2009524. [CrossRef]
90. Fu, Z.; Wang, N.; Legut, D.; Si, C.; Zhang, Q.; Du, S.; Germann, T.C.; Francisco, J.S.; Zhang, R. Rational design of flexible two-dimensional MXenes with multiple functionalities. *Chem. Rev.* **2019**, *119*, 11980–12031. [CrossRef]
91. Pyo, S.; Lee, J.; Kim, W.; Jo, E.; Kim, J. Multi-layered, hierarchical fabric-based tactile sensors with high sensitivity and linearity in ultrawide pressure range. *Adv. Funct. Mater.* **2019**, *29*, 1902484. [CrossRef]
92. Li, L.; Xiang, H.; Xiong, Y.; Zhao, H.; Bai, Y.; Wang, S.; Sun, F.; Hao, M.; Liu, L.; Li, T. Ultrastretchable fiber sensor with high sensitivity in whole workable range for wearable electronics and implantable medicine. *Adv. Sci.* **2018**, *5*, 1800558. [CrossRef]
93. Liu, H.; Li, Q.; Bu, Y.; Zhang, N.; Wang, C.; Pan, C.; Mi, L.; Guo, Z.; Liu, C.; Shen, C. Stretchable conductive nonwoven fabrics with self-cleaning capability for tunable wearable strain sensor. *Nano Energy* **2019**, *66*, 104143. [CrossRef]

94. Yu, P.; Li, X.; Li, H.; Fan, Y.; Cao, J.; Wang, H.; Guo, Z.; Zhao, X.; Wang, Z.; Zhu, G. All-fabric ultrathin capacitive sensor with high pressure sensitivity and broad detection range for electronic skin. *ACS Appl. Mater. Interfaces* **2021**, *13*, 24062–24069. [CrossRef] [PubMed]
95. Xu, H.; Gao, L.; Wang, Y.; Cao, K.; Hu, X.; Wang, L.; Mu, M.; Liu, M.; Zhang, H.; Wang, W. Flexible waterproof piezoresistive pressure sensors with wide linear working range based on conductive fabrics. *Nano Micro Lett.* **2020**, *12*, 159. [CrossRef] [PubMed]
96. Zhang, L.; He, J.; Liao, Y.; Zeng, X.; Qiu, N.; Liang, Y.; Xiao, P.; Chen, T. A self-protective, reproducible textile sensor with high performance towards human-machine interactions. *J. Mater. Chem. A* **2019**, *7*, 26631–26640. [CrossRef]
97. Wu, R.; Ma, L.; Patil, A.; Hou, C.; Zhu, S.; Fan, X.; Lin, H.; Yu, W.; Guo, W.; Liu, X.Y. All-textile electronic skin enabled by highly elastic spacer fabric and conductive fibers. *ACS Appl. Mater. Interfaces* **2019**, *11*, 33336–33346. [CrossRef] [PubMed]
98. Cheng, B.; Wu, P. Scalable fabrication of Kevlar/$Ti_3C_2T_x$ MXene intelligent wearable fabrics with multiple sensory capabilities. *ACS Nano* **2021**, *15*, 8676–8685. [CrossRef]
99. Nie, B.; Huang, R.; Yao, T.; Zhang, Y.; Miao, Y.; Liu, C.; Liu, J.; Chen, X. Textile-based wireless pressure sensor array for human-interactive sensing. *Adv. Funct. Mater.* **2019**, *29*, 1808786. [CrossRef]
100. Zhao, Z.; Huang, Q.; Yan, C.; Liu, Y.; Zeng, X.; Wei, X.; Hu, Y.; Zheng, Z. Machine-washable and breathable pressure sensors based on triboelectric nanogenerators enabled by textile technologies. *Nano Energy* **2020**, *70*, 104528. [CrossRef]
101. Liu, Z.; Zheng, Y.; Jin, L.; Chen, K.; Zhai, H.; Huang, Q.; Chen, Z.; Yi, Y.; Umar, M.; Xu, L. Highly breathable and stretchable strain sensors with insensitive response to pressure and bending. *Adv. Funct. Mater.* **2021**, *31*, 2007622. [CrossRef]
102. Gholami, M.; Rezaei, A.; Cuthbert, T.J.; Napier, C.; Menon, C. Lower body kinematics monitoring in running using fabric-based wearable sensors and deep convolutional neural networks. *Sensors* **2019**, *19*, 5325. [CrossRef]
103. Nasiri, S.; Khosravani, M.R. Progress and challenges in fabrication of wearable sensors for health monitoring. *Sens. Actuators A Phys.* **2020**, *312*, 112105. [CrossRef]
104. Wicaksono, I.; Tucker, C.I.; Sun, T.; Guerrero, C.A.; Liu, C.; Woo, W.M.; Pence, E.J.; Dagdeviren, C. A tailored, electronic textile conformable suit for large-scale spatiotemporal physiological sensing in vivo. *NPG Flex. Electron.* **2020**, *4*, 5. [CrossRef]
105. Homayounfar, S.Z.; Rostaminia, S.; Kiaghadi, A.; Chen, X.; Alexander, E.T.; Ganesan, D.; Andrew, T.L. Multimodal smart eyewear for longitudinal eye movement tracking. *Matter* **2020**, *3*, 1275–1293. [CrossRef]
106. Ma, Y.; Ouyang, J.; Raza, T.; Li, P.; Jian, A.; Li, Z.; Liu, H.; Chen, M.; Zhang, X.; Qu, L. Flexible all-textile dual tactile-tension sensors for monitoring athletic motion during taekwondo. *Nano Energy* **2021**, *85*, 105941. [CrossRef]
107. Ishac, K.; Suzuki, K. Lifechair: A conductive fabric sensor-based smart cushion for actively shaping sitting posture. *Sensors* **2018**, *18*, 2261. [CrossRef]
108. Han, M.; Kwak, J.W.; Rogers, J.A. Soft sign language interpreter on your skin. *Matter* **2020**, *3*, 337–338. [CrossRef]
109. Zhang, Z.; He, T.; Zhu, M.; Sun, Z.; Shi, Q.; Zhu, J.; Dong, B.; Yuce, M.R.; Lee, C. Deep learning-enabled triboelectric smart socks for IoT-based gait analysis and VR applications. *NPG Flex. Electron.* **2020**, *4*, 29. [CrossRef]
110. Shuai, L.; Guo, Z.H.; Zhang, P.; Wan, J.; Pu, X.; Wang, Z.L. Stretchable, self-healing, conductive hydrogel fibers for strain sensing and triboelectric energy-harvesting smart textiles. *Nano Energy* **2020**, *78*, 105389. [CrossRef]
111. Chen, J.; Wen, X.; Liu, X.; Cao, J.; Ding, Z.; Du, Z. Flexible hierarchical helical yarn with broad strain range for self-powered motion signal monitoring and human-machine interactive. *Nano Energy* **2021**, *80*, 105446. [CrossRef]

 materials

Article

Design and 3D Printing of Stretchable Conductor with High Dynamic Stability

Chao Liu [1,2,3], Yuwei Wang [2,3], Shengding Wang [2,3], Xiangling Xia [2,3], Huiyun Xiao [2,3], Jinyun Liu [2,3], Siqi Hu [2,3], Xiaohui Yi [2,3], Yiwei Liu [2,3], Yuanzhao Wu [2,3], Jie Shang [2,3,*] and Run-Wei Li [4,*]

1 School of Materials Science and Chemical Engineering, Ningbo University, Ningbo 315211, China
2 CAS Key Laboratory of Magnetic Materials and Devices, Ningbo Institute of Materials Technology and Engineering, Chinese Academy of Sciences, Ningbo 315201, China
3 Zhejiang Province Key Laboratory of Magnetic Materials and Application Technology, Ningbo Institute of Materials Technology and Engineering, Chinese Academy of Sciences, Ningbo 315201, China
4 College of Materials Science and Opto-Electronic Technology, University of Chinese Academy of Sciences, Beijing 100049, China
* Correspondence: shangjie@nimte.ac.cn (J.S.); runweili@nimte.ac.cn (R.-W.L.)

Abstract: As an indispensable part of wearable devices and mechanical arms, stretchable conductors have received extensive attention in recent years. The design of a high-dynamic-stability, stretchable conductor is the key technology to ensure the normal transmission of electrical signals and electrical energy of wearable devices under large mechanical deformation, which has always been an important research topic domestically and abroad. In this paper, a stretchable conductor with a linear bunch structure is designed and prepared by combining numerical modeling and simulation with 3D printing technology. The stretchable conductor consists of a 3D-printed bunch-structured equiwall elastic insulating resin tube and internally filled free-deformable liquid metal. This conductor has a very high conductivity exceeding 10^4 S cm^{-1}, good stretchability with an elongation at break exceeding 50%, and great tensile stability, with a relative change in resistance of only about 1% at 50% tensile strain. Finally, this paper demonstrates it as a headphone cable (transmitting electrical signals) and a mobile phone charging wire (transmitting electrical energy), which proves its good mechanical and electrical properties and shows good application potential.

Keywords: liquid metal; stretchable conductor; 3D printing; high dynamic stability; wearable devices

Citation: Liu, C.; Wang, Y.; Wang, S.; Xia, X.; Xiao, H.; Liu, J.; Hu, S.; Yi, X.; Liu, Y.; Wu, Y.; et al. Design and 3D Printing of Stretchable Conductor with High Dynamic Stability. *Materials* **2023**, *16*, 3098. https://doi.org/10.3390/ma16083098

Academic Editor: Heung Cho Ko

Received: 14 March 2023
Revised: 10 April 2023
Accepted: 12 April 2023
Published: 14 April 2023

1. Introduction

Flexible electronic devices have mechanical properties that traditional devices do not have, such as bendability [1–3], foldability [4–6] and stretchability [7–9], and are widely used in healthcare [10,11], medicine [12,13], and soft robots [14–16]. Stretchable conductors, as the basic components of flexible electronic devices, have always been a hot topic of research. Compared with traditional rigid conductors, stretchable conductors have more deformability, which greatly expands the applications of wearable devices, and saves a lot of wiring space for robotic arms [17]. Stretchable conductors maintain stable conductivity under large deformation, that is, high dynamic stability, which is crucial for the stable operation of devices such as robotic arms and wearable electronics devices. In order to fully reflect the advantages of stretchable conductors, it is necessary to have both high stretchability and stable conductivity [18].

As a key component of flexible electronic devices, stretchable conductors are usually realized by the structure or material [19–24]. Structure-based stretchable conductors are one of the most widely used methods to achieve scalable interconnections. Some non-stretchable conductive materials are designed as serpentine [25], mesh [26,27], cracks [28–30], and longitudinal waves [31,32]. However, high stretchability usually requires more complex patterns, which increases the wire resistance and process complexity. In terms of materials, conductive composites are mainly prepared by using polymers that are inherently conductive or mixing conductive materials into elastomers. For example, poly (3,4-ethyldioxythiophene):poly (styrenesulfonate) (PEDOT:PSS) is intrinsically conductive. The modified PEDOT:PSS film has more than 50% mechanical stretchability and more than 1000 S cm^{-1} of conductivity [33]. Despite this, intrinsically conductive polymers have disadvantages such as high cost and poor stability [34]. In addition, conductive networks can also be provided by mixing conductive materials in elastic polymers, such as metal nanowires [35,36], graphene [37,38], carbon nanotubes [39,40], and liquid metals [41]. This polymer composite material allows the manufacture of flexible and scalable conductors through a simple mixing process. However, it is difficult for this conductive composite material to maintain a stable resistance during deformation, and in repeated deformation, due to the mutual extrusion of conductive materials or the disconnection of the conductive path. These shortcomings limit the practical application of conductive composites as stretchable conductors [42].

Some researchers have achieved relatively stable resistance changes during stretching by constructing a three-dimensional network of conductive materials. For example, Gao et al. reported a stretchable conductor inspired by the maple leaf shape. A prestrain finishing method was used to make the surface of the conductive polypyrrole coating a layered wrinkle structure, so the resistance change was 66% under 600% stretching and its conductivity was 100 S m^{-1} [43]. Zhang et al. formed a three-dimensional conductive network by welding carbon nanotubes, and then packaged it with PDMS. The resistance change was about 5% and the conductivity was about 132 S m^{-1} under 90% stretching [44]. Hong et al. used the bimodal porous structure made of silver nanowires. At 40% strain, the resistance change was 8% and the conductivity was 42 S cm^{-1} [45]. Nevertheless, most of the reported stretchable conductors have electrical degradation or a conductivity that cannot meet the requirements of practical applications, and the preparation process is complicated so cannot be prepared in batches.

There are other examples of stretchable conductors by using intrinsically flexible liquid metal. Liang et al. filled liquid metal into a porous sponge made of elastomers, the maximum conductivity could reach 10,000 S cm^{-1}, and the resistance change was about 10% at 50% stretch, but it required a large liquid metal consumption and was not easy to package [46]. Ning et al. improved its mechanical properties by mixing liquid metal and elastic matrix PUS and using PDMS for encapsulation. The conductivity was 478 S cm^{-1}, and the resistance change was about 2% at 50% stretch [47]. This highly loaded composite material is prone to liquid metal leakage after multiple stretching and the comparably low conductivity is also problematic. Injecting liquid metal into the elastomer channel also makes it possible to prepare high-conductivity stretchable conductors, but the complex preparation process and the great resistance change after stretching are problematic. It is also difficult to obtain high conductivity by mixing with materials, because it is difficult for micron-sized liquid metal droplets with low aspect ratio to form a dense conductive network.

In this paper, to develop the next generation of wearable flexible electronic devices and improve the efficiency of expensive liquid metals, an innovative linear bunch three-dimensional conductive structure is proposed to achieve stable resistance change within a certain strain range. Stretchable conductors with high dynamic stability are fabricated by designing strain-insensitive elastomeric channels that can be prepared in batches through the 3D printing technology developed in recent years. Liquid metal is selected as the conductive filler of a three-dimensional conductive structure. Because of its inherent deformability and good conductivity, it can respond to external stress and undergo reversible deformation without any hysteresis or mechanical degradation, which avoids the phenomenon of electrical degradation during repeated stretching. The results show that the liquid-metal-based stretchable conductor with a linear bunch three-dimensional conductive network has an excellent conductivity of up to 10,000 S cm^{-1}, and the elongation at break is greater than 50%. More importantly, it has an excellent dynamic stability during stretching. At 50% tensile strain, the relative resistance change ($\Delta R/R_0$) is only about 1%, which shows great prospects as a high-performance stretchable conductor and is of great significance to the development of flexible electronic devices. In subsequent practical application testing, as an elastic headphone line, it has an excellent electrical signal transmission capability and has little effect on the music signal after stretching 50%. As an elastic charging line for mobile phones, it still transmits power stably under dynamic stretching.

2. Materials and Methods

2.1. Materials

Photocurable resin Agilus30 (Stratasys, Eden Prairie, MN, USA), high-purity metal gallium (99.99%; Beijing Founde Star Sci. & Technol. Co., Ltd., Beijing, China), and indium (99.995%; Beijing Founde Star Sci. & Technol. Co., Ltd.) were used. The extra reagents utilized in the experiment were altogether gained from Sinopharm Chemical Reagent Co., Ltd., Shanghai, China.

2.2. Preparation Process

The linear bunch structure was fabricated using a photocuring 3D printer (AUTOCERA-L, Beijing Shiwei Technology Co., Ltd., Beijing, China). The elastic resin Agilus30 was used for molding, the molded sample was then placed in a beaker filled with alcohol, and the ultrasonic machine was used to ultrasound at 25 kHz and 25 °C for 10 min. The object after cleaning was put into the oven for drying for 5 min, and then the sample was post-cured for 20 min by a UV curing machine. After curing completely, the bracket was removed. We mixed gallium and indium in a mass ratio of 3:1, then heated the mixture in a water bath at 60 °C and stirred for 30 min to obtain a liquid metal (gallium indium alloy). A syringe was used to fill the liquid metal with a bunch-structured conductive network, and the elastic resin Agilus30 was used at both ends to fix the copper sheet and seal the port for easy testing.

2.3. Young's Modulus

The Young's modulus of the elastic resin Agilus30 was determined by tensile testing. Standard stretch parts were made using a 3D printer, and the sample was secured on a stretching machine (Instron 5943, Norwood, MA, USA) with an initial length of 2 cm. The sample was stretched at a rate of 50 mm min^{-1}, and the obtained stress–strain curve was linearly fitted to confirm that the slope was Young's modulus.

2.4. Computational Simulations

COMSOL Multiphysics 6.0 was used for finite element simulation. The Yeoh model in hyperelastic materials was used for simulation experiments, the measured material parameters were input, and the density of the material was set to 970 kg m^{-3}. As liquid metal is fluid, it is necessary to adopt a multi-physical field of fluid–solid coupling. In the simulation, one end of the sample was fixed and the other end was stretched. As the stretching proceeded, the shape of the sample changed. The electrical module recorded the change in resistance according to the formula $R = \rho L S^{-1}$ of resistance.

2.5. Mechanical Features

The basic mechanical properties such as the strain and strength of the printed linear-bunch-structured stretchable conductor were tested by a stretcher (Instron 5943).

2.6. Electrical Characteristics

The resistance was measured using the DC current source (Keithley 6221, Cleveland, OH, USA) and the nanovoltmeter (Agilent 34420A, Santa Clara, CA, USA) by the four-wire method. The copper sheets at both ends of the sample were connected to the current source and voltmeter, and the data acquisition system was used to observe the resistance change of the linear-bunch-structured stretchable conductor under the stretching of the stretching machine (Instron 5943).

2.7. Making a Headphone Cable

Part of the earphone cable was replaced with the prepared sample so that the two ends of the earphone wire were bonded to both ends of the sample by light-curing resin, and a stretchable earphone cable was obtained. To test the voltage waveform, an oscilloscope (710 110/DLM2024, Yokogawa, Tokyo, Japan) was connected in parallel with a speaker to record the music signal.

3. Results

3.1. Structural Design

Figure 1a shows the three strain-insensitive structures designed in this paper, and the three shapes are numbered as 1, 2, and 3. When the structure with uneven thickness is stretched, the deformation squeezes the liquid metal in the coarse position so that the fine position is supplemented, which may retain the overall resistance in a relatively stable state during the stretching process. Taking shape 2 as an example, the structural parameters that control its shape are analyzed. The coarse local radius is set to the outer diameter, the fine position is set to the inner diameter, and the distance between the two adjacent inner diameters is set to the length, as shown in Figure 1b. Then, a comparative analysis of the designed structures is conducted to select the most strain-insensitive structure for subsequent analysis. Therefore, an orthogonal test table (Table S1) with four factors and three levels is designed. The shape and structural parameters of the outer diameter, inner diameter, and length are listed as the factors affecting the relative resistance change after stretching. Each factor is given three values. The relative resistance change of the stretchable conductor after 50% stretching in the orthogonal experiment is calculated by the finite element method. Figure 1c shows the estimated marginal mean of each factor. Its role is to control the results of each independent variable to predict the average of the dependent variables when other variables remain unchanged. It is usually used in multivariate analysis. When the estimated marginal mean is lower, the value of ΔR is smaller. It can be seen from the figure that the three-dimensional conductive network structure of shape 2 maintains a more stable resistance during stretching.

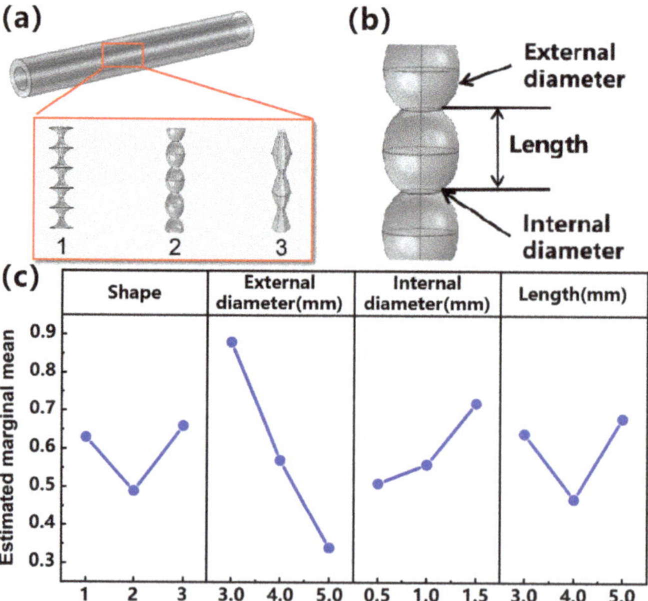

Figure 1. (**a**) Schematic diagram of three structures. (**b**) Structural parameters of the shape. (**c**) Results of orthogonal experiment.

3.2. Simulation and Verification

Therefore, in this paper, the second kind of linear bunch conductive structure is studied in depth. First, two kinds of elastic matrix (Figure 2a) with uniform thickness and equiwall thickness containing linear bunch conductive structures are designed; the enlarged diagram is a schematic diagram of the internal structure of uniform thickness and equiwall thickness. The internal structure in Figure 2a is the bunch structure, while the external structure is of uniform thickness and equiwall thickness. The parameters that determine the elastic matrix of the linear bunch structure are horizontal diameter (a), longitudinal diameter (b), neck length (c), and thickness (d). The Young's modulus of the elastomer material is 0.4 MPa (Figure S1). This value is used for finite element analysis simulation. Using Yeoh's mechanical module and electrical module, a finite element analysis and simulation are carried out under the action of fluid–solid coupling. Based on the model of this structure, the change in resistance in the case of stretching is simulated to match the change in resistance in the actual tensile experiment. As shown in Figure 2b, in the simulation process, the linear bunch elastic matrix is stretched and deformed. It can be seen from the figure that the tensile strain is mainly released by the gourd ball with larger curvature, and there is only slight deformation at the neck connection between the balls. The thin neck is indeed the main contributor to the overall resistance. Therefore, using this string structure, it is possible to suppress the change in resistance to a lower level during the stretching process. Next, we perform a finite element simulation and experiment on two sets of samples of uniform thickness and equiwall thickness and compare their results to verify the accuracy of the finite element simulation model. The horizontal diameter (a) of the two groups of samples is 4 mm, the longitudinal diameter (b) is 5 mm, the neck length (c) is 0.75 mm, and the thickness (d) is 1 mm. The resistance of the uniform-thickness and equiwall-thickness bunch structure made by 3D printing is measured under the stretching of the stretching machine. Figure 2c,d show the resistance change curve of the finite element simulation and the actual experiment under 50% stretching. It can be seen from the figure that the resistance change during the simulated stretching process is in good agreement with the actual experimental results, which shows that the finite element simulation model is valid in this study.

In order to grasp the influence of each structural parameter on the change in resistance during stretching, control experiments of the finite element simulation are carried out on them. Figure 2e shows the effect of the change in horizontal diameter (a) on the change in resistance in stretching (50%) when the longitudinal diameter (b) is 5 mm, the neck length (c) is 0.75 mm, and the thickness (d) is 1 mm; the change in resistance decreases with the increase in the horizontal diameter (a). Figure 2f shows that when the horizontal diameter (a) is 3 mm, the neck length (c) is 0.75 mm, and the thickness (d) is 1 mm, the change in longitudinal diameter (b) has an effect on the change in resistance when stretching (50%). The change in resistance decreases first and then increases with the increase in longitudinal diameter (b). There is an optimal value to minimize the change in resistance. Figure 2g shows that when the horizontal diameter (a) is 3 mm, the longitudinal diameter (b) is 5 mm, and the thickness (d) is 1 mm, the change in resistance during stretching (50%) increases with the increase in the neck length (c). Figure 2h shows that when the horizontal diameter (a) is 3 mm, the longitudinal diameter (b) is 5 mm, and the neck length (c) is 0.75 mm, the change in the thickness (d) during stretching (50%) has little effect on the change in resistance, which is basically negligible compared to several other parameters. It can be seen from the diagram that the structure of equiwall thickness is more stable than that of the uniform-thickness structure under the same structural parameters, so when preparing samples for practical application, the linear bunch three-dimensional conductive network structure with equiwall thickness is selected.

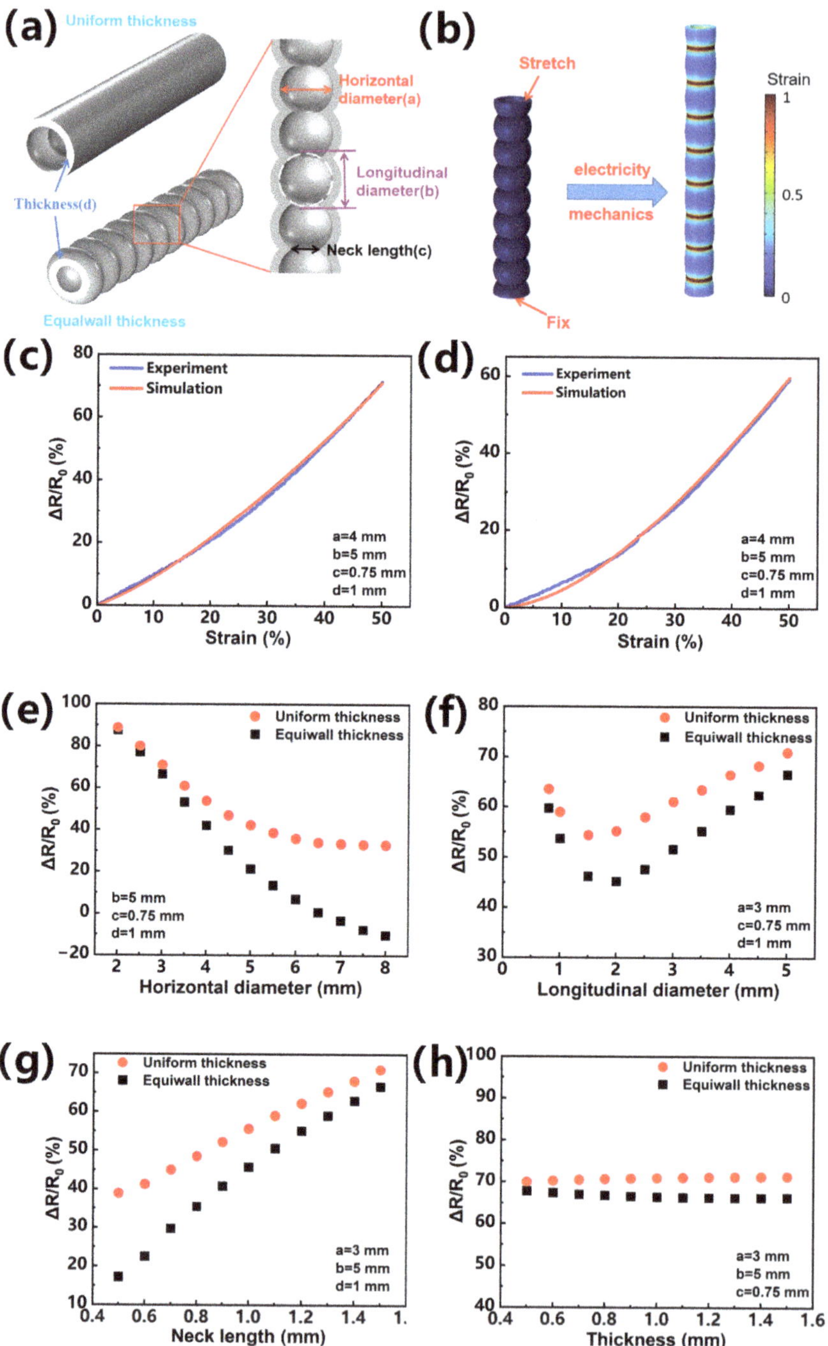

Figure 2. Parameters and electrical properties of 3D-printed bunch structures. (**a**) Schematic diagram of linear bunch conductive network with two elastic matrix shapes. (**b**) Simulation diagram of linear bunch conductive network. Color map represents strain. (**c**) The electrical properties of the experimental

and simulated tensile tests of the uniform-thickness bunch structure. a = 4 mm, b = 5 mm, c = 0.75 mm, d = 1 mm. (**d**) The electrical properties of the experimental and simulated tensile tests of the equiwall-thickness bunch structure. a = 4 mm, b = 5 mm, c = 0.75 mm, d = 1 mm. (**e–h**) The control experiment of each structural parameter of uniform-thickness bunch structure and equiwall-thickness bunch structure under 50% tensile strain.

3.3. *Performance Characterization*

According to the influence of the parameters of the bunch structure on the resistance change during stretching summarized by the control experiment, a stretchable conductor with basically stable resistance under 50% stretching is further produced by 3D printing. Figure 3a shows the fabrication process of the stretchable conductor, first printing an elastic matrix with a bunch structure of equiwall thickness through a 3D printer, then rinsing the sample in alcohol, removing the support, and filling the channel with liquid metal using the injection method. According to the influence law of the parameters obtained by the finite element simulation, the horizontal diameter (a) is 2.5 mm, the longitudinal diameter (b) is 2 mm, the neck length (c) is 0.25 mm, and the thickness (d) is 0.5 mm. The equiwall-thickness bunch structure has a resistance change of only about 1% when stretching 50%, while the resistance of the ordinary cylindrical pipe changes by about 126% (Figure 3b) under the same stretching condition, which greatly improves the strain stability of the stretchable conductor. In addition, the stability of the bunch structure and the ordinary cylindrical structure during twisting and bending is also tested (Figure S2). It can be seen that when bending to 180 degrees, the resistance of the ordinary cylinder changes by 6.5%, while the resistance of the bunch structure only changes by 0.3%. Similarly, when twisted to 360 degrees, the resistance of the ordinary cylinder changes by 9.3%, while the resistance of the bunch structure changes by only 0.1%. Under different deformation, the bunch structure still maintains excellent stability performance compared with the ordinary cylindrical structure. It indirectly reflects that the design of the equiwall-thickness linear-bunch-structured conductive network plays an important role in obtaining excellent strain insensitive conductivity. The quality factor Q, defined as the percent strain divided by the percent relative resistance change, is an important evaluation factor for the strain-insensitive conductivity of stretchable conductors. As depicted in Table 1, the equiwall-thickness linear bunch structure exhibits a prominent Q value (33) even though it is in a state of strain (50%) compared to recently reported stretchable conductors based on the porous structure [45], wrinkle structure [48], and others [44,47,49–53], suggesting the outstanding strain-insensitive conductivity.

Table 1. Comparison table of equiwall-thickness bunch structure and other representative structure stretchable conductors.

Conductive Structure	Quality Factor (Q)	Electrical Conductivity	Durability (Number of Cycles)	Reference
welded CNT structure	18 (90% strain)	1.32 S cm^{-1}	1000	[44]
porous structure of AgNWs	5 (40% strain)	42 S cm^{-1}	100	[45]
wrinkle-crumple rGo structure	23 (70% strain)	/	200	[48]
AgNWs spongy structure	2.5 (50% strain)	27.78 S cm^{-1}	1000	[49]
liquid metal corrugated structure	2 (100% strain)	/	500	[50]
AgNWs buckled structure	0.28 (130% strain)	21 S cm^{-1}	1000	[51]
micro-wrinkled rGo structure	1.48 (80% strain)	0.256 S cm^{-1}	500	[52]
liquid metal spongy structure	25 (50% strain)	478 S cm^{-1}	1000	[47]
gold thin-film crack structure	1 (50% strain)	/	500	[53]
liquid metal bunch structure	33 (50% strain)	10,000 S cm^{-1}	1000	this work

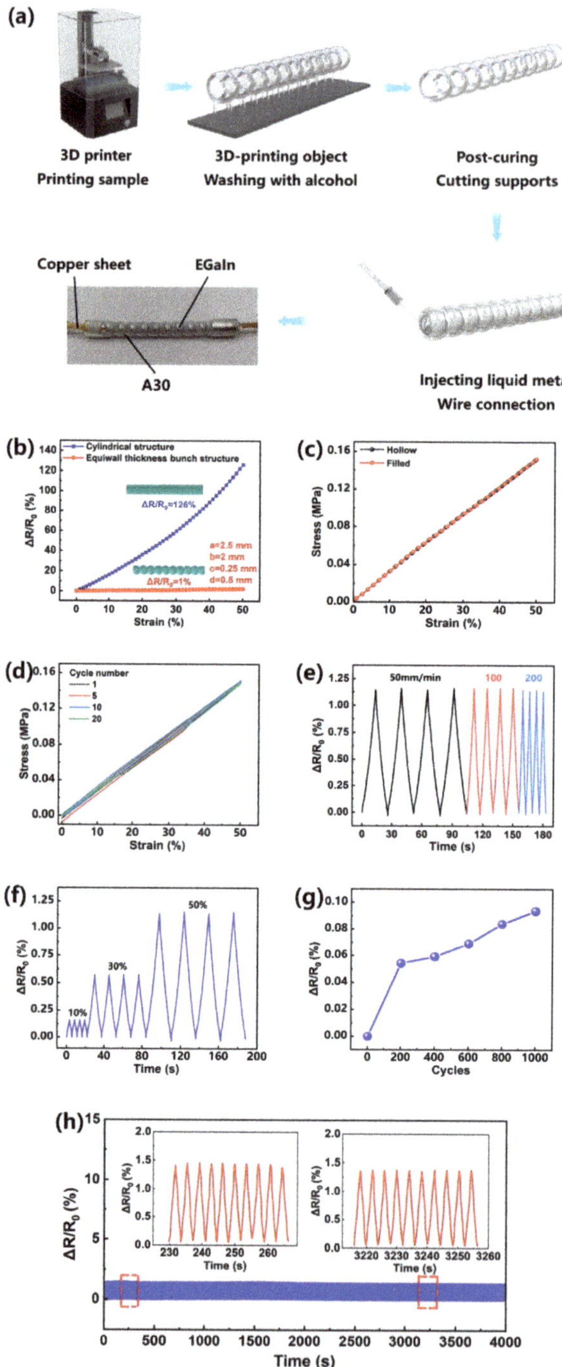

Figure 3. Three-dimensional-printing process and performance of string structure wire. (**a**) Fabrication process. The elastic matrix of the equiwall-thickness linear bunch structure was made by a 3D printer, and the structure was filled with liquid metal. (**b**) Comparison of the resistance change of the

strain-insensitive equiwall-thickness bunch structure conductor with the ordinary cylindrical pipe during stretching, through the optimization of structural parameters. (**c**) Comparison of mechanical properties of bunch-structured stretchable conductors filled with liquid metal and unfilled with liquid metal. (**d**) Cyclic diagram of conductor under 50% stretch. (**e**) Variation in resistance with a strain of 50% at different strain rates. (**f**) The resistance change under different strain cycles at a strain rate of 200 mm min^{-1}. (**g**) The change in inherent resistance after different stretching cycles. (**h**) Tensile stretching/releasing cycle of 50% strain for 1000 cycles.

The reason for choosing liquid metal as a filler is that it has very little effect on the mechanical properties of the elastic material, and this low-viscosity liquid easily flows in response to the applied strain [54]. The mechanical properties (Figure 3c) of stretchable conductors with hollow and filled bunch structures are measured by a stretching machine. The bunch structures with and without liquid metal have almost the same mechanical properties, indicating that the effect of liquid metal on mechanical properties is negligible. Under 20 large strain cycles, the change in tensile properties is negligible (Figure 3d). This shows that the elastomer material has a relatively high mechanical durability. In order to further test the stability of the stretchable conductor, Figure 3e shows that we apply different stretching rates (50−200 mm min^{-1}), which have little effect on the change in resistance, which is of great significance for the different rates of actual use in different parts of the manipulator. Figure 3f illustrates the change in resistance of the equiwall bunch structure under cyclic extending and relaxing with various strains by exerting the same stimulus rate of 20 mm min^{-1}. The influence shown in Figure 3f shows the resistance change apparently under various strain states (10, 30, and 50%), and the response is found to be stable. Finally, in order to evaluate the stability and durability of the equiwall-bunch-structured stretchable conductor, the sample is subjected to 1000 cycles of 50% strain and a 200 mm min^{-1} rate. Figure 3g shows that in 1000 cycles, the resistance change of the sample after cyclic stretching can be neglected relative to the change in the initial resistance. The resistance increase is less than 0.06% after 200 stretching cycles and is less than 0.1% after 1000 stretching cycles, which proves the high dynamic stability of the bunch-structured stretchable conductor. Figure 3h shows that the change in resistance is still stable in 1000 cycles, and the illustration of Figure 3h is the enlarged area in 230−265 s and 3215−3255 s. For most stretchable conductors, repeated stretching often leads to a sharp decline in conductivity due to damage to the internal conductive channel. Due to the good recoverability of the conductive network, this bunch structure can be almost restored to the initial value by releasing the applied strain. It is further shown that this bunch structure has a long working life and reliability.

3.4. Principal Analysis

Figure 4 shows the structural change of the designed bunch-structured conductive network before and after stretching. An enlarged view of Figure 4a shows the shape of the neck of the conductor and the ball before and after stretching, and it can be seen that the liquid metal at the ball releases strain and squeezes into the neck during stretching. The ball of this bunch structure is similar to a reservoir, which is supplemented at the neck during stretching, so the resistance is always maintained in a relatively stable state during stretching. Figure 4b shows a schematic diagram of the structural changes of this bunch-structured conductive network and ordinary cylindrical conductive network during stretching, and it is obvious that the ordinary cylindrical conductive network becomes elongated in the middle after stretching, while the neck of the bunch-structured conductive network does not change much [43]. The results of finite element simulations further confirm this conclusion, and when stretched, the neck of the bunch structure resistance will be supplemented by liquid metal from the ball, and even increase the diameter of the neck (Figure S3). Therefore, this bunch-structured stretchable conductor with equiwall thickness has significant durability and stable strain-insensitive performance, and it has great application prospects in flexible bodies and wearable systems.

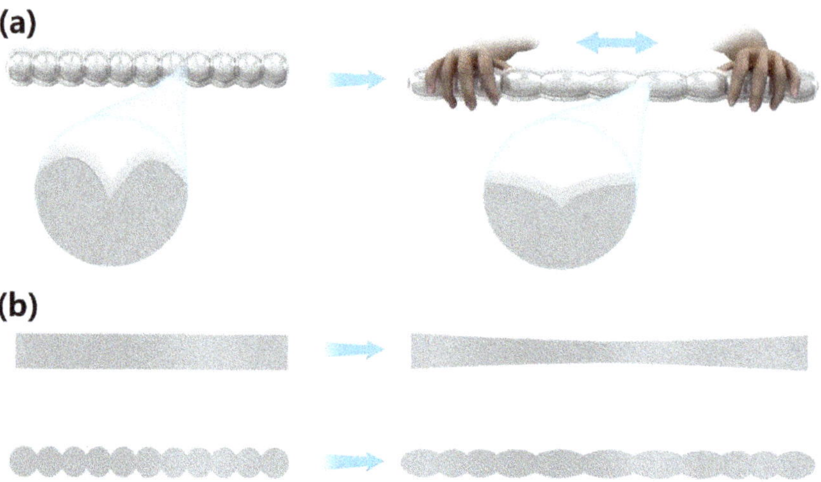

Figure 4. (**a**) Structure diagram of bunch conductive network with equiwall thickness and elastic matrix. (**b**) Schematic diagram of structural changes of equiwall-thickness bunch conductive network and ordinary cylindrical conductive network.

3.5. Application

Now, more and more people use headphones, which can be seen everywhere in our daily lives. While wireless headphones are more widely recognized for their flexibility, cable headphones still have a wide audience because of their music's conductivity and accuracy. However, the length and breakage of the headphone cord have been troubling users. Therefore, a stretchable headphone line can ensure the stability of signal transmission while maintaining good tensile properties, which can be an optimization solution to this problem. The bunch-structured stretchable conductor designed in this article can be used as a headphone cable with good tensile properties (Figure 5a). To prove its electrical performance, the transmission ability of its musical signal (Chinese national anthem) is tested, which is compared after being unstretched and stretched. By comparing the amplitude data before and after stretching 50% (Figure 5b), we find that the maximum amplitude deviation is only 0.042 V, which hardly affects the quality of the music signal. In addition, the effect of the applied tensile strain on the frequency characteristics of the music signal (Figure S4) is analyzed. The spectrogram is obtained by the Fourier transform of the voltage waveform in Figure 5b. It can be seen that after applying a 50% stretching, its frequency characteristics are similar to those not stretched. In addition, this bunch-like stretchable conductor is used in the charging cable of the mobile phone (Supporting Movie S1), and it is found that before and after stretching, it does not affect the charging of the mobile phone at all. Figure 6 directly shows the dynamic stability of the equiwall-bunch-structured stretchable conductor (Figure S5 is the cross section of the structure); with a 100 mA current connected, the voltage change after 50% stretching is small, reflecting that the resistance barely changes before and after stretching, and Supporting Movie S2 shows the real-time change in resistance during dynamic stretching.

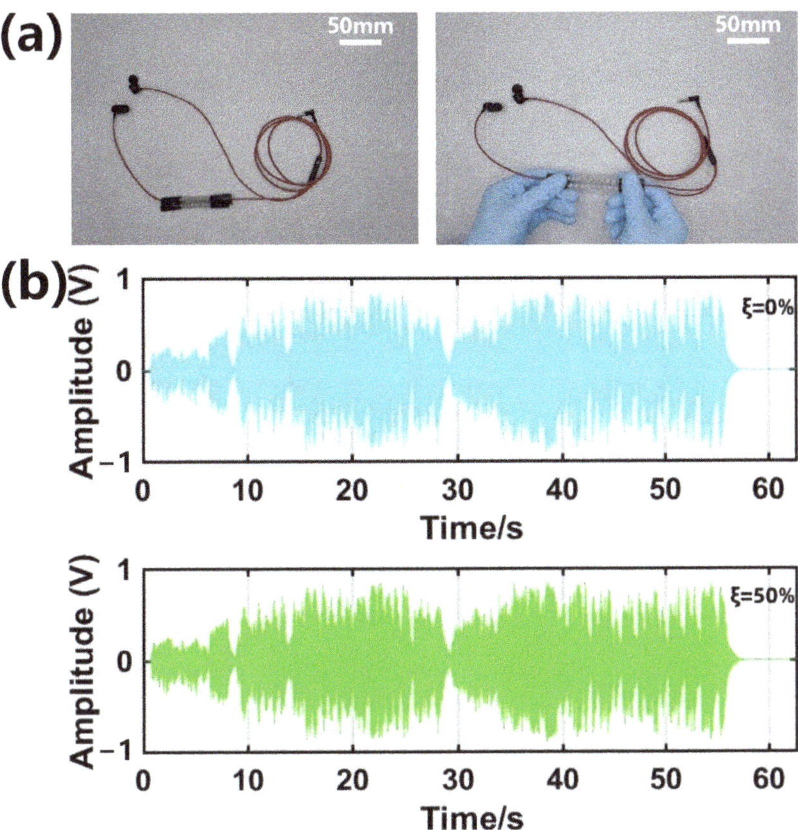

Figure 5. (**a**) Image of stretchable conductor as earphone wire. (**b**) Amplitude–time curve of complex music signals from retractable headphones.

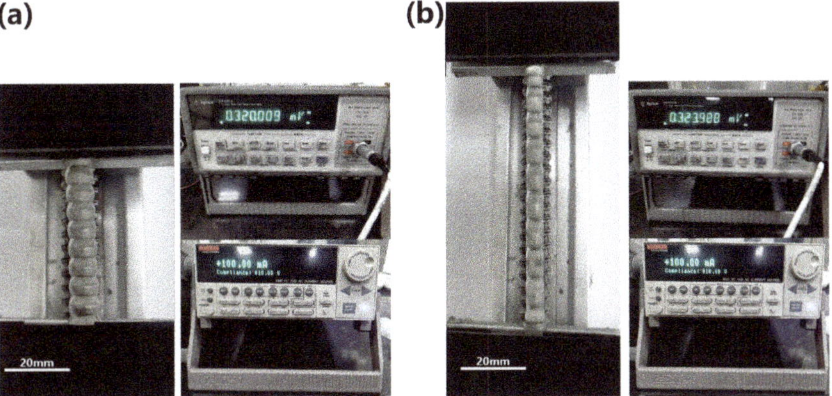

Figure 6. (**a**) Voltage of the initial state of stretchable conductor at 100 mA. (**b**) Voltage of 50% tensile strain of stretchable conductor at 100 mA.

4. Conclusions

In summary, this paper designs a conductive network with an equiwall-thickness bunch structure, uses 3D printing technology to print an elastic matrix, and then injects liquid metal to form a stretchable conductor. Through experiments and simulations, the influence of the structural parameters of the linear bunch conductive network with equiwall thickness on the resistance change during stretching is summarized, and a stretchable conductor with high dynamic stability is produced according to the summarized rules. When this bunch structure is stretched, the liquid metal at the ball is squeezed, so the neck is replenished, similar to the reservoir structure, which allows the conductor to maintain a stable resistance when stretched. As a stretchable conductor, the resistance changes by only about 1% at 50% tensile strain. In addition, the conductor exhibits high durability and excellent electromechanical stability in 1000 cycles at 50% tensile strain. Due to the high conductivity and superior stretchability of this equiwall-thickness linear bunch conductive network, it has broad application prospects in the field of wearable electronics. It is used in the charging line and headphone line of mobile phones to verify the applicability of the structure as a stretchable wire in practical applications. This work provides inspiration for the development of stretchable conductors based on three-dimensional conductive structures through structural design that can be used in various applications of next-generation wearable devices.

Supplementary Materials: The following supporting information can be downloaded at: https://www.mdpi.com/article/10.3390/ma16083098/s1: Four-factor three-level table of orthogonal test (Table S1). Testing Young's Modulus of standard drawing parts of photocurable resins (Figure S1). Simulation experiment of radius change of neck cross-sectional area of linear bunch structure (Figure S2). Effect of tensile strain on frequency characteristics of music signals (Figure S3) (PDF). (a) Spectrum of initial state voltage waveform. (b) Spectrogram of voltage waveform at 50% tensile stress (Figure S4). Cross-sectional section diagram of equiwall thickness bunch structure (Figure S5). The linear-bunch-structured stretchable conductor is used as an elastic mobile phone charging line (Movie S1) (MP4). The voltage change of a stretchable conductor with a linear string structure is measured when it is stretched by 50% at 100 mA (Movie S2) (MP4).

Author Contributions: Conceptualization, J.S., X.Y., Y.L and Y.W. (Yuanzhao Wu); Methodology, C.L. and H.X.; Software, C.L. and X.X.; Validation, C.L.; Formal analysis, C.L. and S.W.; Investigation, C.L.; Resources, C.L., J.L. and S.H.; Data curation, C.L., S.W. and J.S.; Writing—original draft, C.L.; Writing—review & editing, Y.W. (Yuwei Wang); Supervision, X.Y., Y.L., Y.W. (Yuanzhao Wu) and J.S.; Project administration, R.-W.L.; Funding acquisition, J.S. and R.-W.L. The manuscript was written through contributions of all authors. All authors have read and agreed to the published version of the manuscript.

Funding: This research was partially funded by the National Natural Science Foundation of China (U22A20248, 52127803, 51931011, 51971233, 62174165, M-0152, U20A6001, U1909215 and 52105286, 52201236, 62204246, 92064011, 62174164), the External Cooperation Program of the Chinese Academy of Sciences (174433KYSB20190038, 174433KYSB20200013), the Instrument Developing Project of the Chinese Academy of Sciences (YJKYYQ20200030), the K.C. Wong Education Foundation (GJTD-2020-11), the Chinese Academy of Sciences Youth Innovation Promotion Association (2018334), the "Pioneer" and "Leading Goose" R&D Program of Zhejiang (2022C01032), the Zhejiang Provincial Key R&D Program (2021C01183), the Natural Science Foundation of Zhejiang Province (LD22E010002), the Zhejiang Provincial Basic Public Welfare Research Project (LGG20F010006), the Ningbo Scientific and Technological Innovation 2025 Major Project (2019B10127, 2020Z022), and Ningbo Natural Science Foundations (20221JCGY010312).

Institutional Review Board Statement: Not applicable.

Informed Consent Statement: Informed consent was obtained from all subjects involved in the study.

Data Availability Statement: The data presented in this study are available on request from the corresponding author.

Acknowledgments: This research was partially supported by the National Natural Science Foundation of China (U22A20248, 52127803, 51931011, 51971233, 62174165, M-0152, U20A6001, U1909215 and 52105286, 52201236, 62204246, 92064011, 62174164), the External Cooperation Program of the Chinese Academy of Sciences (174433KYSB20190038, 174433KYSB20200013), the Instrument Developing Project of the Chinese Academy of Sciences (YJKYYQ20200030), the K.C. Wong Education Foundation (GJTD-2020-11), the Chinese Academy of Sciences Youth Innovation Promotion Association (2018334), the "Pioneer" and "Leading Goose" R&D Program of Zhejiang (2022C01032), the Zhejiang Provincial Key R&D Program (2021C01183), the Natural Science Foundation of Zhejiang Province (LD22E010002), the Zhejiang Provincial Basic Public Welfare Research Project (LGG20F010006), the Ningbo Scientific and Technological Innovation 2025 Major Project (2019B10127, 2020Z022), and Ningbo Natural Science Foundations (20221JCGY010312).

Conflicts of Interest: The authors declare no conflict of interest.

References

1. Zhao, J.; He, C.; Yang, R.; Shi, Z.; Cheng, M.; Yang, W.; Xie, G.; Wang, D.; Shi, D.; Zhang, G. Ultra-sensitive Strain Sensors Based on Piezoresistive Nanographene Films. *Appl. Phys. Lett.* **2012**, *101*, 063112. [CrossRef]
2. Zhao, J.; Wang, G.; Yang, R.; Lu, X.; Cheng, M.; He, C.; Xie, G.; Meng, J.; Shi, D.; Zhang, G. Tunable Piezoresistivity of Nanographene Films for Strain Sensing. *ACS Nano* **2015**, *9*, 1622–1629. [CrossRef] [PubMed]
3. Tee, B.C.; Wang, C.; Allen, R.; Bao, Z. An Electrically and Mechanically Self-healing Composite with Pressure and Flexion-sensitive Properties for Electronic Skin Applications. *Nat. Nanotechnol.* **2012**, *7*, 825–832. [CrossRef]
4. Suo, Z.; Ma, E.Y.; Gleskova, H.; Wagner, S. Mechanics of Rollable and Foldable Film-on-foil Electronics. *Appl. Phys. Lett.* **1999**, *74*, 1177–1179. [CrossRef]
5. Han, M.J.; Khang, D.Y. Glass and Plastics Platforms for Foldable Electronics and Displays. *Adv. Mater.* **2015**, *27*, 4969–4974. [CrossRef] [PubMed]
6. Ren, H.; Cui, N.; Tang, Q.; Tong, Y.; Zhao, X.; Liu, Y. High-Performance, Ultrathin, Ultraflexible Organic Thin-Film Transistor Array Via Solution Process. *Small* **2018**, *14*, 1801020. [CrossRef]
7. Rogers, J.A.; Someya, T.; Huang, Y. Materials and Mechanics for Stretchable Electronics. *Science* **2010**, *327*, 1603–1607. [CrossRef]
8. An, B.W.; Gwak, E.J.; Kim, K.; Kim, Y.C.; Jang, J.; Kim, J.Y.; Park, J.U. Stretchable, Transparent Electrodes as Wearable Heaters Using Nanotrough Networks of Metallic Glasses with Superior Mechanical Properties and Thermal Stability. *Nano Lett.* **2016**, *16*, 471–478. [CrossRef]
9. Kim, D.H.; Xiao, J.; Song, J.; Huang, Y.; Rogers, J.A. Stretchable, Curvilinear Electronics Based on Inorganic Materials. *Adv. Mater.* **2010**, *22*, 2108–2124. [CrossRef]
10. Wang, Y.; Yin, L.; Bai, Y.; Liu, S.; Wang, L.; Zhou, Y.; Hou, C.; Yang, Z.; Wu, H.; Ma, J.; et al. Electrically Compensated, Tattoo-like Electrodes for Epidermal Electrophysiology at Scale. *Sci. Adv.* **2020**, *6*, eabd0996. [CrossRef]
11. Liu, H.; Li, M.; Ouyang, C.; Lu, T.J.; Li, F.; Xu, F. Biofriendly, Stretchable, and Reusable Hydrogel Electronics as Wearable Force Sensors. *Small* **2018**, *14*, 1801711. [CrossRef] [PubMed]
12. Liu, Y.; Liu, J.; Chen, S.; Lei, T.; Kim, Y.; Niu, S.; Wang, H.; Wang, X.; Foudeh, A.M.; Tok, J.B.H.; et al. Soft and Elastic Hydrogel-based Microelectronics for Localized Low-voltage Neuromodulation. *Nat. Biomed. Eng.* **2019**, *3*, 58–68. [CrossRef] [PubMed]
13. Morikawa, Y.; Yamagiwa, S.; Sawahata, H.; Numano, R.; Koida, K.; Ishida, M.; Kawano, T. Ultrastretchable Kirigami Bioprobe. *Adv. Healthc. Mater.* **2018**, *7*, 1701100. [CrossRef]
14. Li, W.; Matsuhisa, N.; Liu, Z.; Wang, M.; Luo, Y.; Cai, P.; Chen, G.; Zhang, F.; Li, C.; Liu, Z.; et al. An on-demand Plant-based Actuator Created Using Conformable Electrodes. *Nat. Electron.* **2021**, *4*, 134–142. [CrossRef]
15. Larson, C.; Peele, B.; Li, S.; Robinson, S.; Totaro, M.; Beccai, L.; Mazzolai, B.; Shepherd, R. Highly Stretchable Electroluminescent Skin for Optical Signaling and Tactile Sensing. *Science* **2016**, *351*, 1071. [CrossRef]
16. Roh, Y.; Kim, M.; Won, S.M.; Lim, D.; Hong, I.; Lee, S.; Kim, T.; Kim, C.; Lee, D.; Im, S.; et al. Vital Signal Sensing and Manipulation of A Microscale Organ with A Multifunctional Soft Gripper. *Sci. Rob.* **2021**, *6*, eabi6774. [CrossRef]
17. Liu, X. The More and Less of Electronic-skin Sensors. *Science* **2020**, *370*, 910. [CrossRef] [PubMed]
18. Lei, Z.Y.; Wu, P.Y. Double-slit Photoelectron Interference in Strong- field Ionization of The Neon Dimer. *Nat. Commun.* **2019**, *10*, 1.
19. Yao, S.; Zhu, Y. Nanomaterial-enabled Stretchable Conductors: Strategies, Materials and Devices. *Adv. Mater.* **2015**, *27*, 1480–1511. [CrossRef]
20. He, H.; Zhang, L.; Guan, X.; Cheng, H.; Liu, X.; Yu, S.; Wei, J.; Ouyang, J. Biocompatible Conductive Polymers with High Conductivity and High Stretchability. *ACS Appl. Mater. Interfaces* **2019**, *11*, 26185–26193. [CrossRef]
21. Vosgueritchian, M.; Lipomi, D.J.; Bao, Z. Highly Conductive and Transparent PEDOT:PSS Films with a Fluorosurfactant for Stretchable and Flexible Transparent Electrodes. *Adv. Funct. Mater.* **2012**, *22*, 421–428. [CrossRef]
22. Kayser, L.V.; Lipomi, D.J. Stretchable Conductive Polymers and Composites Based on PEDOT and PEDOT:PSS. *Adv. Mater.* **2019**, *31*, e1806133. [CrossRef]

23. Qi, D.; Zhang, K.; Tian, G.; Jiang, B.; Huang, Y. Stretchable Electronics Based on PDMS Substrates. *Adv. Mater.* **2021**, *33*, e2003155. [CrossRef]
24. McLellan, K.; Yoon, Y.; Leung, S.N.; Ko, S.H. Recent Progress in Transparent Conductors Based on Nanomaterials: Advancements and Challenges. *Adv. Mater. Technol.* **2020**, *5*, 1900939. [CrossRef]
25. Kim, D.H.; Liu, Z.; Kim, Y.S.; Wu, J.; Song, J.; Kim, H.S.; Huang, Y.; Hwang, K.C.; Zhang, Y.; Rogers, J.A. Optimized Structural Designs for Stretchable Silicon Integrated Circuits. *Small* **2009**, *5*, 2841–2847. [CrossRef] [PubMed]
26. Miyamoto, A.; Lee, S.; Cooray, N.F.; Lee, S.; Mori, M.; Matsuhisa, N.; Jin, H.; Yoda, L.; Yokota, T.; Itoh, A.; et al. Inflammation-free, Gas-permeable, Lightweight, Stretchable on-skin Electronics with Nanomeshes. *Nat. Nanotechnol.* **2017**, *12*, 907–913. [CrossRef] [PubMed]
27. Someya, T.; Kato, Y.; Sekitani, T.; Iba, S.; Noguchi, Y.; Murase, Y.; Kawaguchi, H.; Sakurai, T. Conformable, Flexible, Large-area Networks of Pressure and Thermal Sensors with Organic Transistor Active Matrixes. *Proc. Natl. Acad. Sci. USA* **2005**, *102*, 12321–12325. [CrossRef]
28. Lacour, S.P.; Chan, D.; Wagner, S.; Li, T.; Suo, Z. Mechanisms of Reversible Stretchability of Thin Metal Films on Elastomeric Substrates. *Appl. Phys. Lett.* **2006**, *88*, 4103. [CrossRef]
29. Graz, I.M.; Cotton, D.P.J.; Lacour, S.P. Extended Cyclic Uniaxial Loading of Stretchable Gold Thin-films on Elastomeric Substrates. *Appl. Phys. Lett.* **2009**, *94*, 1902. [CrossRef]
30. Liu, Z.; Wang, X.; Qi, D.; Xu, C.; Yu, J.; Liu, Y.; Jiang, Y.; Liedberg, B.; Chen, X. High-Adhesion Stretchable Electrodes Based on Nanopile Interlocking. *Adv. Mater.* **2017**, *29*, 1603382. [CrossRef]
31. Drack, M.; Graz, I.; Sekitani, T.; Someya, T.; Kaltenbrunner, M.; Bauer, S. An Imperceptible Plastic Electronic Wrap. *Adv. Mater.* **2015**, *27*, 34–40. [CrossRef]
32. Qi, D.; Liu, Z.; Yu, M.; Liu, Y.; Tang, Y.; Lv, J.; Li, Y.; Wei, J.; Liedberg, B.; Yu, Z.; et al. Highly Stretchable Gold Nanobelts with Sinusoidal Structures for Recording Electrocorticograms. *Adv. Mater.* **2015**, *27*, 3145–3151. [CrossRef] [PubMed]
33. He, H.; Zhang, L.; Yue, S.; Yu, S.; Wei, J.; Ouyang, J. Enhancement in the Mechanical Stretchability of PEDOT:PSS Films by Compounds of Multiple Hydroxyl Groups for Their Application as Transparent Stretchable Conductors. *Macromolecules* **2021**, *54*, 1234–1242. [CrossRef]
34. Fan, X.; Nie, W.; Tsai, S.H.; Wang, N.; Huang, H.; Cheng, Y.; Wen, R.; Ma, L.; Yan, F.; Xia, Y. PEDOT:PSS for Flexible and Stretchable Electronics: Modifications, Strategies, and Applications. *Adv. Sci.* **2019**, *6*, 1900813. [CrossRef] [PubMed]
35. Lee, S.; Shin, S.; Lee, S.; Seo, J.; Lee, J.; Son, S.; Cho, H.J.; Algadi, H.; Al-Sayari, S.; Kim, D.E.; et al. Ag Nanowire Reinforced Highly Stretchable Conductive Fibers for Wearable Electronics. *Adv. Funct. Mater.* **2015**, *25*, 3114–3121. [CrossRef]
36. Matsuhisa, N.; Inoue, D.; Zalar, P.; Jin, H.; Matsuba, Y.; Itoh, A.; Yokota, T.; Hashizume, D.; Someya, T. Printable Elastic Conductors by in Situ Formation of Silver Nanoparticles from Silver Flakes. *Nat. Mater.* **2017**, *16*, 834–840. [CrossRef] [PubMed]
37. Kil, M.S.; Kim, S.J.; Park, H.J.; Yoon, J.H.; Jeong, J.M.; Choi, B.G. Highly Stretchable Sensor Based on Fluid Dynamics-Assisted Graphene Inks for Real-Time Monitoring of Sweat. *ACS Appl. Mater. Interfaces* **2022**, *14*, 48072–48080. [CrossRef]
38. Sun, F.; Tian, M.; Sun, X.; Xu, T.; Liu, X.; Zhu, S.; Zhang, X.; Qu, J. Stretchable Conductive Fibers of Ultrahigh Tensile Strain and Stable Conductance Enabled by a Worm-Shaped Graphene Microlayer. *Nano Lett.* **2019**, *19*, 6592–6599. [CrossRef]
39. Yang, C.H.; Wu, Y.; Nie, M.; Wang, Q.; Liu, Y. Highly Stretchable and Conductive Carbon Fiber/Polyurethane Conductive Films Featuring Interlocking Interfaces. *ACS Appl. Mater. Interfaces* **2021**, *13*, 38656–38665. [CrossRef]
40. Gao, Y.; Guo, F.; Cao, P.; Liu, J.; Li, D.; Wu, J.; Wang, N.; Su, Y.; Zhao, Y. Winding-Locked Carbon Nanotubes/Polymer Nanofibers Helical Yarn for Ultrastretchable Conductor and Strain Sensor. *ACS Nano* **2020**, *14*, 3442–3450. [CrossRef]
41. Wang, H.; Yao, Y.; He, Z.; Rao, W.; Hu, L.; Chen, S.; Lin, J.; Gao, J.; Zhang, P.; Sun, X.; et al. A Highly Stretchable Liquid Metal Polymer as Reversible Transitional Insulator and Conductor. *Adv. Mater.* **2019**, *31*, 1901337. [CrossRef] [PubMed]
42. Matsuhisa, N.J.; Chen, X.D.; Bao, Z.N.; Someya, T. Materials and Structural Designs of Stretchable Conductors. *Chem. Soc. Rev.* **2019**, *48*, 2946. [CrossRef] [PubMed]
43. Gao, Y.; Yu, L.; Li, Y.; Wei, L.; Yin, J.; Wang, F.; Wang, L.; Mao, J. Maple Leaf Inspired Conductive Fiber with Hierarchical Wrinkles for Highly Stretchable and Integratable Electronics. *ACS Appl. Mater. Interfaces* **2022**, *14*, 49059–49071. [CrossRef]
44. Zhang, F.; Ren, D.; Huang, L.; Zhang, Y.; Sun, Y.; Liu, D.; Zhang, Q.; Feng, W.; Zheng, Q. 3D Interconnected Conductive Graphite Nanoplatelet Welded Carbon Nanotube Networks for Stretchable Conductors. *Adv. Funct. Mater.* **2021**, *31*, 2107082. [CrossRef]
45. Oh, J.Y.; Lee, D.; Hong, S.H. Ice-Templated Bimodal-Porous Silver Nanowire/PDMS Nanocomposites for Stretchable Conductor. *ACS Appl. Mater. Interfaces* **2018**, *10*, 21666–21671. [CrossRef]
46. Liang, S.; Li, Y.; Chen, Y.; Yang, J.; Zhu, T.; Zhu, D.; He, C.; Liu, Y.; Handschuh-Wang, S.; Zhou, X. Liquid Metal Sponges for Mechanically Durable, All-Soft, Electrical Conductors. *J. Mater. Chem. C* **2017**, *5*, 1586. [CrossRef]
47. Huang, Y.; Yu, B.; Zhang, L.; Ning, N.; Tian, M. Highly Stretchable Conductor by Self-Assembling and Mechanical Sintering of a 2D Liquid Metal on a 3D Polydopamine-Modified Polyurethane Sponge. *ACS Appl. Mater. Interfaces* **2020**, *11*, 48321–48330. [CrossRef]
48. Chang, T.H.; Tian, Y.; Li, C.; Gu, X.; Li, K.; Yang, H.; Sanghani, P.; Lim, C.M.; Ren, H.; Chen, P.Y. Stretchable Graphene Pressure Sensors with Shar-Pei-like Hierarchical Wrinkles for Collision-Aware Surgical Robotics. *ACS Appl. Mater. Interfaces* **2019**, *11*, 10226–10236. [CrossRef]
49. Li, L.; Zhu, C.; Wu, Y.; Wang, J.; Zhang, T.; Liu, Y. A Conductive Ternary Network of A Highly Stretchable AgNWs/AgNPs Conductor Based on A Polydopamine-modified Polyurethane Sponge. *RSC Adv.* **2015**, *5*, 62905–62912. [CrossRef]

50. Chen, J.; Zhang, J.; Luo, Z.; Zhang, J.; Li, L.; Su, Y.; Gao, X.; Li, Y.; Tang, W.; Cao, C.; et al. Superelastic, Sensitive, and Low Hysteresis Flexible Strain Sensor Based on Wave-Patterned Liquid Metal for Human Activity Monitoring. *ACS Appl. Mater. Interfaces* **2020**, *12*, 22200–22211. [CrossRef]
51. Weng, C.; Dai, Z.; Wang, G.; Liu, L.; Zhang, Z. Elastomer-Free, Stretchable, and Conformable Silver Nanowire Conductors Enabled by Three-Dimensional Buckled Microstructures. *ACS Appl. Mater. Interfaces* **2019**, *11*, 6541–6549. [CrossRef] [PubMed]
52. Feng, C.; Yi, Z.; Dumée, L.F.; Garvey, C.J.; She, F.; Lin, B.; Lucas, S.; Schütz, J.; Gao, W.; Peng, Z.; et al. Shrinkage Induced Stretchable Micro-wrinkled Reduced Graphene Oxide Composite with Recoverable Conductivity. *Carbon* **2015**, *93*, 878–886. [CrossRef]
53. Zhang, B.; Lei, J.; Qi, D.; Liu, Z.; Wang, Y.; Xiao, G.; Wu, J.; Zhang, W.; Huo, F.; Chen, X. Stretchable Conductive Fibers Based on a Cracking Control Strategy for Wearable Electronics. *Adv. Funct. Mater.* **2018**, *28*, 1801683. [CrossRef]
54. Dickey, M.D.; Chiechi, R.C.; Larsen, R.J.; Weiss, E.A.; Weitz, D.A.; Whitesides, G.M. Eutectic Gallium-Indium (EGaIn): A Liquid Metal Alloy for the Formation of Stable Structures in Microchannels at Room Temperature. *Adv. Funct. Mater.* **2008**, *18*, 1097–1104. [CrossRef]

Article

A Novel Measurement Approach to Experimentally Determine the Thermomechanical Properties of a Gas Foil Bearing Using a Specialized Sensing Foil Made of Inconel Alloy

Adam Martowicz [1,*], Jakub Roemer [1], Paweł Zdziebko [1], Grzegorz Żywica [2], Paweł Bagiński [2] and Artur Andrearczyk [2]

1 Department of Robotics and Mechatronics, AGH University of Science and Technology, al. Mickiewicza 30, 30-059 Krakow, Poland
2 Institute of Fluid-Flow Machinery, Polish Academy of Sciences, Department of Turbine Dynamics and Diagnostics, Fiszera 14 Street, 80-231 Gdansk, Poland
* Correspondence: adam.martowicz@agh.edu.pl

Abstract: Modern approaches dedicated to controlling the operation of gas foil bearings require advanced measurement techniques to comprehensively investigate the bearings' thermal and thermomechanical properties. Their successful long-term maintenance with constant operational characteristics may be feasible only when the allowed thermal and mechanical regimes are rigorously kept. Hence, an adequate acquisition of experimental readings for the critical physical quantities should be conducted to track the actual condition of the bearing. The above-stated demand has motivated the authors of this present work to perform the thermomechanical characterization of the prototype installation of a gas foil bearing, applying a specialized sensing foil. This so-called top foil is a component of the structural part of the bearing's supporting layer and composed of a superalloy, Inconel 625. The strain and temperature distributions were identified based on the readings from the strain gauges and integrated thermocouples mounted on the top foil. The measurements' results were obtained for the experiments that represent the arbitrarily selected operational conditions of the tested bearing. Specifically, the considered measurement scenario relates to the operation at a nominal rotational speed, i.e., during the stable process, as well as to the run-up and run-out stages. The main objectives of the work are: (a) experimental proof for the described functionalities of the designed and manufactured specialized sensing foil that allow for the application of a novel approach to the bearing's characterization, and (b) qualitative investigation of the relation between the mechanical and thermal properties of the tested bearing, using the measurements conducted with the newly proposed technical solution.

Keywords: gas foil bearing; Inconel alloy; sensing foil; strain field; temperature field; thermomechanical coupling; thermomechanical characterization; rotor dynamics; turbomachinery

Citation: Martowicz, A.; Roemer, J.; Zdziebko, P.; Żywica, G.; Bagiński, P.; Andrearczyk, A. A Novel Measurement Approach to Experimentally Determine the Thermomechanical Properties of a Gas Foil Bearing Using a Specialized Sensing Foil Made of Inconel Alloy. *Materials* **2023**, *16*, 145. https://doi.org/10.3390/ma16010145

Academic Editors: Bing Wang, Tung Lik Lee and Yang Qin

Received: 26 October 2022
Revised: 13 December 2022
Accepted: 20 December 2022
Published: 23 December 2022

1. Introduction

Bearings are one of the critical components of any machine. Very often, they determine the durability and reliability of machinery parts, and their limitations in terms of rotational speed and transmitted loads directly impact the permissible operating range of the entire mechanical system. This is particularly important in modern turbomachines, such as microturbines, whose components are subjected to extreme mechanical and thermal loads. In the case of bearings operating at very high rotational speeds and high temperatures, unique design solutions and materials must be used [1]. Among various types of high-speed fluid film bearings (slide bearings), gas foil bearings (GFBs), also known as air foil bearings (AFBs), play an increasingly important role and exhibit unique properties, which are impossible to obtain with other bearing systems [2,3]. Specifically, GFBs can operate at high temperatures, reaching 600 °C. The high mechanical compliance of the GFBs' foils makes these bearings able to operate with the shaft misalignment of 0.01 mm at high rotational speeds, i.e., 100,000 rpm. Moreover, the enhanced damping properties of the supporting layer exhibited by the mentioned type of bearings, compared to the classic gas bearings, improve the dynamics of the rotor. Unlike typical slide bearings, where the lubricating medium is primarily oil, GFBs can use almost any gas to operate [3]. They are self-acting (hydrodynamic) bearings, and their geometry changes depending on operating conditions [4]. There are also developed and utilized axial versions of GFBs, known as gas foil thrust bearings [5]. The geometry of a typical radial GFB is shown schematically in Figure 1.

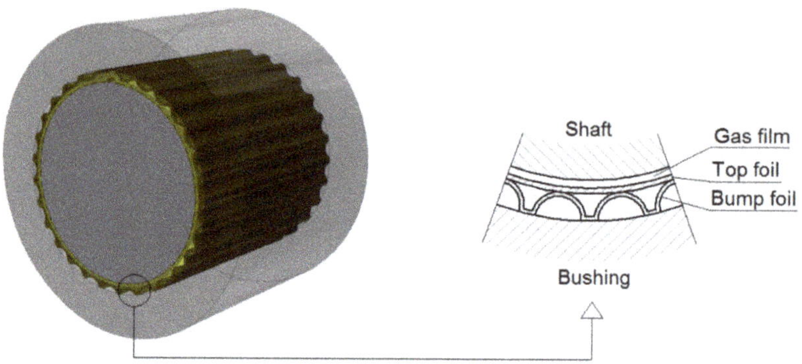

Figure 1. Typical geometry of a bump-type radial GFB. The top and bump foils compose a structural part of the bearing's supporting layer, whereas the gas film creates a fluidic layer in addition. Both layers ensure the required suspension and elevation capability for the rotating shaft's journal in a GFB while its operation.

The most characteristic component of a GFB is a group of thin foils that form a compliant structure with variable geometry. When the shape of the top foil, supported in many small-sized areas by the bump foil, is adjusted to the current position of the journal, the load is transferred to a large area, and the unit pressures are low. When the bearing is working and the journal rotates at a certain speed, it is lifted by a gaseous lubricating film, which is formed due to the hydrodynamic effect. Thus, above a certain speed (called the "lift-off speed"), the journal is no longer in direct metal-to-metal type contact with the top foil and no wear occurs in the bearing. Figure 2 shows an example of a hydrodynamic pressure profile that develops during the operation of a GFB. For clarity, this profile does not consider local changes in the locations of top foil support regions.

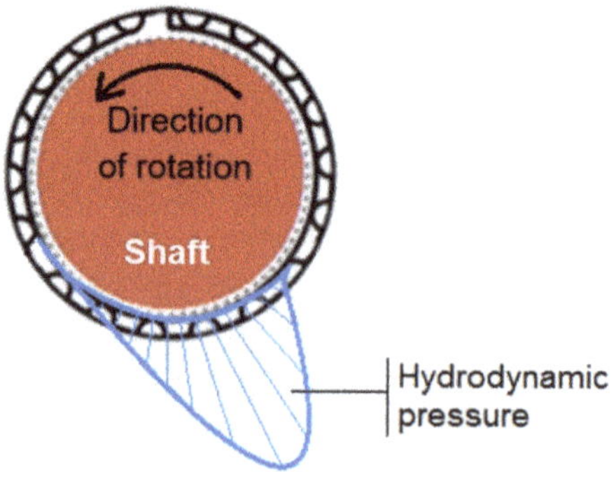

Figure 2. Approximate representation of the hydrodynamic pressure in a GFB.

Thanks to their specific design, GFBs can self-adapt to variable operating conditions [3,4,6]. However, it should be mentioned that GFBs are only suitable for lightly loaded and high-speed rotors due to the relatively low load capacity, which results from the low viscosity of the gases [6,7]. Among the various applications of GFBs in oil-free turbomachines, in which rotational speeds can exceed 100,000 r/min (100 krpm), their use in gas/vapor microturbines and turbocompressors is also known [3,8–10].

There are many design variations of gas foil bearings, and they can be modified to achieve desirable mechanical properties or a better heat transfer [7,11]. GFBs can therefore differ significantly in the geometry and layout of the bump foil, and other flexible elements composed of different materials or a metal mesh can be used instead [11–14]. In addition, the modified contact conditions between the bumps and the sleeve and the use of additional springs change the stiffness and damping properties of the whole bearing [15,16]. In order to actively change the properties of foil bearings during their operation, more advanced designs are proposed, where smart materials, such as shape-memory materials, piezoelectric materials, or thermoelectric materials, are used [17–19]. Other structural modifications of GFBs have been presented in a number of papers [20–24].

GFBs also have some disadvantages. One of the major drawbacks is their sensitivity to changes in operating temperatures. Changes in foil geometry due to the thermal effects can affect the stable operation of the bearing. Thermal deformations of the top foil and bump foils can also interfere with the formation of a gaseous lubrication film, leading to accelerated wear or even damage of the bearing. These phenomena are very complex and are still required to be studied when thermal, flow, and structural behavior, as well as the interactions between them, occur. Changes in the mechanical characteristics of GFBs, which were caused by temperature shifts, have been confirmed experimentally by many researchers [6,25–27] and concerned bearing parameters, such as stiffness, damping, load capacity, and coefficient of friction. The relationships between temperature, rotational speed, assembly preload, lift-off speed, and torque are also discussed in articles [28,29]. In reference to the above-cited works, it is worth mentioning, that the preload affects both the lift-off speed and the thickness of the gaseous film. These parameters, in turn, effectively influence the damping and stiffness of the foil assembly, as well as the friction torque of the bearing. The control of preload can be performed in several ways, e.g., by selecting the nominal installation clearance for the shaft's journal (within the range from -0.02 mm up to 0.02 mm), with the shape and number of bumps of the bump foils or via selection of the materials used to construct the components undergoing thermally-controlled geometric expansion. However, there is still a lack of publications on thermomechanical couplings observed in GFBs. This results, among other things, form the difficulty of accurately measuring the performance of thin flexible foils when bearings operate. Although such studies can be performed using advanced numerical models, a direct measurement of the entire deformation zone and foil temperature still presents technical difficulties. An overview of GFB modeling methods, which take into account temperature distribution, can be found in papers [2,7,30]. Examples of thrust GFB models that allow the analysis of thermal, flow, and structural phenomena are also presented in papers [31,32].

In the available literature, there are many publications in which the numerical methods used to analyze the displacements, deformations, and stresses of foils are discussed. For this purpose, the finite element method (FEM) is most often used, which is implemented in an in-house developed computer program or commercially available software. In this way, all the operational parameters of the foils, including their displacements, deformations, and stresses, can be determined at any point. Specifically, among other applications, FEM has been successfully employed to investigate the multiphysics properties of GFBs, including thermo-hydrodynamic analysis [28], characterize the bearing's adaptive and non-linear structural properties [4,6], and conduct study on an active bump-type foil bearing [24], as well as GFB dynamics [33,34].

Authors of experimental studies carried out on foil bearings most often focus their attention on determining more general working conditions and the bearing's properties or the entire rotating system, such as the bearing's structural stiffness, shaft position, journal and bushing orbits, gas film thickness, or critical rotor speeds [34–36]. The mechanical parameters that determine the foils' operation are not the subject of such studies, as this would require very advanced measurement techniques—such as, for example, vision methods [37]—or the integration of measuring elements with the examined object [38,39]. Therefore, new solutions and measurement methods are constantly being sought in this field of research.

The great importance of thermally induced phenomena in GFBs and their impact on the mechanical properties motivated the authors of this article to develop a new measurement method that uses a specialized sensing top foil composed of a superalloy, Inconel 625, and, then, perform the thermomechanical characterization of the prototype installation of a gas foil bearing to the extent not yet taken into account. Particularly, extraordinary measuring properties were sought for the GFB's structural components acting as the hosts for the integrated sensors. As successfully experimentally proven, a standard structural component, i.e., the bearing's top foil, gained new functionality and, therefore, become a functional component as well. This part enables the acquisition of temperature and strain field readings and characterizes their spatial distribution. The identification of the temporal courses for the above-mentioned critical physical quantities makes it possible to track the actual condition of the bearing.

The temperature measuring methods currently used for GFBs are mainly supported by a thermal imaging camera or several, preferably industrial, thermocouples rarely placed at different locations of the bearing [7,32,40,41]. However, more reliable measurement techniques are needed, which would allow a denser arrangement of sensors. The main objectives of this study are to experimentally validate the proposed measurement approach, which is based on data obtained from thermocouples and strain gauges integrated with the top foil, as well as qualitative investigation of the relation between the mechanical and thermal properties of the tested bearing, making use of the measurements conducted with the newly proposed technical solution.

The temperature and deformation measurements, which were performed using a prototype system developed by the authors, should allow for further studies of the thermomechanical effects occurring in the GFB, including, for example, the identification of new multiphysical relationships. This will make it possible to determine the conditions necessary to maintain long-term stable bearing operation at different rotational speeds and loads.

The paper is composed as follows. After the introductory section (Section 1), Section 2 discusses the novelty of the developed sensing foil, which makes use of the components dedicated to accurate temperature and deformation measurements, and describes a prototype installation of the tested GFB. The presentation of both the measurement system and investigated cases declared within the assumed measurement scenario is addressed in Section 3. The considered scenario relates to all stages of the operating bearing, including: run-up, steady-state run at a nominal rotational speed, and run-out. This is followed by Section 4, which reports the results of the experiments conducted for the operation scenario selected for the examined GFB. The final sections (Sections 5 and 6) discuss the obtained outcomes and a summary of the research conducted, complemented by the key conclusions.

2. Novel Prototypes of GFB and Specialized Sensing Top Foil

In the following, the novel features of the utilized GFB's test stand, equipped with the developed sensing top foil composed of a super alloy, Inconel 625, are briefly presented and discussed to highlight application areas available for the proposed measurement approach. As mentioned in the introductory part of the paper, new components, i.e., thermocouples and strain gauges, that were boned to the GFB's structural part, have advantageously enabled accurate temperature and deformation measurements, and, hence, allow for the assessment of the bearing's operational conditions in the mechanical and thermal domains, simultaneously.

Figure 3 presents the CAD model of the elaborated prototype installation. As visualized, several changes were introduced into the classical construction of a GFB to allow for demanded extraordinary measurement capabilities in the investigated bearing [1–3]. First, the geometry of the bump foils was adapted to provide sufficient space for the sensing components installed on the outer surface of the top foil. It is worth mentioning, that even though the area of the bump foils was significantly reduced, the bearing operated correctly providing carrying load capability. Advantageously, there was not experienced any effect that would significantly disturb both the process of developing the air film during the run-up stage and stable operation of the tested bearing. No sagging effect was indicated as well. The number of the circumferentially distributed supporting areas for the bump foils was not changed compared to the initial GFB's design; however, the authors are aware of the presence of significant changes in the foils' stiffness and damping properties induced by the bump foils' incisions. Moreover, the bushing has undergone considerable remodeling, including such additional machining operations as: removal of a significant part to become a spool-shaped component, and milling 18 holes divided into three tierces. All the above-listed operations eventually allowed for the installation of the sensing top foil as well as for radial guidance of all wiring coming out of the thermocouples and strain gauges. Again, the experimental tests performed to complete the run-in process of the top foil confirmed sufficiently high stiffness of the newly-constructed bushing and correct structural behavior of the developed GFB. Specifically, the expected temporal courses of the measured friction torque and temperature were registered during experiments [42], as shown in Section 4. After the running-in process was completed, the values of the measured quantities expectedly dropped. Following the authors' previous experience, it should be expected that any significant change of the bearing's elasto-damping properties registered during the running-in process should lead to the considerable variations of the friction torque and temperature.

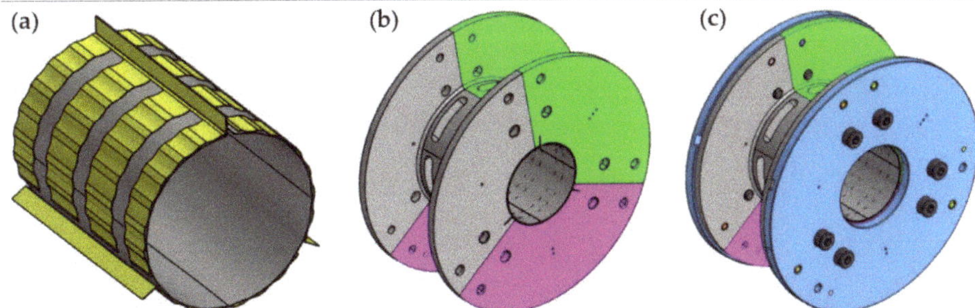

Figure 3. CAD model of the developed prototype a GFB: (**a**) top foil (seen as the inner grey cylindrical component) covered by the set of three bump foils (marked in yellow), (**b**) spool-shaped bushing composed of joined tierces (visualized with three different colors), (**c**) complete assembly equipped with two side flanges (marked in blue).

Figure 4 presents prototypes of novel constructions of both GFB and the specialized sensing top foil. As previously mentioned, additional circumferentially and axially distributed openings provide access to all sensing nodes and ensure enough space for cabling guidance. The developed specialized top foil makes use of the sensing components installed on its out surface and is dedicated to accurate temperature and strain measurements. The thermocouples are constructed using electrofusion welding with a 0.1 mm-diameter platinum-rhodium alloy. A single piece of Inconel 625 alloy, being a part of the top foil that comes out of the bearing, was used as a common electrode for all thermocouples. Strain gauges, in turn, were glued on the top foil with a dedicated adhesive to provide the necessary mechanical coupling for force and deformation transfer and protection against over-stiffening that would prevent the bearing from normal operation. The geometric pattern used to distribute all the sensing components on the top foil is shown in Figure 5. The main geometric parameters of the tested bearing are collected in Table 1.

Figure 4. Assembled prototype GFB's bushing (on the left) and sensing top foil before installation in the bearing (on the right). The integrated thermocouples and strain gauges are visible through the openings in the spool-shaped tierces of the GFB's bushing that were machined to allow radial guidance of the sensors' cabling. On the sensing top foil, there were visible spots caused by the gluing process of the strain gauges. However, the excessive amount of glue was removed from the area of contact between the top and bump foils with chemical agents right after completion of gluing process. Effectively, no unexpected behavior of the bearing was indicated due to both the possible change of friction properties and local over-stiffening.

Table 1. Geometric characteristics of the tested bearing.

Parameter/Characteristics	Value/Description
Shaft's journal nominal diameter	30 mm
Bushing's inner nominal diameter	31 mm
Bearing width	40 mm
Foil thickness	0.1 mm
Characteristics of the bump foils:	
Bump pitch	4.8 mm
Bump height	0.4 mm
Bump length	3.4 mm
Number of bumps in segment	7
Number of segments	4
Number of foils	3
Theoretical nominal clearance	0.02 mm

Figure 6 presents a complete assembly of the constructed GFB prototype mounted on the electrospindle's shaft. It is worth noting that the inner surface of the top foil was covered with a protective polymer layer AS20, which is based on Teflon with additives. More detailed description of the layer's properties and application can be found in [42]. Similarly, the shaft's journal was subjected to a hardening operation, making use of chromium oxide Cr_2O_3. The described surface thermal treatment reduced the friction torque and led to a smaller wear of the cooperating structural metallic parts that appeared during the run-up and run-out stages of GFB's operation.

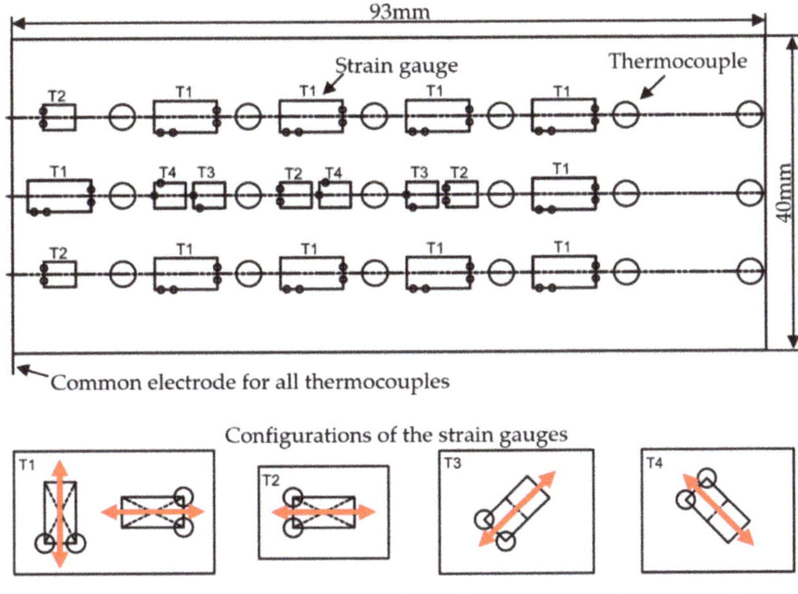

Figure 5. View of the flattened prototype sensing top foil to visualize the circumferential and axial distribution of the integrated thermocouples and strain gauges. Four configurations of the strain gauges were considered to allow multiaxial strain acquisition. However, in the current study, the authors only investigate the temperature and strain change based on the centrally axially located sensors to provide preliminary results and general thermomechanical characterization for the inspected bearing, as announced in Section 3.

Figure 6. Complete GFB prototype assembly mounted on the electrospindle's shaft. A simple 125 mm-long lever structure allowed for indirect friction torque measurement via a force sensor. Dedicated 3D-printed clamps, seen at both sides of the bearing, held the sensor cables. The fully assembled and equipped bearing weighed 1165 g. Apart from the gravitational force generated by the bearing itself, no additional load was considered in the test. Hence, the total load for the GFB was estimated to equal 11.43 N, considering the gravitational acceleration 9.81 m/s^2. The digits handwritten on the bearing close to the shaft approximate the circumferential locations for the temperature sensors.

3. Measurement System and Investigated GFB's Operation Scenario

This section provides the details of the measurement system utilized for temperature, strain, friction torque, and rotational speed readings acquisition, and presents an overview on the bearing's operation scenario arbitrarily selected by the authors to investigate the GFB's thermal and mechanical behavior for various operational conditions experienced within a single run of the bearing, including the stages: run-up, steady-state run at a nominal rotational speed, and run-out.

3.1. Experimental Test Stand

A freely-suspended configuration of the tested GFB was considered during measurements, as visualized in Figure 7. A photograph of the prototype installation is presented in Figure 8. A close-up view of the selected components of the test stand, including the GFB, is shown in Figure 9.

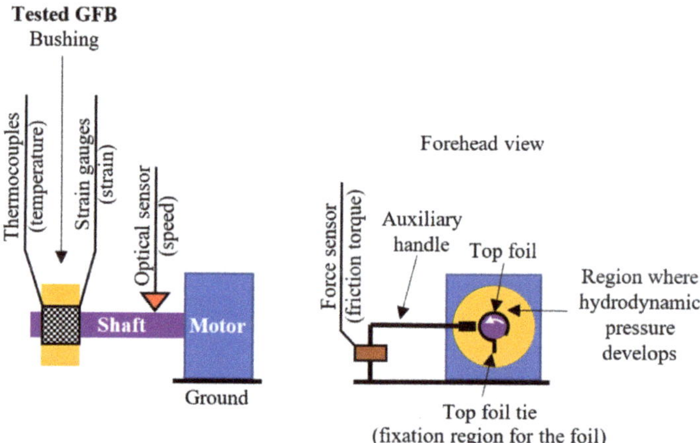

Figure 7. Schematic view of the freely-suspended configuration used for thermal and mechanical characterization of the tested GFB. The specialized sensing top foil—marked with the chessboard pattern—was mounted in the tested GFB and enabled temperature and strain measurements. The white arrow drawn on the shaft indicates its direction of rotation.

This above-mentioned operational configuration, also known as floating configuration, is usually applied for the initial inspection of the bearing's characteristics. It does not use the housing, nor is another bearing support considered in a test stand. Effectively, the lack of rigid fixation allows a GFB to adjust the orientation and position of its structural parts in the supporting layer and, hence, complete the process of running-in for the top foil safely. To ensure proper operation of the bearing, i.e., the generation of the air film that elevates the GFB over the shaft's journal, the top foil tie must lie at the bottom.

To conduct the current research on the thermomechanical characterization of GFBs, the authors have extended the measurement capabilities of the previously investigated bearing's prototype installation, which is described in the work [43]. In the present work, the strain bridge modules were considered for the deformation measurements. Due to the limitations regarding the number of measurement channels available in the utilized hardware, not all strain gauges—among the sensors presented in Figure 5 that cover the entire surface of the top foil—were active to provide strain readings simultaneously. The authors arbitrarily selected the groups of the sensors used in the performed measurement campaigns. Specifically, in the present study, only the strain and temperature sensors lying in the middle column, i.e., centrally axially located in the top foil, were considered to provide preliminary results and general thermomechanical characterization for the inspected bearing. Hence, for the sake of clarity regarding data interpretation, the cross-sectional views for the circumferential distributions of the two above-mentioned quantities identified only within the GFB's central transversal plane are analyzed.

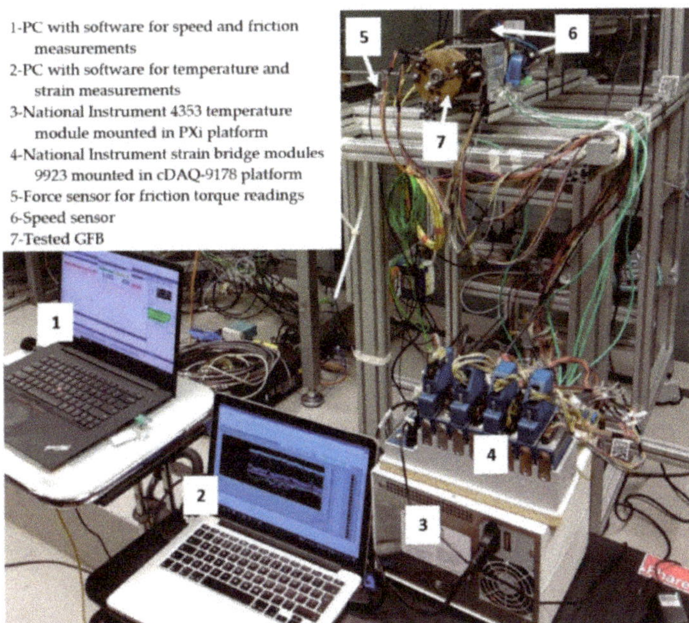

Figure 8. Measurement system used for the acquisition of temperature, strain, friction torque, and rotational speed readings.

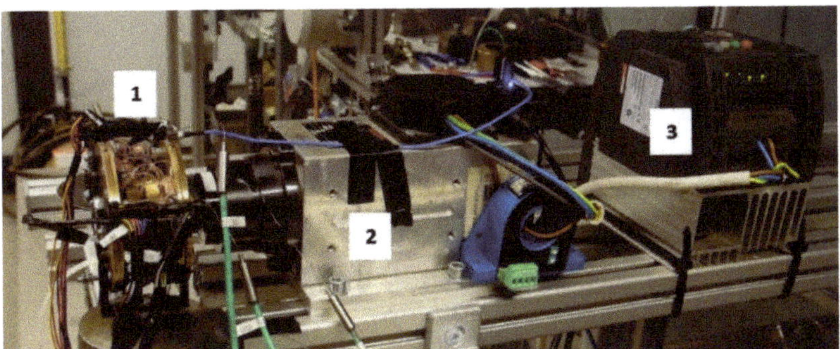

Figure 9. Close-up view of the bearing (**1**), electrospindle (**2**) and inverter (**3**).

Characteristics of the temperature and strain sensors used to provide the experimental results in the current study are presented in Table 2. In the case of the thermocouples, regularly distributed sensing nodes were available every 60 deg. along the top foil's circumference to uniformly cover the angular domain visible in a cross-sectional view. Contrarily, the circumferential distribution of the strain gauges employed for the strain readings was irregular. On one hand, not all sensors were available for the considered measurement scenario due to the limitations resulting from the number of strain bridge modules or identified strain gauges malfunctions. On the other hand, by default, the sensors providing the strain measurement in the axial direction (i.e., the ones of the configuration T1) were eliminated as not providing useful information for the circumferential distribution of the analyzed quantity. Nonetheless, almost the entire angular domain was covered with the strain readings to allow for further interpolation and extrapolation of this quantity along the top foil's circumference.

Table 2. Thermocouples and strain gauges centrally axially localized in the top foil used to collect measurement data for the current study.

Type of Sensor	Angular Position with Respect to the Localization of the Top Foil Tie [deg.]	Orientation of the Strain Component with Respect to Circumference [deg.] (Row along Circumference from the Top Foil Tie/Sensor Configuration)	Comments
Thermocouples	42.5 102.5 162.5 222.5 282.5 342.5	Not applicable	All thermocouples have a common electrode at the top foil tie
Strain gauges (sequentially denoted in Figure 10: SG1, ... SG5)	62.5 82.5 182.5 202.5 267.5	−30 (Row 2/Configuration T4) 30 (Row 2/Configuration T3) 30 (Row 4/Configuration T3) 0 (Row 4/Configuration T2) 0 (Row 5/Configuration T1)	Skipped sensors: • Row 1/Configuration T1—angular position 7.5 deg., orientation 90 deg. • Row 5/Configuration T1—angular position 247.5 deg., orientation 90 deg. Not available sensors: • Row 1/Configuration T1—angular position 27.5 deg., orientation 0 deg. • Row 3/Configuration T2—angular position 122.5 deg., orientation 0 deg. • Row 3/Configuration T4—angular position 142.5 deg., orientation −30 deg.
Please see Figure 5 for reference			

It should be noted that only circumferential components of the measured strains were analyzed by the authors in the current study. They are considered dominant ones in terms of the top foil's behavior and the capability of developing the air film [44,45]. Specifically, the circumferential curvature of the top foil is much more prone to both the external load and the hydrodynamic effect. In contrast, axial deformations of the top foil require a higher level of mechanical and thermal excitation. Similarly, a general assumption was made by the authors that the circumferential and axial directions for the measured strains coincide with the directions of the principal strains following the specificity of the top foil's behavior. Finally, initial non-zero strains, registered for all strain gauges before the initialization of the measurement session, were canceled. Hence, the changes in the measured strains were primarily studied in the present research to infer the GFB's operation.

A precise 24-bit module PXi-NI4353 was utilized for temperature measurement. However, the built-in procedure of automatic compensation, making use of the integrated thermistor that is dedicated to standard industrial thermocouples, could not be applied in the case of the welded sensors. Consequently, the authors identified their properties with a calibration procedure. A constant value of 25 °C, with its experimentally confirmed variation not greater than +/−1 °C, was assumed at the cold-junction, i.e., at the common electrode for all thermocouples. Nonetheless, the estimated maximum error regarding relative temperature measured with the integrated sensors equals 0.5 °C for the temperature range 0 °C up to 100 °C.

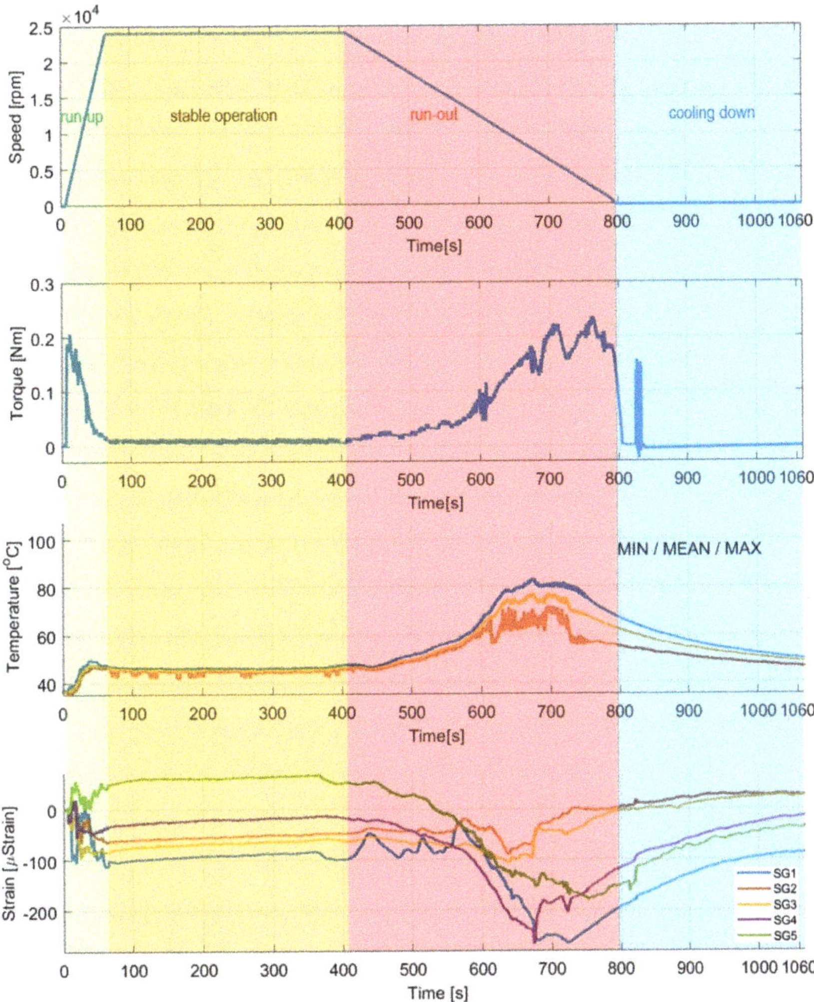

Figure 10. Temporal courses for the investigated operational characteristics of the GFB's prototype. From the top down to bottom: rotational speed of the shaft's journal, friction torque, temperature, and strain for the circumferentially distributed measurement points in the sensing top foil. Localization of the strain gauges SG1 up to SG5 is provided in Table 2.

3.2. Measurement Scenario Selection

The considered measurement scenario (i.e., GFB's operation scenario) was arbitrarily selected to allow the investigation of all stages of operation of the tested prototype bearing. This scenario assumed that the tested bearing undergo a single 1070-s-long cycle of its complete run, including:

- Measurement process initialization at non-rotating shaft—lasting for 7 s;
- Run-up stage—for 60 s with the rotational acceleration 6.67 Hz/s (6.67 r/s^2);
- Stage of stable operation at 24,000 r/min (24 krpm, i.e., 400 Hz)—for 340 s;
- Run-out stage—for 390 s with the rotational acceleration –1.03 Hz/s (–1.03 r/s^2);
- Cooling down at a non-rotating shaft and the process of completing the measurements—for 273 s.

Among all the temperature and strain readings experimentally acquired during the considered GFB's operation scenario, the authors have intentionally selected some of the measurement results for discussion in the current paper due to their significant contribution to the understanding of the GFB's thermomechanical behavior. Specifically, the experimental outcomes obtained while changing the bearing's operational conditions, i.e., during the run-up and run-out stages for the processes of both developing and losing the air film are presented (in Section 4) and then analyzed (in Section 5). Moreover, the stage of the GFB's stable operation, i.e., its nominal rotational speed, considered to be 24 krpm for the performed tests, is investigated. This speed guaranteed the presence of the air film continuously elevating the bearing over the shaft's journal as desired for the configuration of a freely-suspended GFB. Finally, the cooling down stage is also addressed to identify a plastic deformation of the top foil. Adequately, several case studies were selected for more comprehensive analysis based on the temporal plots drawn for the experimental readings of the temperature, as discussed in detail in Section 4. As previously experienced by the authors, the temperature may be considered a reliable indicator of the current operational state of the bearing [43].

4. Experimental Results

In the current section, the outcomes of the experiments conducted for the considered scenario of the bearing's operation are reported. Temperature and strain results visualization was performed using both temporal plots (shown in Section 4.1) and transversal cross-sectional views to enable the presentation of circumferential distributions of the above-mentioned quantities being caught at given time moments (presented in Section 4.2). The transversal cross-section located at the center of the top foil was taken into account to characterize the GFB and formulate adequate conclusions in Section 5.

Specifically, the time moments at which the air film developed and was lost are of the authors' particular concern. This approach was intentionally considered to primarily allow for an inference about the respective changes of the GFB's thermal and mechanical characteristics for the most critical stages of its operation.

4.1. Temporal Courses for the GFB's Operational Characteristics

Figure 10 presents the temporal plots for the quantities that characterize the bearing's operation identified during the experiments, including: speed of rotation, friction torque, temperature, and strain readings on the top foil.

As seen in Figure 10, the characteristic non-monotonic changes of the operational parameters accompany the air film development and its losing processes. Certain time periods are required for these processes to take place during both: (a) complete formulation of the thin air layer above the shaft's journal, and (b) the final return to a dry friction contact, as the run-out stage proceeds and the shaft stops, eventually.

In Section 4.2, specific cases are declared by the authors with reference to the results presented in Figure 10, to allow for a more detailed analysis of the bearing's behavior in Section 5. It should be noted that, although the experiments were conducted for the run-in top foil, a progressive process of its mechanical adjustment is still present in the bearing that continuously affects the measured strain and torque. In fact, no externally induced changes of the measured strains are seen within the last part of the stage of stable operation. Moreover, the stick-slip phenomenon was observed at various time moments, including the stage of cooling down, i.e., when the GFB does not operate. Specifically, there was a significant thermal shrinking appearing at no air film a dry friction contact is brought back during the run-out stage and then maintained while there is a break of the shaft. This results in both sudden and gradual changes in the measured friction torque. The greatest shift regarding the above-mentioned quantity is registered within the approximate time interval [820,835] seconds. Effectively, even though the value of the friction torque returns to zero immediately after the shaft stops, it still undergoes a considerable change later. This phenomenon, which is identified for a non-rotating shaft, is, however, considered to be out of the scope of the present study. Nonetheless, it does not significantly interfere with the general trends of the temperature and strain plots for the stage of cooling down. For the presented strain courses, an offset removal procedure was performed only once, at the measurement initialization, to visualize the total change of the identified strain gauges' readings. As seen, a single run may cause small-amplitude elastic deformations of the top foil that are not completely canceled or, possibly, plastic ones may contribute.

4.2. Circumferential Distributions of Temperature and Strain in the Top Foil

Taking into consideration the registered temporal plots (Figure 10) and the overall assumptions regarding the investigated measurement scenario mentioned in Section 3.2, the following cases at various GFB's operation stages were selected for a further, more comprehensive study on the thermomechanical properties of the tested bearing:

- Case A—phase of development of an air film during the run-up stage of the bearing's operation at the measurement time moment 20 s, i.e., 13 s after the run-up stage was initiated (at 5650 rpm), specifically at the moment when the rates of the measured temperatures began to decrease. From that moment, the measured temperatures did not grow faster and faster anymore, until they finally reached the maximum level during the run-up stage of approximately 50 °C at 15,900 rpm. The results obtained for Case A are presented in Figure 11.

For the sake of clarity, having introduced Figure 11 as an example, the meaning of all cross-sectional views in the presented plots, as well as the assumed convention regarding the presentation of the signs of the visualized quantities is explained briefly here. The full color scale (i.e., from dark blue to dark red) is used to represent all the scatter of the visualized quantities, irrespective of their absolute values.

First, the two upper plots of the set of four circumferential views seen in Figure 11, respectively, present the temperature (on the left) and strain (on the right) distributions in the top foil. These plots make use of the equivalent (redundant) color mapping and shape scaling to visualize the values of the considered quantity. The exact measured values of these quantities are represented with black dots at given circumferential positions—following the prototype's description found in Section 3. The spatial courses for both the temperature and strain along the curved top foil were found using linear interpolation and extrapolation. This approximation technique was also considered outside the angular range corresponding to the localizations of the integrated sensors to provide rough estimates of the mentioned quantities at the top foil-free end, as well as in the area where this foil is clamped to the bushing. However, the estimate for the strain at the top foil tie (i.e., outside the region covered by the strain gauges close to the foil's tie) is not identified due to the expected, by the authors, complicated nature of the phenomena of the mechanical load carrying in this area. In fact, a total mechanical interaction between the bushing and the shaft's journal includes force reactions that are completed via the foil's tie. The lack of sensors in the mentioned region makes it impossible to reliably conclude a strain distribution in that place.

As explained in Section 3, the strain gauges are mounted on the outer surface of the top foil. Hence, the positive values of the measured strain (specifically, the positive change of the strain found after initial offset level removal) indicate proceeding with a clamping of the foil around the shaft's journal. In reference, the behavior of the top foil is represented by the reduced radius of the colored curve visualizing the strain distribution. Consequently, the negative strains are represented with the opposite change of the plotted curve's radius. The positive strains generated due to thermal expansion (visualized in the bottom left plot in Figure 11) are modeled in a consistent way to provide the same interpretation of the behavior of the top foil—actually wrapping the shaft during the foil's elongation in a continuous presence of the supporting bump foils and the caused contact limits. The thermal expansion coefficient applied to find the thermally induced strain in the top foil equals 12.8×10^{-6}.

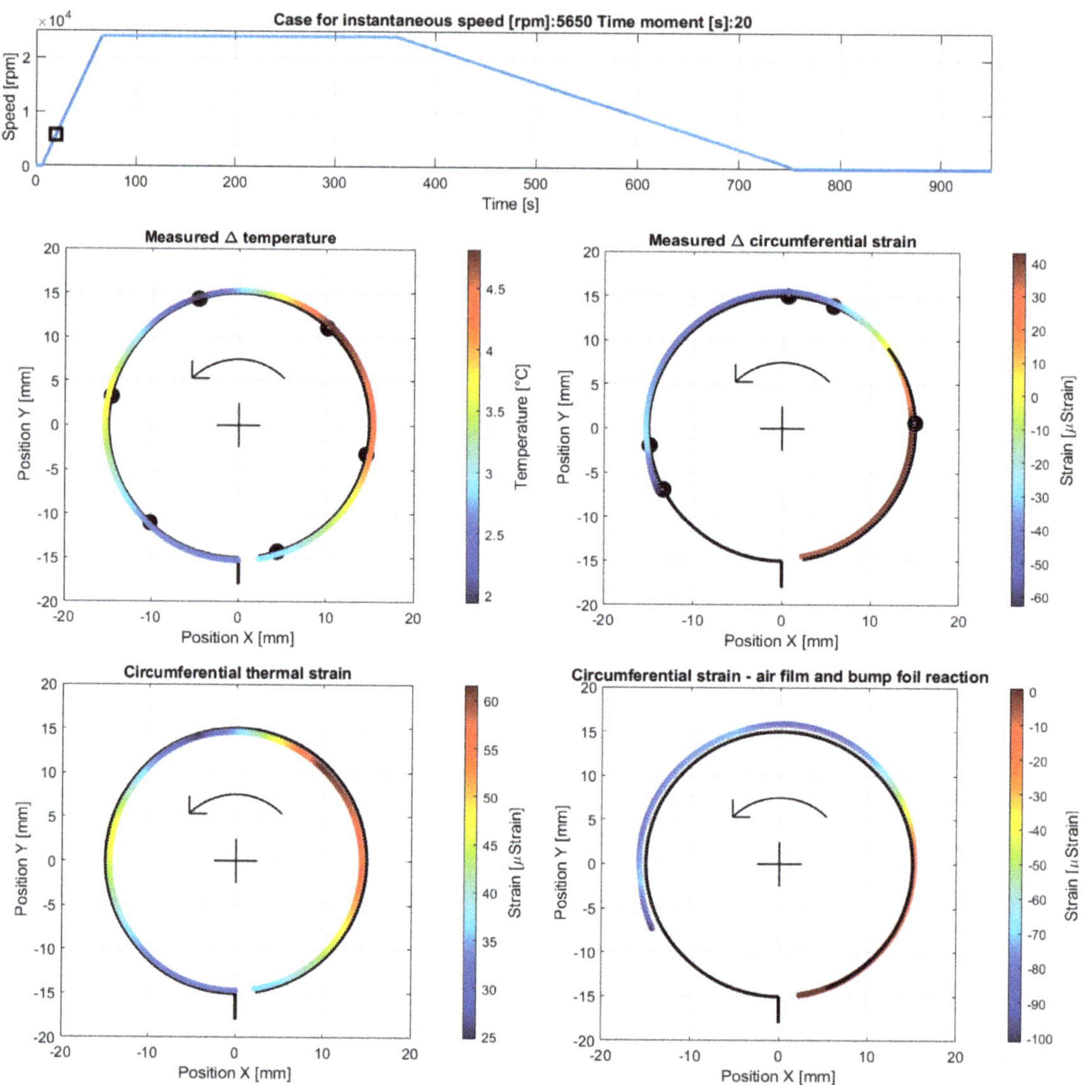

Figure 11. Cross-sectional views for the circumferential distribution of the temperature and strain identified in the top foil at 5650 rpm during the run-up stage of the GFB's operation. The referential course of the rotational speed is provided on top. Its current value is visualized with a black square marked on the entire speed profile. The arrows indicate the direction of rotation of the shaft. The data presented in the cross-sectional views are explained in the current section.

The results shown in the bottom right plot in Figure 11 visualize the circumferential strain approximating the hydrodynamic action of the air film complemented with a presumable contribution of the bump foils' interaction. The presented strains were found by subtracting the calculated thermally induced strains (bottom left curve in Figure 11) from the resultant strains measured with the strain gauges (upper right curve in Figure 11). However, it should be noted that an additional assumption was made regarding the value of the strain visualized in the bottom right plot in Figure 11. Specifically, its value at the free end of the top foil was arbitrarily fixed to zero, which indicates the assumed realistic lack of a non-thermally induced contribution to the deformation of the mentioned part of the foil, irrespective of its actual displacement. Consequently, an adequate value of the strain was considered in the upper right curve in Figure 11 to proceed with the approximation procedure, of the magnitude equal to the thermally induced strain estimated for the mentioned localization and shown in the bottom left curve in Figure 1. The above-described correction was made to follow the assumption, considered by the authors, that only strains originating from the thermal expansion are found at the free end of the foil.

- Case B—operation with a fully developed air film during the run-up stage, at the measurement time moment 47 s, i.e., 40 s after the run-up stage was initiated (at 16,615 rpm) before the temperature along the top foil became homogenized; Figure 12 presents the results obtained for the current case.

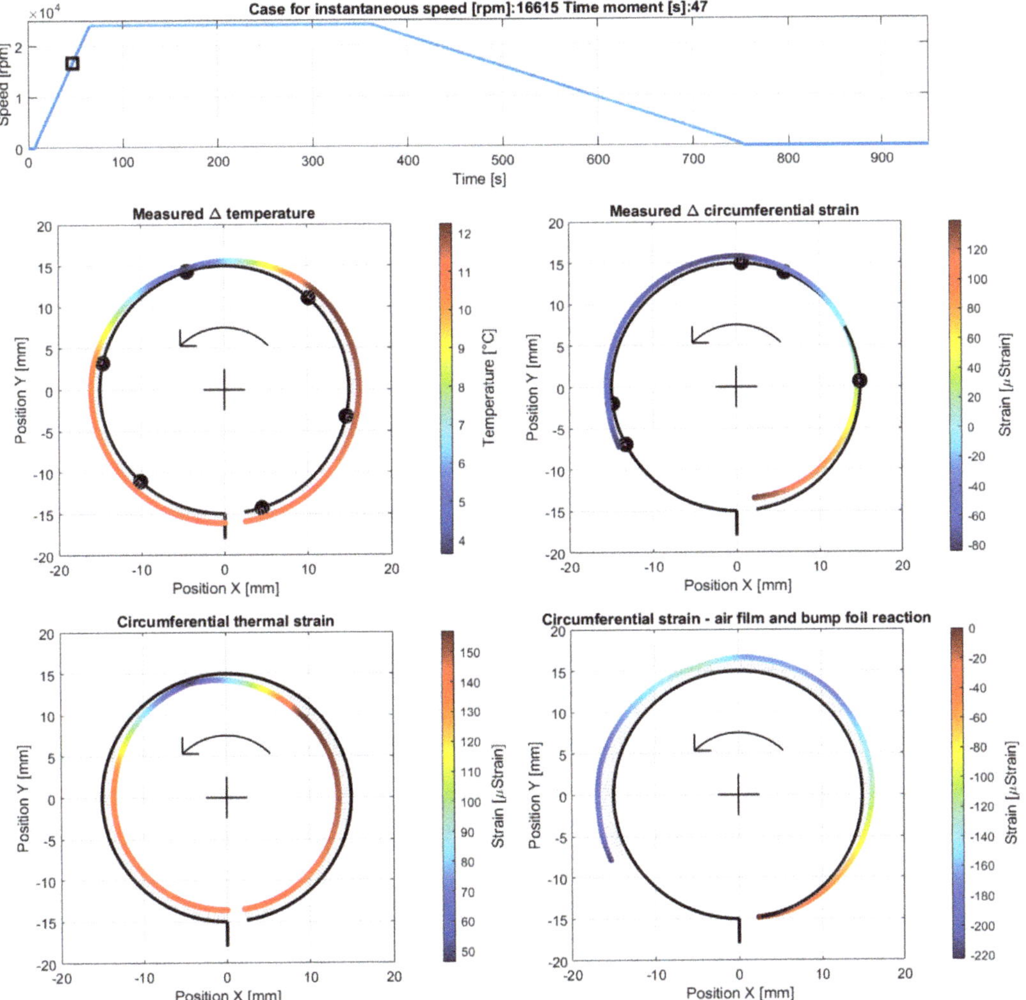

Figure 12. Cross-sectional views for the circumferential distribution of the temperature and strain identified in the top foil at 16,615 rpm during the run-up stage of the GFB's operation before the homogenization of the temperature. The black square and arrows respectively indicate the actual rotational speed of the shaft and its direction.

- Case C—operation with a fully developed air film during the run-up stage, at the measurement time moment 57 s, i.e., 50 s after the run-up stage was initiated (at 20,641 rpm) after the temperature in the top foil became homogenized (Figure 13).

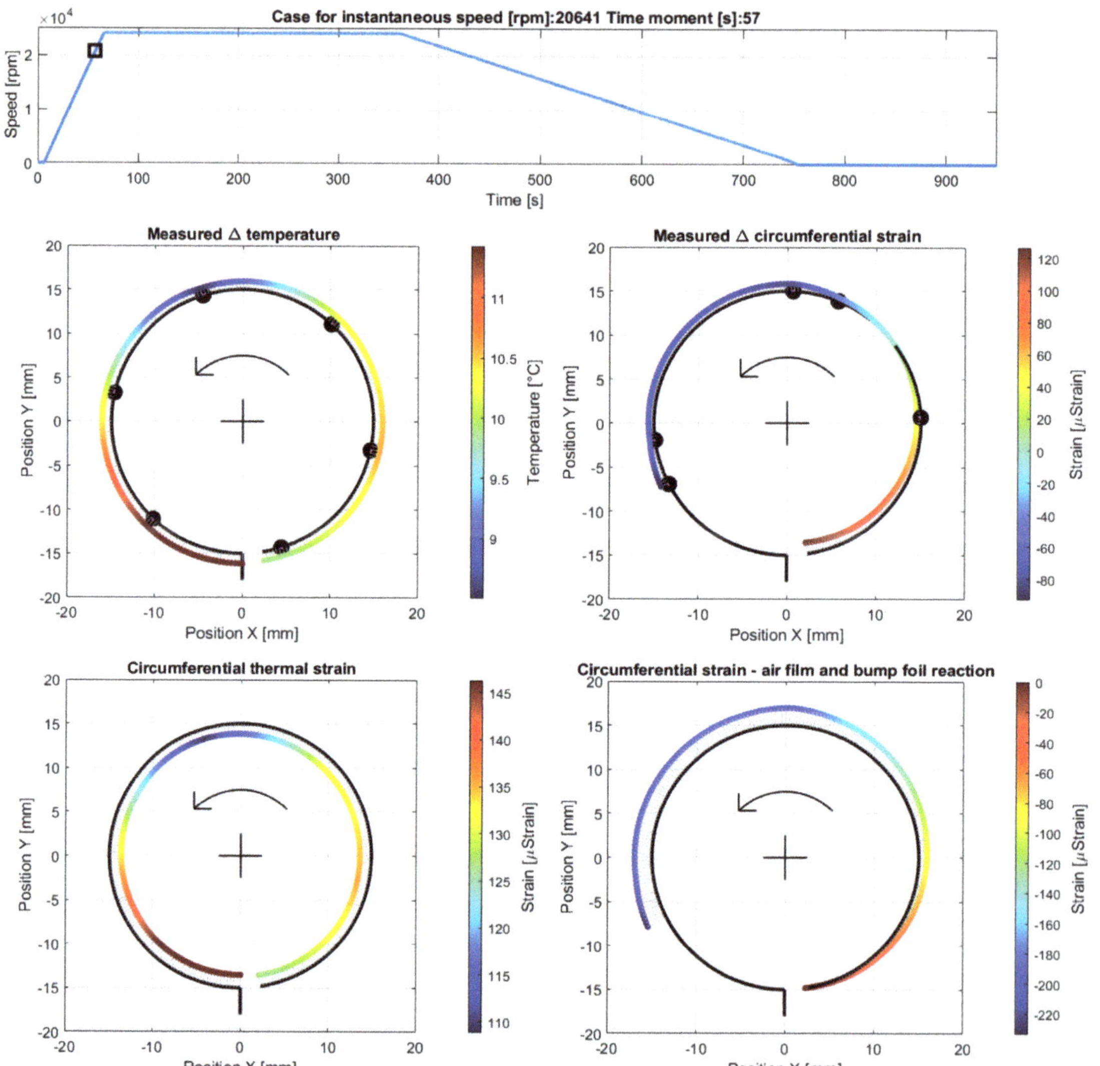

Figure 13. Cross-sectional views for the circumferential distribution of the temperature and strain identified in the top foil at 20,641 rpm during the run-up stage of the GFB's operation after the homogenization of the temperature.

- Case D—operation at the end of the stage of stable operation, at the measurement time moment 350 s, i.e., 283 s after the run-up stage was initiated (at 24,032 rpm); The results are presented in Figure 14.

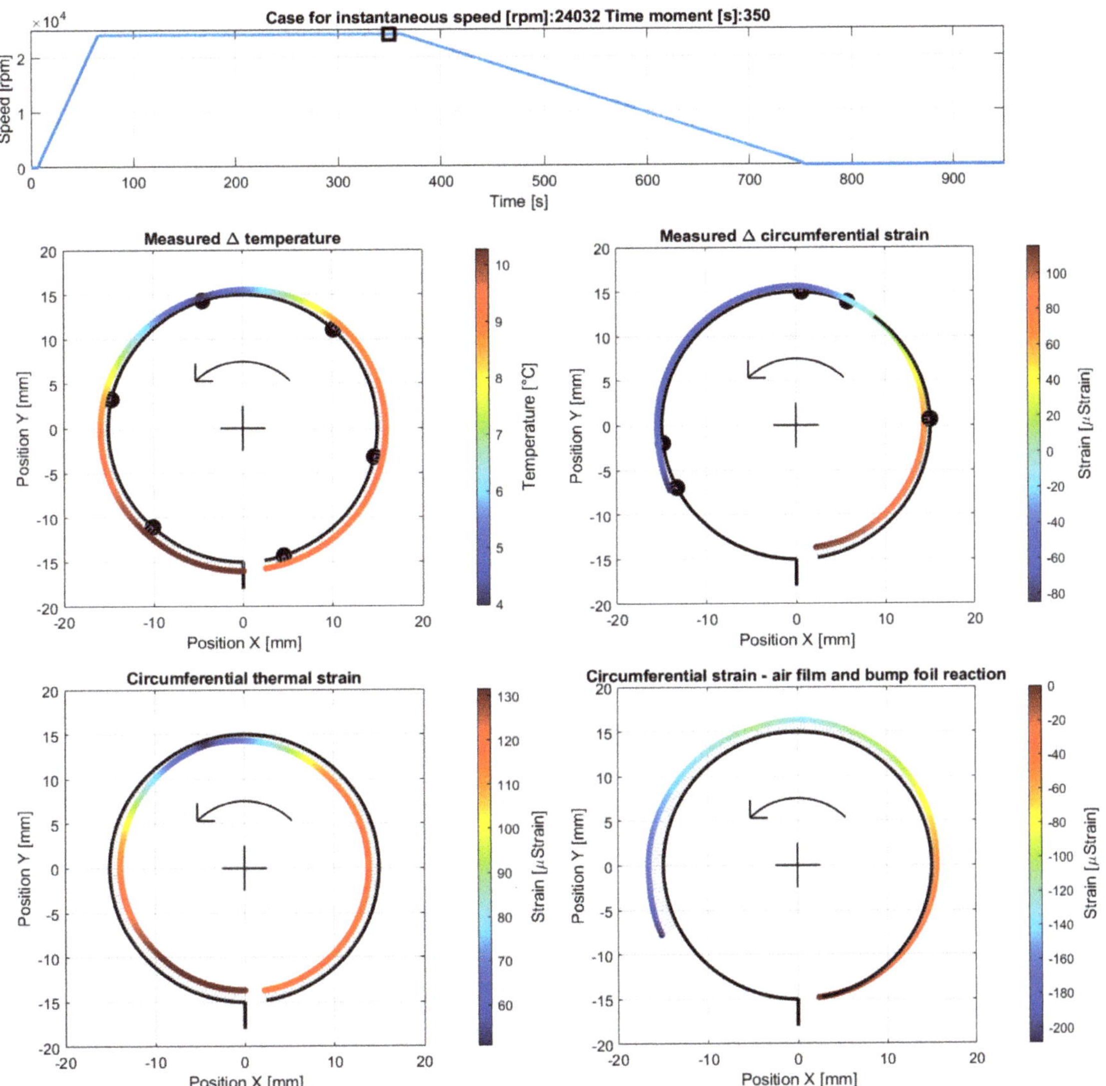

Figure 14. Cross-sectional views for the circumferential distribution of the temperature and strain identified in the top foil at 24,032 rpm during the stage of stable operation of the GFB.

- Case E—operation at the run-out stage before a complete loss of the air film, at the measurement time moment 600 s, i.e., 193 s after the run-out stage was initiated (at 9614 rpm); The results are presented in Figure 15.

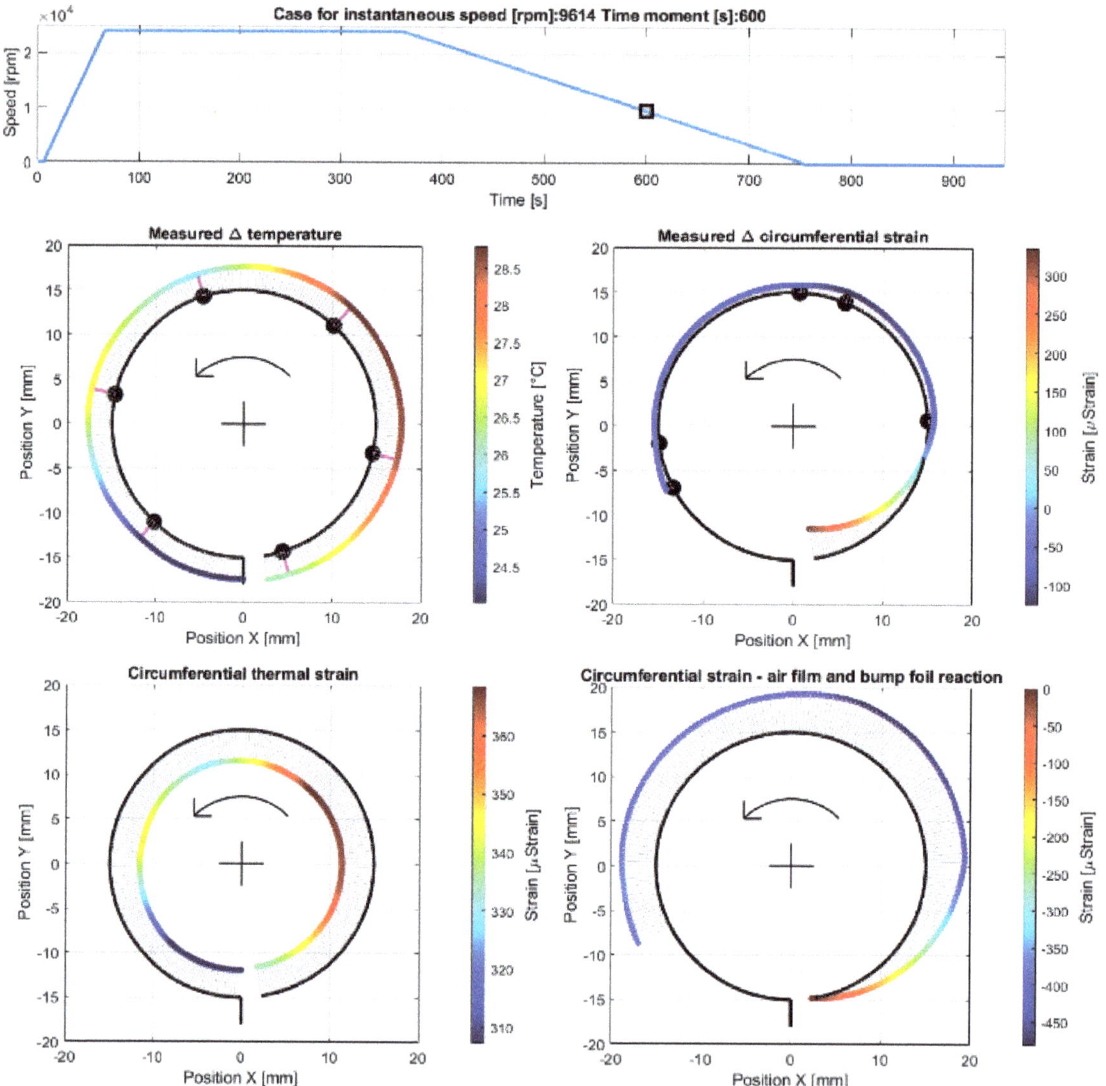

Figure 15. Cross-sectional views for the circumferential distribution of the temperature and strain identified in the top foil at 9614 rpm during the run-out stage of the GFB's operation before the air film was completely lost.

- Case F—operation at the end of the run-out stage after a complete loss of the air film, at the measurement time moment 700 s, i.e., 293 s after the run-out stage was initiated (at 3528 rpm). The results are presented in Figure 16.

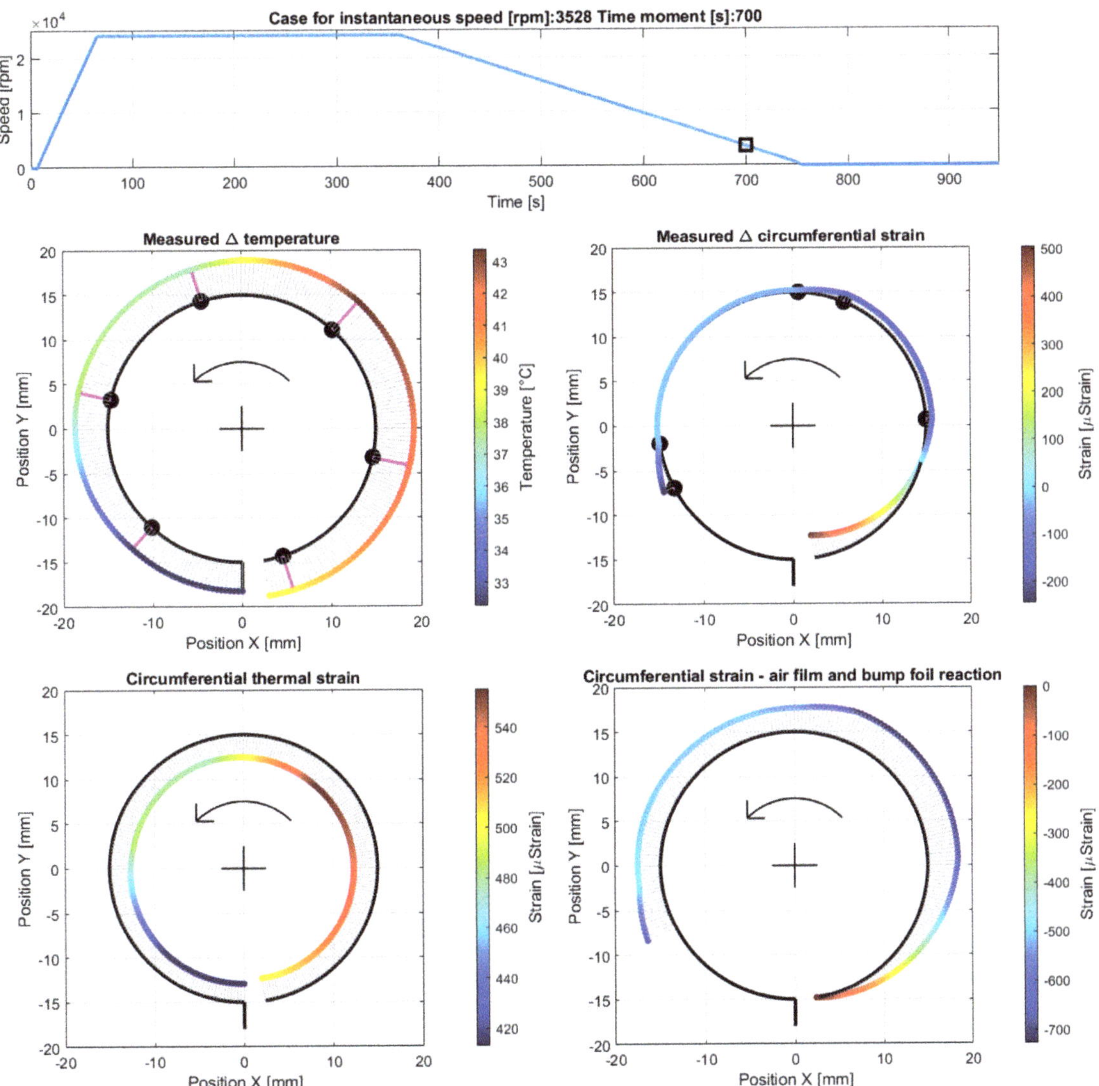

Figure 16. Cross-sectional views for the circumferential distribution of the temperature and strain identified in the top foil at 3528 rpm during the end of the run-out stage of the GFB's operation after the air film was completely lost. The curves used to visualize the strains are rescaled by the factor 0.5 with respect to the previously shown cases.

- Case G—initiation of the cooling down stage after the shaft's stop, at the measurement time moment 760 s. The results are presented in Figure 17.

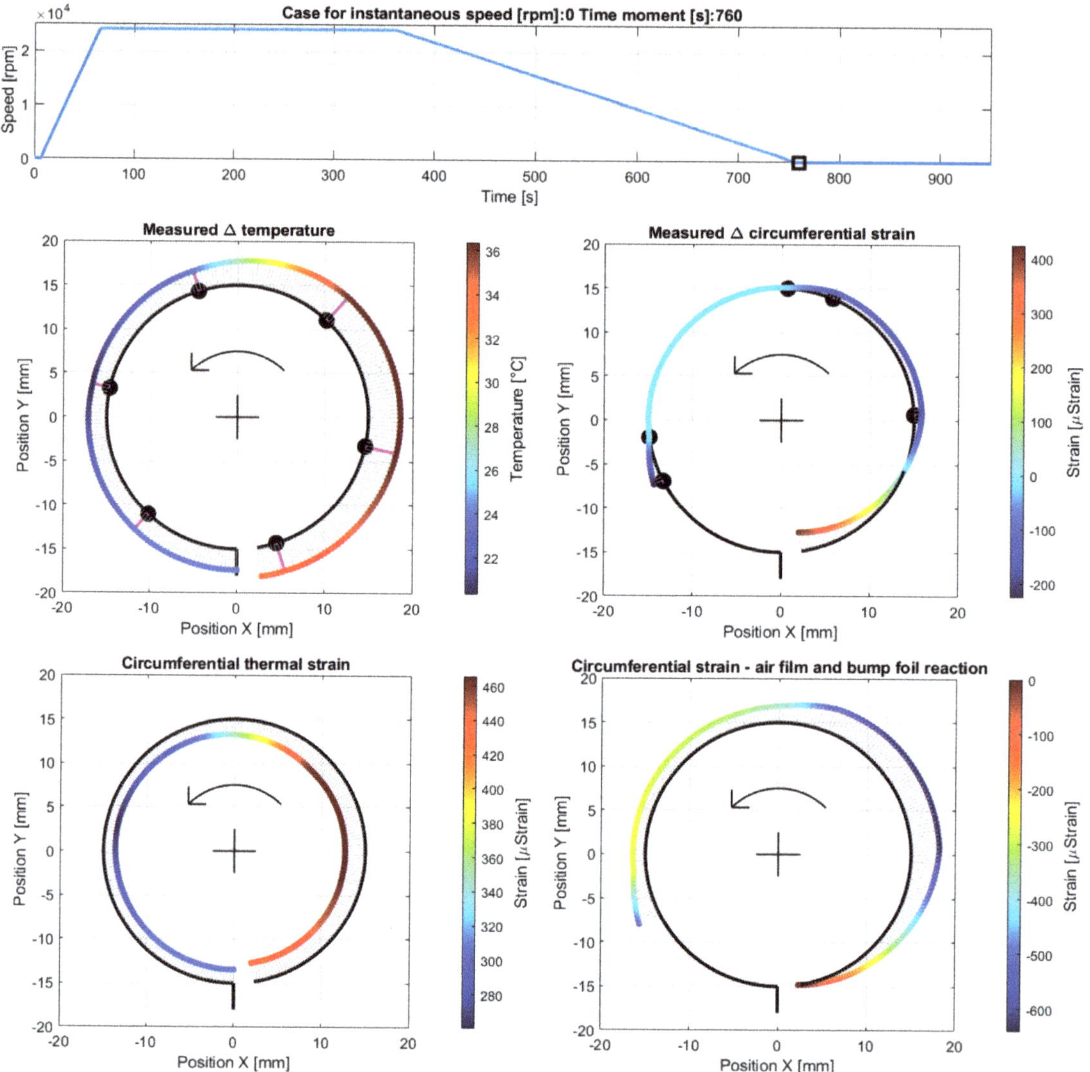

Figure 17. Cross-sectional views for the circumferential distribution of temperature and strain identified in the top foil at the stopped shaft. The curves used to visualize the strains are rescaled by the factor 0.5 compared to Cases A to F.

- Case H—end of the measurements during the cooling down stage at the time moment 1070 s. The results are presented in Figure 18.

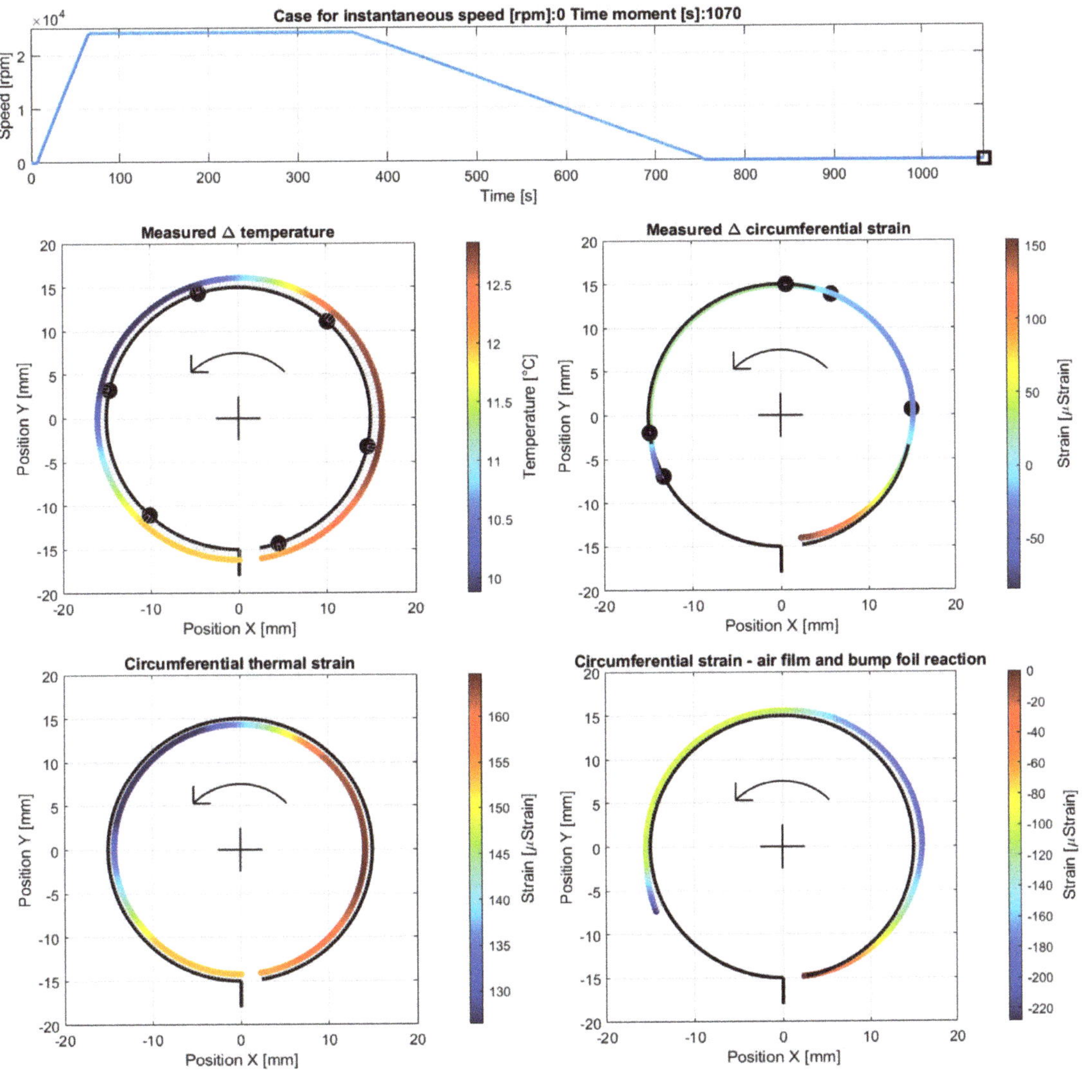

Figure 18. Cross-sectional views for the circumferential distribution of the temperature and strain identified in the top foil at the stopped shaft at the end of the measurements. The curves used to visualize the strains are rescaled by the factor 0.5 compared to Cases A to F.

The aggregated data related to the circumferential distribution of the measured temperature and strain for all the above-introduced cases are shown in Figure 19.

As previously mentioned, an analysis of the experimental outcomes is performed in Section 5.

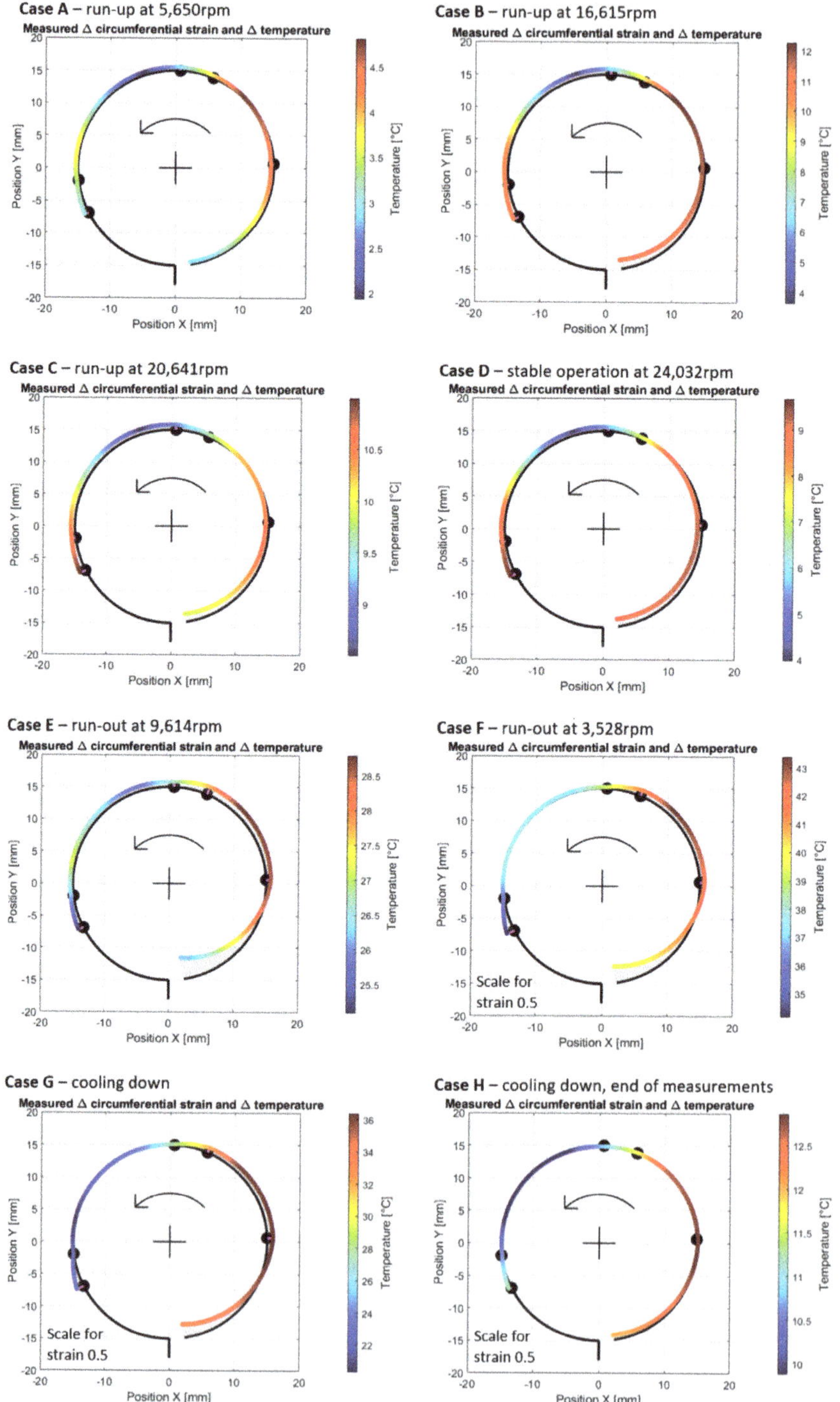

Figure 19. Compacted cross-sectional views for the circumferential distribution of the temperature and strain identified in the top foil for the studied cases (**A–H**), referencing the various stages of the GFB's operation.

5. Discussion

The following provides a discussion of the experimental outcomes obtained for the cases introduced in Section 4. These cases address, in the authors' opinion, the most interesting phenomena appearing in the tested bearing during the subsequent stages of its operation.

5.1. Cases A–C—Development of an Air Film during Run-Up Stage

After the bearing's shaft starts rotating, a sudden growth of the friction torque was identified (shown in Figure 10). This initial behavior of the GFB is natural, as it results from the dry friction present between the surfaces of the components which are in mechanical contact, i.e., the rotating shaft's journal and the top foil. It should be highlighted that the presence of the protective layers on both cooperating elements prevented damage to the bearing. Otherwise, bearing malfunction inevitably occurs before the air film development is complete. Similarly, a positive rate of the temperature readings confirmed a not yet successfully finished process of the gaseous film generation. However, starting from the approximate speed of 5 krpm (Case A) the temperature gradually gets stabilized until it reaches its maximum value at approximately 15 krpm (Case B). Further increase of the rotational speed does not lead to any significant change of all of the measured parameters (Case C) that confirm the stable state of the previously generated air film.

It is worth mentioning that during the entire process of the air film development (Cases A–C) sudden and chaotic-looking changes of the strain registered by all strain gauges were observed. The continuous adjustment of the bump and top foils occurred in response to both film generation (via hydrodynamic effect) and variation of the friction torque (due to dry friction between the metallic parts). Moreover, the top foil underwent the phenomenon of a micro slip over the bump foils due to the progressive thermal expansion.

Before the gaseous film entirely develops (as addressed in case A), the highest temperatures, which are seen in Figure 11, were recorded in the upper right localization in the top foil, where it keeps contact with the rotating shaft. In fact, following the counter-clockwise direction of the rotation of the shaft, the GFB's bushing tried to lift slightly and move to the left. The registered increase of the temperature along the GFB's circumference, found with the applied approximation procedure, varied within the range from 2 °C up to 4.7 °C. The resulting thermal strain (in the order of 25–60 microstrains) amplified the effect of wrapping the top foil around the shaft.

The strain measured for Case A indicated the resultant wrapping effect at the free end of the top foil. As the speed increased (Case B), which meant simultaneous completion of the generation of the air film and further temperature change, the wrapping effect grew, which confirmed the hydrodynamic action. The maximum temperature was found at the top right zero-strain localization on the cross-sectional view in Figure 12. The dominant foil flattening effect was visible at its tie region. After a stable air film was generated (Case C), the temperature became homogenized along the circumference, even though it still reached relatively high values in the order of 11.5 °C above its ambient counterpart with the scatter equal to 3 °C. Moreover, for Case C, the maximum temperature was no longer found in the region where the air film was the thinnest. Contrarily, as shown in Figure 13, the raised temperature was observed within the region that squeezed the air before it entered the neck of the gaseous film. Accordingly, a sudden temperature drop occurred where the air expanded on top of the circumferential view. Next, i.e., moving counterclockwise, the air again became heated on the bottom—near the top foil tie.

5.2. Case D—Continuous Operation of the Bearing with a Stable Air Film

As seen in Figure 10, the values of all measured quantities remain unchanged. Only a slightly rising trend regarding the measured strains, which is found equal and common for all sensors, was observed. The GFB's behavior did not change significantly for the constant rotational speed and non-changing external load. Adequately, no further deformation of the top foil was identified for the considered stage of operation. The amplitude of the friction torque was approximately ten times smaller than its counterpart, measured during both the run-up and run-out stages of the GFB's operation.

The above-discussed effect of squeezing and expanding the air film around its neck is visible in Figure 14. Similar local changes in the temperature distribution were found. Moreover, again, no resultant in-plane strain was identified around the neck of the gaseous film which confirmed the expected presence of radial forces (originated from the hydrodynamic effect) as dominant ones for the considered angular localization.

5.3. Cases E and F—Run-Out Stage and Gradual Loss of the Air Film

First, during the initial phase of the run-out stage of operation (i.e., during Case E and earlier), continuous growth of the friction torque and temperature was observed. In fact, since the rotational speed decreased, the air film underwent gradual destruction. The occurrence of a dry friction in the area where the bushing and shaft cooperate led to a sudden growth of both heat generation and temperature, as visualized in Figure 15. These symptoms clearly indicate a change in the bearing's operational parameters, specifically regarding the state of the air film. Finally, when the rotational speed became insufficient to continuously maintain the gaseous part of the supporting layer, a dry friction contact appeared and remained until the shaft stopped (Case F).

Low deceleration of the rotational speed resulted in a relatively high temperature of the top foil, from 33 °C up to 43 °C above the ambient temperature, which caused significant thermal expansion. The highest temperatures for Cases E and F were found in the dry friction contact region. The analysis of the strains led to observations similar to those found for Cases A–C. Specifically, the top foil tended to wrap around the shaft at its free end and, contrarily, got flattened on top and on the left of the cross-sectional data presentation, as shown in Figure 16. The amplitudes of strains, however, were approximately four times greater during the run-out stage.

5.4. Cases G and H—Cooling Down

The strains registered for Case G did not converge to their initial values, i.e., at the moment of measurement initialization. The large scatter of the strains, which are shown in Figure 17 and were observed at the beginning of the cooling down stage, resulted from both high temperatures (i.e., due to thermal expansion, which is a reversible phenomenon) and the progressive process of the foil's shape adjustment. A single run of the bearing led to the irreversible change of the locations of all foils, and, possibly, to the plastic deformation of the top foil. These phenomena naturally occurred during the bearing's operation due to the process of losing the air film, progressive thermal expansion and, finally, resulted from the force response of the bump foils.

The identified gradual temperature drops followed the expected change of the generated heat energy. In fact, when the shaft was stopped, no source of heat originating from the evolving friction conditions was present in the bearing any longer. At the end of measurements (which is addressed by Case H), the registered temperature scatters decreased to achieve the interval from 10 °C up to 13 °C above the ambient temperature (Figure 18).

6. Summary and Concluding Remarks

The conducted research experimentally confirmed the demanded measurement functionality for the designed and constructed specialized sensing top foil that was successfully installed in a normally operating GFB prototype. Specifically, the novel approach and the newly proposed technical solution, which makes use of the integrated thermocouples and bonded strain gauges, allowed for the acquisition of the temperature and strain readings at selected localizations in the outer surface of the top foil.

The present study is considered by the authors as an attempt of a comprehensive characterization of the bearing which may help to identify both the causes of the top foil's deformations, i.e., the change of the top foils' geometry and the thermomechanical conditions at which an air film develops due to the hydrodynamic action. The conducted, primarily qualitative, investigation deals with the two interacting physical domains, having considered circumferential distribution of the temperature and strain.

With the spatial courses of the measured quantities, the critical regions that contribute to the phenomena important for the GFB's operation may be clearly indicated. Specifically, the angular position of the region exhibiting the raised temperature during both the run-up and run-out stages showed the part of the top foil's surface where a dry friction contact was present. Moreover, a characteristic local decrease of the temperature on top of the rotating shaft found during the stage of a stable operation (as shown in Figure 14) indicated the large flow of the air being expanded when coming out of the area of the thinnest gaseous film. The adequate greater curvature of the top foil (i.e., local wrapping of the shaft), which is seen on the right, and its simultaneous flattening on the left with respect to the above-mentioned region confirmed the resultant shift of that foil due to the hydrodynamic interaction.

The authors are aware of the existing limitations of the present study. First, the temperature and strain readings were acquired and analyzed only for the central cross-section of the tested bearing. Moreover, the floating configuration without the external load was investigated, which stood for the first step of the analysis—prior to the planned use of a two-node support configuration of a GFB [43]. Next, significant geometric modifications of the foils' geometry, with the intentionally introduced gaps (incisions) in the bump foils, as well as the measurement method used to identify temperature and strain, made the proposed technical solution not yet ready to be implemented in commercial cases. The obtained preliminary results, however, allow for the inspection of the relation between the mechanical and thermal properties of the tested bearing while it runs at all subsequent stages of operation.

There are also several issues that need to be addressed during further studies. First, the existing technical limitations related to the strain measurements at the top foil's free end and its tie make it difficult to reliably estimate the foil's behavior (deformation) at these localizations. Hence, other measurement techniques should be possibly proposed and then used to fill in the existing gap and complete the circumferential distribution of the strain in the top foil. Next, the installed strain gauges caused local changes in the top foil's stiffness (overstiffening) which may influence the process of developing the air film and proper bearing's operation in the case of other configurations of GFBs, characterizing various shaft's diameters, bearing's widths, loads, and rotational speeds. The authors admit that each modification of the bearing should be subjected to an additional study to reliably formulate general conclusions on a given generation of GFBs. In the authors' opinion, however, the proof of concept for the investigated measurement technique was successful. Moreover, the repeatability of the temperature and strain measurements should be assessed. Finally, even though the strains were measured accurately, it is not a trivial task to decide when their offset values should be cancelled and, in turn, when the tracking of the temporal courses for the strains should begin. In fact, the respective temporal force (pressure) contributions of both the air film and the bump foil to the measured resultant strains in the top foil are not yet experimentally investigated.

Author Contributions: Conceptualization, A.M. and J.R.; methodology, A.M., J.R., P.Z. and G.Ż.; software, J.R., P.Z. and P.B.; validation, A.M., J.R., P.Z., G.Ż., P.B. and A.A.; formal analysis, A.M. and G.Ż.; investigation, A.M., J.R., P.Z., G.Ż., P.B. and A.A.; resources, A.M. and G.Ż.; writing—original draft preparation, A.M.; writing—review and editing, G.Ż.; visualization, P.Z. and P.B.; supervision, A.M.; project administration, A.M.; funding acquisition, A.M. All authors have read and agreed to the published version of the manuscript.

Funding: This research was funded by the National Science Center, Poland, within the project grant number 2017/27/B/ST8/01822 entitled "Mechanisms of stability loss in high-speed foil bearings—modeling and experimental validation of thermomechanical couplings". Additionally, the third author (P.Z.) would like to thank for his personal support provided by the Department of Robotics and Mechatronics affiliated at the AGH University of Science and Technology, Krakow, Poland.

Data Availability Statement: The data presented in this study are available on request from the corresponding author.

Conflicts of Interest: The authors declare no conflict of interest.

References

1. DellaCorte, C. Oil-Free shaft support system rotordynamics: Past, present and future challenges and opportunities. *Mech. Syst. Signal Process.* **2012**, *29*, 67–76. [CrossRef]
2. Samanta, P.; Murmu, N.C.; Khonsari, M.M. The evolution of foil bearing technology. *Tribol. Int.* **2019**, *135*, 305–323. [CrossRef]
3. DellaCorte, C.; Bruckner, R.J. Remaining technical challenges and future plans for oil-free turbomachinery. *ASME J. Eng. Gas Turbines Power* **2011**, *133*, 042502. [CrossRef]
4. Żywica, G.; Bagiński, P. Investigation of gas foil bearings with an adaptive and non-linear structure. *Acta Mech. Et Autom.* **2019**, *13*, 5–10. [CrossRef]
5. Yu, T.-Y.; Wang, P.-J. Simulation and experimental verification of dynamic characteristics on Gas Foil Thrust Bearings based on Multi-Physics Three-Dimensional Computer Aided Engineering methods. *Lubricants* **2022**, *10*, 222. [CrossRef]
6. Żywica, G.; Bagiński, P.; Bogulicz, M.; Martowicz, A.; Roemer, J.; Kantor, S. Numerical identification of the dynamic characteristics of a nonlinear foil bearing structure: Effect of the excitation force amplitude and the assembly preload. *J. Sound Vib.* **2022**, *520*, 116663. [CrossRef]
7. Martowicz, A.; Roemer, J.; Lubieniecki, M.; Żywica, G.; Bagiński, P. Experimental and numerical study on the thermal control strategy for a gas foil bearing enhanced with thermoelectric modules. *Mechanical Syst. Signal Process.* **2020**, *138*, 106581. [CrossRef]
8. Mikolajczak, P. Vibration analysis of reconditioned high-speed electric motors. *J. Vibroeng.* **2019**, *21*, 1917–1927. [CrossRef]
9. Weber, M.; Weigold, M. High speed synchronous reluctance drives for motor spindles. *MM Sci. J.* **2019**, *4*, 3323–3329. [CrossRef]
10. Kim, T.H.; San Andres, L. Limits for high-speed operation of gas foil bearings. *ASME J. Tribol.* **2006**, *128*, 670–673. [CrossRef]
11. Zhang, K.; Zhao, X.; Feng, K.; Zhao, Z. Thermohydrodynamic analysis and thermal management of hybrid bump-metal mesh foil bearings: Experimental tests and theoretical predictions. *Int. J. Therm. Sci.* **2018**, *127*, 91–104. [CrossRef]
12. Zhang, C.; Ao, H.; Jiang, H.; Zhou, N. Investigations on start-up performances of novel hybrid metal rubber-bump foil bearings. *Tribol. Int.* **2021**, *154*, 106751. [CrossRef]
13. San Andrés, L.; Chirathadam, T.A. Performance characteristics of metal mesh foil bearings: Predictions versus measurements. *ASME J. Eng. Gas Turbines Power* **2013**, *135*, 122503. [CrossRef]
14. Feng, K.; Liu, Y.; Zhao, X.; Liu, W. Experimental evaluation of the structure characterization of a novel hybrid bump-metal mesh foil bearing. *ASME J. Tribol.* **2016**, *138*, 021702. [CrossRef]
15. Hoffmann, R.; Liebich, R. Experimental and numerical analysis of the dynamic behaviour of a foil bearing structure affected by metal shims. *Tribol. Int.* **2017**, *115*, 378–388. [CrossRef]
16. Feng, K.; Liu, W.; Yu, R.; Zhang, Z. Analysis and experimental study on a novel gas foil bearing with nested compression springs. *Tribol. Int.* **2017**, *107*, 65–76. [CrossRef]
17. Martowicz, A.; Roemer, J.; Kantor, S.; Zdziebko, P.; Żywica, G.; Bagiński, P. Gas foil bearing technology enhanced with smart materials. *Appl. Sci.* **2021**, *11*, 2757. [CrossRef]
18. Breńkacz, Ł.; Witanowski, Ł.; Drosińska-Komor, M.; Szewczuk-Krypa, N. Research and applications of active bearings: A state-of-the-art review. *Mech. Syst. Signal Process.* **2021**, *151*, 107423. [CrossRef]
19. Park, J.; Sim, K. A feasibility study of controllable gas foil bearings with piezoelectric materials via rotordynamic model predictions. *J. Eng. Gas Turbines Power* **2019**, *141*, 021027. [CrossRef]
20. Kikuchi, H.; Ibrahim, M.D.; Ochiai, M. Evaluation of lubrication performance of foil bearings with new texturing. *Tribology Online* **2019**, *14*, 339–344. [CrossRef]
21. Lyu, P.; Feng, K.; Zhu, B.; Zhang, K.; Sun, D. The performance evaluation of the promising high-stability foil bearings basing with flexure pivot tilting pads. *Mech. Syst. Signal Process.* **2019**, *134*, 106313. [CrossRef]
22. Pattnayak, M.R.; Pandey, R.K.; Dutt, J.K. Performance behaviours of a self-acting gas journal bearing with a new bore design. *Tribol. Int.* **2020**, *151*, 106418. [CrossRef]
23. Guan, H.Q.; Feng, K.; Yu, K.; Cao, Y.L.; Wu, Y.H. Nonlinear dynamic responses of a rigid rotor supported by active bump-type foil bearings. *Nonlinear Dyn.* **2020**, *100*, 2241–2264. [CrossRef]
24. Guan, H.Q.; Feng, K.; Cao, Y.L.; Huang, M.; Wu, Y.H.; Guo, Z.Y. Experimental and theoretical investigation of rotordynamic characteristics of a rigid rotor supported by an active bump-type foil bearing. *J. Sound Vib.* **2020**, *466*, 115049. [CrossRef]
25. Howard, S.; Dellacorte, C.; Valco, M.J.; Prahl, J.M.; Heshmat, H. Dynamic stiffness and damping characteristics of a high-temperature air foil journal bearing. *Tribol. Trans.* **2008**, *44*, 657–663. [CrossRef]
26. Kim, T.H.; Breedlove, A.W.; San Andres, L. Characterization of foil bearing structure for increasing shaft temperatures: Part I-Static load performance. In Proceeding of the ASME Turbo Expo GT2008, Berlin, Germany, 9–13 June 2008; p. 50567. [CrossRef]
27. Kim, T.H.; Breedlove, A.W.; San Andres, L. Characterization of foil bearing structure at increasing temperatures: Static load and dynamic force performance. *J. Tribol.* **2009**, *131*, 041703. [CrossRef]
28. San Andres, L.; Kim, T.H. Thermohydrodynamic analysis of bump type gas foil bearings: A model anchored to test data. *ASME J. Tribol.* **2010**, *132*, 011701. [CrossRef]

29. Mahner, M.; Bauer, M.; Lehn, A.; Schweizer, B. An experimental investigation on the influence of an assembly preload on the hysteresis, the drag torque, the lift-off speed and the thermal behavior of three-pad air foil journal bearings. *Tribol. Int.* **2019**, *137*, 113–126. [CrossRef]
30. Ghalayini, I.; Bonello, P. Nonlinear and linearised analyses of a generic rotor on single-pad foil-air bearings using Galerkin Reduction with different applied air film conditions. *J. Sound Vib.* **2022**, *525*, 116774. [CrossRef]
31. Rieken, M.; Mahner, M.; Schweizer, B. Thermal optimization of air foil thrust bearing using different foil materials. *J. Turbomach.* **2020**, *142*, 101003. [CrossRef]
32. Liu, X.; Li, C.; Du, J.; Nan, G. Thermal characteristics study of the bump foil thrust gas bearing. *Appl. Sci.* **2021**, *11*, 4311. [CrossRef]
33. Kozanecki, Z.; Tkacz, E.; Łagodziński, J.; Miazga, K. Theoretical and experimental investigations of oil-free bearings and their application in diagnostics of high-speed turbomachinery. *Key Eng. Mater.* **2014**, *588*, 302–309. [CrossRef]
34. Zhou, Y.; Shao, L.; Zhang, C.; Ji, F.; Liu, J.; Li, G.; Ding, S.; Zhang, Q.; Du, F. Numerical and experimental investigation on dynamic performance of bump foil journal bearing based on journal orbit. *Chin. J. Aeronaut.* **2021**, *34*, 586–600. [CrossRef]
35. Yan, J.; Liu, Z.; Zhang, G.; Yu, X.; Xu, L. Feasibility study of a turbocharger rotor supported by air foil bearings with diameter of 17 mm focusing on rotordynamic performance. *Proc. Inst. Mech. Eng. Part D J. Automot. Eng.* **2018**, *233*, 1331–1344. [CrossRef]
36. LaTray, N.; Kim, D.; Song, M. Static performance of a hydrostatic thrust foil bearing for large scale oil-free turbomachines. *J. Eng. Gas Turbines Power* **2021**, *143*, 041017. [CrossRef]
37. Breńkacz, Ł.; Bagiński, P.; Żywica, G. Experimental research on foil vibrations in a gas foil bearing carried out using an ultra-high-speed camera. *Appl. Sci.* **2021**, *11*, 878. [CrossRef]
38. Roemer, J.; Zdziebko, P.; Martowicz, A. Multifunctional bushing for gas foil bearing-test rig architecture and functionalities. *Int. J. Multiphys.* **2021**, *15*, 73–86. [CrossRef]
39. Zdziebko, P.; Martowicz, A. Study on the temperature and strain fields in gas foil bearings–measurement method and numerical simulations. *Eksploat. I Niezawodn.-Maint. Reliab.* **2021**, *23*, 540–547. [CrossRef]
40. Żywica, G.; Bagiński, P.; Banaszek, S. Experimental studies on foil bearing with a sliding coating made of synthetic material. *J. Tribol.* **2016**, *138*, 011301. [CrossRef]
41. Lubieniecki, M.; Roemer, J.; Martowicz, A.; Wojciechowski, K.; Uhl, T. A multi-point measurement method for thermal characterization of foil bearings using customized thermocouples. *J. Electron. Mater.* **2016**, *45*, 1473–1477. [CrossRef]
42. Bagiński, P.; Żywica, G. Experimental study of various low-friction coatings for high-temperature gas foil bearings under cold-start conditions. *ASME J. Eng. Gas Turbines Power* **2022**, *144*, 081007. [CrossRef]
43. Martowicz, A.; Zdziebko, P.; Roemer, J.; Zywica, G. ; Baginski, P. Thermal characterization of a Gas Foil Bearing—A novel method of experimental identification of the temperature field based on integrated thermocouples measurements. *Sensors* **2022**, *22*, 5718. [CrossRef] [PubMed]
44. Carpino, M.; Medvetz, L.A.; Peng, J.P. Effects of Membrane Stresses in the Prediction of Foil Bearing Performance. *Tribol. Trans.* **1994**, *37*, 43–50. [CrossRef]
45. Hryniewicz, P.; Wodtke, M.; Olszewski, A.; Rzadkowski, R. Structural Properties of Foil Bearings: A Closed-Form Solution Validated with Finite Element Analysis. *Tribol. Trans.* **2009**, *52*, 435–446. [CrossRef]

 *materials*

Article

Rapid-Response and Wide-Range pH Sensors Enabled by Self-Assembled Functional PAni/PAA Layer on No-Core Fiber

Gang Long [1], Liang Wan [1], Binyun Xia [1], Chao Zhao [1], Kunpeng Niu [1], Jianguo Hou [1], Dajuan Lyu [2], Litong Li [2], Fangdong Zhu [3] and Ning Wang [1,*]

1 National Engineering Research Center of Fiber Optic Sensing Technology and Networks, Wuhan University of Technology, Wuhan 430070, China
2 State Key Laboratory of Optical Fiber and Cable Manufacture Technology, Yangtze Optical Fibre and Cable Joint Stock Limited Company, Wuhan 430073, China
3 Ningbo Lianghe Road & Bridge Technology Co., Ltd., Ningbo 315201, China
* Correspondence: ningwang23@whut.edu.cn; Tel.: +86-27-8765-1850-8201

Abstract: The measurement of pH has received great attention in diverse fields, such as clinical diagnostics, environmental protection, and food safety. Optical fiber sensors are widely used for pH sensing because of their great advantages. In this work, an optical fiber pH sensor is fabricated, by combining the merits of the multimode interference configuration and pH-sensitive polyaniline/polyacrylic acid (PAni/PAA) coatings, which was successfully in situ deposited on the no-core fiber (NCF) by the layer-by-layer (LBL) self-assembly method. The sensors' performance was experimentally characterized when used for pH detection. It has a high sensitivity of 0.985 nm/pH and a great linear response in a universal pH range of 2–12. The response time and recovery time is measured to be less than 10 s. In addition, its temperature sensitivity is tested to be about 0.01 nm/°C with a low temperature crosstalk effect, which makes it promising for detecting pH in the liquid phase with temperature variation. The sensors also demonstrated easy fabrication, good stability, and repeatability, which are adapted to pH detection in most practical applications.

Keywords: no-core fiber; polyaniline; polyacrylic acid; pH sensor; multimode interference

Citation: Long, G.; Wan, L.; Xia, B.; Zhao, C.; Niu, K.; Hou, J.; Lyu, D.; Li, L.; Zhu, F.; Wang, N. Rapid-Response and Wide-Range pH Sensors Enabled by Self-Assembled Functional PAni/PAA Layer on No-Core Fiber. *Materials* **2022**, *15*, 7449. https://doi.org/10.3390/ma15217449

Academic Editors: Wiesław Stręk and Marcel Poulain

Received: 30 August 2022
Accepted: 21 October 2022
Published: 24 October 2022

1. Introduction

As known, pH-conditions regulation is of significance in the fields of biochemical industry [1], food safety [2], disease diagnosis [3], environmental engineering [4], and so on. A slight variation in pH conditions will cause the failure of crucial processes within the above domains [5]. Therefore, how to detect various environmental pH conditions accurately and quickly is essential for guiding their precise regulation. A variety of pH detection techniques have been developed, such as pH test paper [6], glass electrodes [7], biological [8], and quantum dots fluorescent [9]. Conventional pH test paper has been widely used, despite its low precision and large errors due to subjective visual judgment. Although glass electrodes have demonstrated high precision, they still show many shortcomings, such as low sensitivity, bulk size, insensitive response, etc. They also need to be recalibrated for each measurement, which limits their development toward online monitoring applications.

Optical fiber sensors have attracted considerable attention due to their small size [10], light weight [5], high sensitivity [11], remote operation for real-time sensing [12], etc., which have been studied and used widely for biochemical sensing applications. Among them, many optical fiber pH sensors have been extensively reported [12–15]. For example, Mishra et al. [16] proposed a fast-response, wide-range pH sensor by coating a smart hydrogel on a long-period fiber grating (LPFG) surface. Li et al. [17] coated the PAA/chitosan (CS)-sensitive film on the surface of the gold-coated optical fiber through the LBL self-assembly technique. The surface plasmon resonance (SPR) sensor realizes the high sensitivity detection of pH by utilizing a characteristic of the sensitive film: expanding/contracting with the change of pH. Yan et al. [18] reported a pH sensor with a Mach–Zehnder interferometer (MZI) cascaded with a fiber Bragg grating (FBG), by coating the graphene oxide/polyvinyl alcohol (GO/PVA)-sensitive film for pH detection, while the FBG was used for temperature calibration, which realized the temperature-self-calibrating pH monitoring. However, the disadvantages of complex optical fiber structures and fabrication processes have hindered their practical application. However, using some inexpensive polymer-sensitive materials combined with simple optical fiber architectures seems to be a promising pathway.

Various pH-sensitive materials have been used as coatings on the surface of optical fiber sensors to achieve highly sensitive detection of pH [19–24]. PAni is used for pH sensing because of its unique variable pH optical properties [25,26]. PAni was gradually deprotonated and converted from ES (emeraldine salt) to EB (emeraldine base) with increasing pH [27]. This results in a change in the doping and conformation of the polymer chain and, thus, a change in its refractive index (RI). PAni has excellent stability, physicochemical properties, and rapid and reversible adsorption or desorption kinetics. For the above reasons, PAni is widely used in various optical or electrical sensors. However, when PAni was deposited on the sensors' surface using in situ oxidation polymerization, the PAni coating was completely not uniform for thicknesses lower than 2 μm [25]. When PAni powder was used, the coating was normally performed by the dip-coating method, and it tended to peel off. PAA is also an excellent pH-sensing material [28,29]. The polymer chain of PAA is rich in carboxyl groups. The degree of ionization of carboxyl groups varies with the changing of the pH condition. However, with the environmental pH increasing, PAA ionizes and expands into a fully solvated open-coil conformation, especially under a strong alkaline condition [29]. Therefore, PAni and PAA are unfavorable for pH sensing individually.

In this work, optical fiber pH sensors with a rapid response and a wide range are proposed by depositing the pH-sensitive polyaniline/polyacrylic acid (PAni/PAA) composite film on the fibers configured with multimode interference, composed of two pieces of single-mode fiber (SMF) and one section of no-core fiber (NCF). Here, benefiting from their differing electronegativity, the PAni/PAA composite coatings were in situ fabricated onto the NCF easily by using the typical layer-by-layer (LBL) self-assembly method. In addition, the combination of PAA and PAni makes up for the each other's defects mentioned above, when used for developing the optical fiber pH sensors. It was experimentally demonstrated that the proposed pH sensors possess high sensitivity, rapid response, and low temperature crosstalk under universal pH conditions.

2. Experimental Section

2.1. Materials

Aniline (C_6H_7N, 99.5%), ammonium persulphate (APS, 98%), PAA (Mw = 100,000, 35 wt% solution in water), sodium hydroxide (NaOH, 96%), hydrochloric acid (HCl, 36%), sodium chloride (NaCl, 99.5%), N, and N–dimethylformamide (DMF, 99.5%) were all purchased from Sinopharm Chemical Reagent Co., Ltd., Shanghai, China. All the reagents were used without any further purification. The NCF (OD: 125 μm) and SMF (G.652) were purchased from Yangtze Optical Fiber and Cable Co., Ltd., Wuhan, China.

PAni powder was first synthesized by the typical chemical oxidative polymerization method [30,31]. During the synthesis processes, 10 mL of 0.4 M HCl solution was prepared and divided into two equal volumes. Subsequently, 0.2 g of aniline and 0.189 g of APS were, respectively, added to the two HCl solutions and thoroughly stirred to mix them uniformly. After that, the two solutions were mixed and placed in a refrigerator (BCD–196DK) at 0 °C for 30 min. Then, they were laid at about 4 °C overnight. Finally, the reaction solution was taken out and centrifuged. The precipitate was then dried in a vacuum oven (DZF–6050ABF) at 80 °C overnight, and the PAni powder was obtained.

When preparing the PAA precursor and the PAni precursor, 300 mg of PAA was first dissolved in 20 mL of deionized (DI) water to obtain the PAA solution with a 15 mg/mL concentration. In addition, 40 mg of the synthesized PAni powder was added into 4 mL of DMF solvent, while sonicated for 10 min and then vigorously stirred overnight. The PAni–DMF solution was then diluted with 16 mL DI water to 2 mg/mL.

2.2. Fabrication of the pH–Sensing Probe

Before coating the PAni/PAA–sensitive film onto the NCF, the SNS (SMF–NCF–SMF) was fabricated first, and a section of NCF (length ~3 cm) was spliced with two SMF by a fusion splicer (FSM–100P, Fujikura, Shanghai, China). The NCF length was designed to be about 3 cm for the probe by considering the self–imaging effect of the NCF. To explain that more clearly, the simulation was conducted using the simulation software Rsoft 2013 BeamPRO, and the simplified mode was built. The dimensions of the simulated model are comparable to those of the actual sensor. The refractive index of the NCF is 1.463, and the length is 3 cm. The core refractive index of the SMFs is 1.4504, and the cladding refractive index is 1.4447. The background refractive index is set to 1. The incident light from the light source (λ = 1550 nm) was coupled into the NCF from a single–mode fiber that excites higher–order modes due to the mode–field mismatch effect, and the higher–order modes undergo multimode interference in the NCF. During the simulation, normalization was first conducted to make the maximum light intensity to be 1 at the center of the NCF. Figure 1a shows the internal optical field of the 3 cm NCF for the SNS probe. As the length of the NCF varies, the optical field is periodically in the present, which is the self–imaging effect. In addition, it was found that the light density reached the maximum when the light was transmitted to a point about 15 mm in the NCF. Thus, the self–imaging distance in the NCF could be regarded to be about 15 mm. If the length of the NCF is exactly equal or multiple to its self–imaging distance, the maximum output light would be obtained, which will be better for pH sensing. Figure 1b also shows the optical intensity distribution in the NCF. On the one hand, the longer the no–core fiber coated with sensitive materials is, the stronger the interaction between the light transmitted in the fiber and the material, which improves the sensitivity and accuracy of the sensing probe [32]. On the other hand, if the probe is too long, its actual application will be affected. Thus, the probe length was finally designed to be 3 cm.

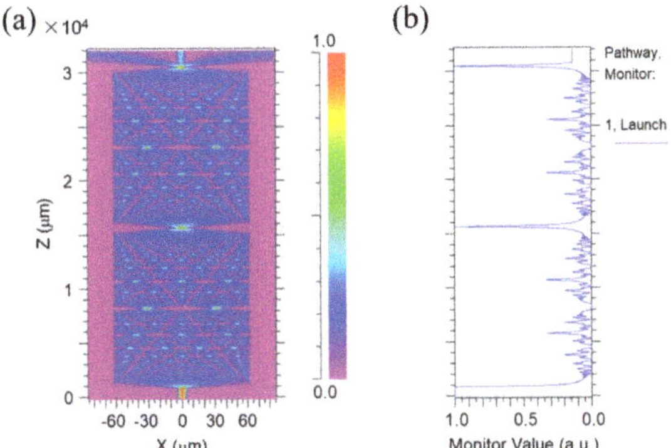

Figure 1. (**a**) Diagram of the optical field distribution inside a 3 cm NCF; (**b**) light–intensity values monitored on the Z–axis.

The PAni/PAA film was coated onto the NCF surface by the LBL self–assembly method, and the details are shown in Figure 2. The NCF section was treated using the oxygen plasma cleaner (power: 80 w; time: 300 s), which made the probe surface rich in hydroxyl groups and, thus, negatively charged. To ensure its smooth execution, the sensitive materials should be self–assembly deposited immediately after the plasma treatment, and the SNS probe was alternately immersed in cationic PAni–DMF solution and anionic PAA solution for 15 min, and the positive–charged PAni particles had the priority for naturally bonding with the hydroxylated glass fiber surface. The SNS probe was then rinsed with DI water for 3 min in between the immersions in cationic and anionic solutions, to remove the excess adsorbed components. The above steps were then repeated several times, and the self–assembled multilayers of PAni/PAA were obtained on the NCF. Finally, the coated SNS probe was delivered to the oven and dried at 60 °C overnight.

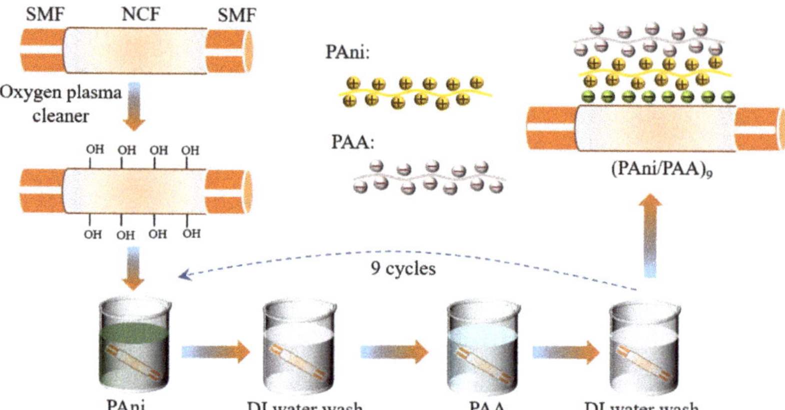

Figure 2. Fabrication of the SNS pH sensors by the LBL self–assembly method.

2.3. Characterization

The pH–sensitive films were characterized by scanning electron microscopy (SEM) (Zeiss Merlin Compact, Carl Zeiss, Germany, 15 kV resolution: 1.0 nm, 1 kV resolution: 1.7 nm, acceleration voltage: 1–20 kV, magnification: 20–300,000×). Fourier transform infrared spectroscopy (FTIR, wavenumber region: 4000–400 cm^{-1}, number of scans: 32, resolution: 4 cm^{-1}) spectra of the PAni, PAA, and PAni/PAA were recorded by the Thermo Scientific Nicolet 6700 spectrophotometer. The RI of different salt solutions was tested using an Abbe refractometer (WAY–2W, range of RI: 1.300–1.700, accuracy: 0.003).

2.4. Setup for pH Detection

Figure 3a shows the experimental setup for pH detection, which consists of the broadband light source (BBS, 1030–1660 nm), the V–groove, the lifting platform, the optical spectrum analyzer (OSA, YOKOGAWA AQ6370B, wavelength range: 600–1700 nm, high wavelength accuracy: ±0.01 nm, high wavelength resolution: 0.02 nm), and the SNS–sensing probe. The NCF coated with the pH–sensitive composite films were immersed in the V–groove filled with solution samples or rinsing regents, while connecting with the light source and the spectrometer by two sections of SMFs, which were fixed on both sides of the V–groove by rubber clamps to obtain the stable output of optical signals. The V–groove mounted on the lifting platform was employed to flexibly insert or remove the probe from the solution samples.

In the experiment, NaCl solutions with the same pH and different RI were prepared with RI of 1.3316, 1.3405, 1.3496, 1.3588, 1.3682, and 1.3773, respectively.The RI sensitivity of the bare SNS sensor was measured by immersing the sensing probe in NaCl solutions with different RI. To examine the pH detection performance of the sensor, the standard pH solutions were prepared by HCl or NaOH in DI water, with pH values from 2 to 12, and were calibrated by a commercial pH meter (HANNA, HI98103, range: 0.0 to 14.0 pH, resolution: 0.1 pH, accuracy: ±0.2 pH). To ensure data reliability, the sensor probe was immersed into the solution samples for 10 min, and the transmission spectrums were continuously monitored and recorded for each sample. Moreover, the response time, stability, and repeatability of this pH sensor were tested. To avoid the influence of temperature fluctuations on the measurement results, the temperature of the solution samples was kept at a constant room temperature (~25 °C). Afterwards, the influence of temperature on the pH sensor was investigated. The pH of the solution is 7, while the temperature is adjusted from 25 °C to 55 °C with an increment interval of 5 °C.

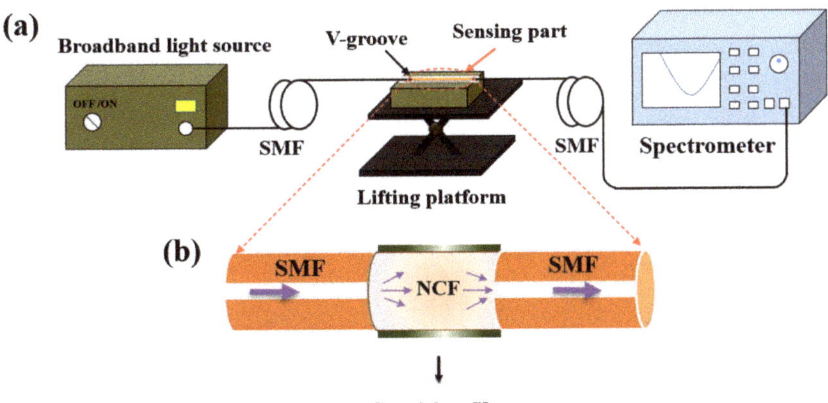

Figure 3. (**a**) Experimental setup for pH detection; (**b**) schematic diagram of the pH sensing probe.

Figure 3b exhibits the schematic diagram of the sensor probe in detail. The NCF with a diameter of 125 μm was fused with two sections of SMFs, and the dark green layer representing the pH–sensitive composite films was coated just outside the NCF surface. When used for RI sensing, the NCF acts as the core, and the surrounding medium acts as the cladding. Multiple modes will be excited when the light enters the NCF from the input SMF, resulting in multimode interference. The interference spectrum dip wavelength of the mth order can be expressed:

$$\lambda_m = \frac{2\Delta n_{eff} L}{2m+1} \tag{1}$$

where Δn_{eff} is the effective RI difference between the fundamental and higher–order modes, L is the length of the NCF.

The pH variations of the samples will cause the RI of the coated pH–sensitive film to change on the surface of the SNS probe. At this point, the effective RI difference between the higher–order mode and the fundamental mode will change. This will eventually lead to a drift in the interference spectrum. The increase in RI of the PAni/PAA composite film reflects the increase in pH, which results in a red shift of the resonance wavelength. The interaction mechanism between the PAni/PAA composite film and the pH solution samples and their effect on the penetration in the SNS probe will be described in the next section.

3. Results

3.1. Characterization of the Sensing Probe

To confirm the successful attachment of the self–assembled film on the fiber surface, the pH–sensitive film on the fiber surface was characterized by FTIR. For comparison, the FTIR spectra of pure PAni and pure PAA were also measured. As shown in Figure 4a, the characteristic peaks of PAni present near the wavenumbers of 1558 cm^{-1} are ascribed to the stretching of the quinoid. In addition, the peaks around 1292 and 1240 cm^{-1} come from the aromatic C–N and $C–N^+$ stretching vibration [33]. This demonstrates the successful synthesis of PAni. Pure PAA presented stretching–vibration bands of the carbonyl groups at 1717 cm^{-1} [34]. The peak of the PAA carbonyl group in the self–assembled film was centered at 1702 cm^{-1}. Compared with pure PAA, the frequencies of C=O stretching in the self–assembled film were red–shifted by 15 cm^{-1}. This indicates that strong intermolecular hydrogen bonds were formed between PAni and PAA during the self–assembly process [35]. The characteristic peaks of multilayers were present near the wavenumbers of 1454 cm^{-1} (C–C stretching mode for the quinoid ring) and 1240 cm^{-1}

(C–N$^+$ stretching vibration) [33,36,37]. This confirms the presence of the PAni in the coating. These results indicated that PAni and PAA were successfully attached to the fiber surface by self–assembly.

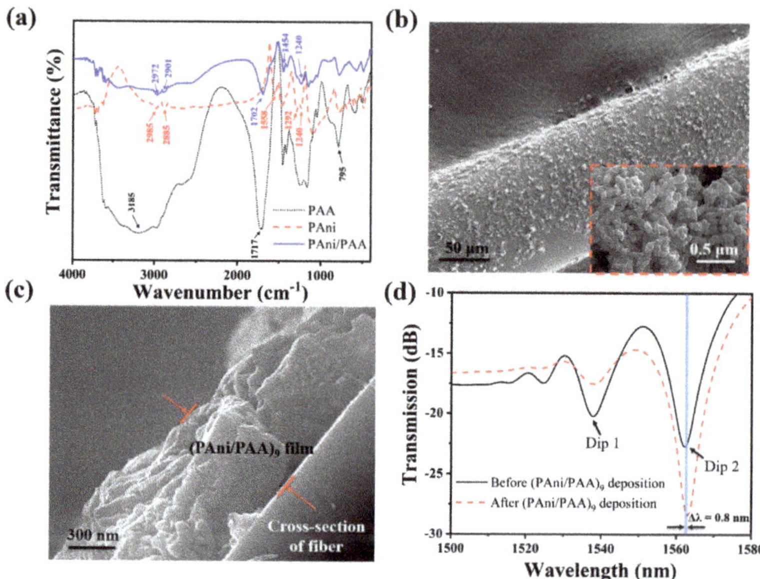

Figure 4. (**a**) FTIR spectra of PAni, PAA, and PAni/PAA; (**b**) SEM photos of the NCF with the PAni/PAA coatings; the inset is the scale–up image; (**c**) the cross–section of the NCF with the PAni/PAA coatings; (**d**) transmission spectrum of the SNS before and after the $(PAni/PAA)_9$ deposition.

Figure 4b shows the scaled–up SEM image of the NCF coated with $(PAni/PAA)_9$. Its original smooth surface becomes rough with noncontinuous coatings, which were composed of many nanoparticles observed from the scaled–up image in the Figure 4b inset, and it looks like a porous structure. Figure 4c exhibits the cross–sectional view of the sensitive coating on the NCF, and its thickness is about 1 μm. Figure 4d plotted the transmission spectrum of the SNS before and after $(PAni/PAA)_9$ deposition in air. The shift of Dip 1 and Dip 2 are different after coating. Dip 1 did not shift, while Dip 2 red shifted by 0.8 nm from 1562.2 nm to 1563 nm. This is because the interference spectrum here is caused by multimode interference, and the response of each peak to the changing external environment reflectivity index is not consistent, which has also been reported and explained in detail [3,38].

3.2. *RI Sensitivity of the Bare SNS Sensor*

The transmission spectrum of the bare SNS is plotted in Figure 5a by immersing it in solutions with a different RI. It can be found that the resonance wavelength experiences a red shift when the RI increased. The relationship between the resonance wavelength and the RI exhibits a great linear fitting, and the RI sensitivity of this SNS sensor is 114.153 nm/RIU. The linear fitting is shown in Figure 5b. Its functional relationship is y = 114.153x + 1420.165 (R^2 = 0.990).

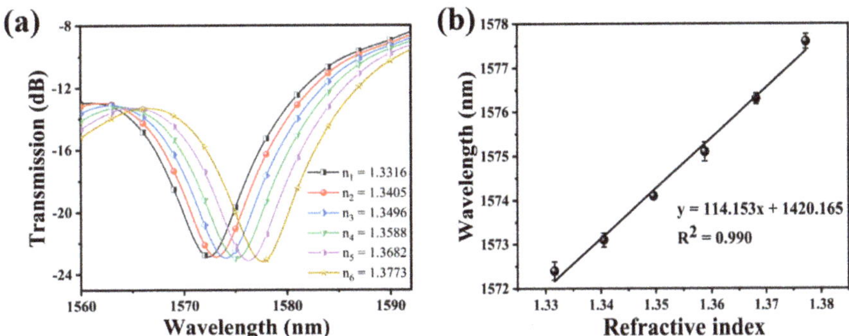

Figure 5. (**a**) Variation of the resonance wavelength of the bare SNS sensor with different RI; (**b**) the relationship between wavelength variation and the RI.

3.3. *pH Detection*

The transmission spectrums of the sensor in solution samples of different pH are shown in Figure 6a. When the pH value of the solution samples was adjusted from 2 to 12, the resonance peak shifted from 1568.6 nm to 1578.3 nm, and the total shift was about 9.7 nm. The relationship between the resonance wavelength and pH showed a great linear fitting of $R^2 = 0.990$ in Figure 6b, and the function could be expressed by $y = 0.982x - 1.818$. Thus, its pH sensitivity can be defined as 0.982 nm/pH, in the pH range between 2 and 12. Here, the error bars are obtained by repeating the experiments five times. To further clarify the effect of the sensitive coating, a control experiment was also conducted, and the bare SNS without any coating was immersed into the solutions with different pH values (2, 4, 7, 10, and 12). There was no obvious variation for the resonant wavelength with the changing pH, indicating that the RI of the configured pH solutions is very close to that of the neutral DI water (1.3330). Therefore, the pH response of the PAni/PAA–coated SNS sensor is mainly caused by the RI change of the sensitive film induced by the different pH solutions.

The pH response characteristics of the sensor are consistent with the dissolution characteristics of the PAni/PAA–sensitive membranes for pH samples. As shown in Figure 6c, the imine (=N–) sites of PAni are protonated in an acidic medium during electrostatic self–assembly. The cation radicals ($-NH_3^+$) generated after PAni protonation are bound to the PAA network, containing anion radicals ($-COO^-$) through strong electrostatic interactions and intermolecular hydrogen bonds, which results in a homogeneous sensitive film. With the increase in pH, the resonance wavelength underwent a red shift. This is because the thickness of the PAni/PAA coating changes with the pH. When the pH increased from 2 to 12, the PAA's ionization degree continued to increase. That is, the number of $-COO^-$ continued to increase. Although PAni gradually deprotonates and gradually changes from ES to EB, the overall degree of ionic cross–linking between the two still increases with increasing pH. This leads to the shrinkage of the multilayer membrane. As the thickness of the sensitive film decreases, the water molecules contained in the film gradually decrease, which causes the effective RI of the sensitive film to increase. This eventually leads to a red shift in the resonance wavelength.

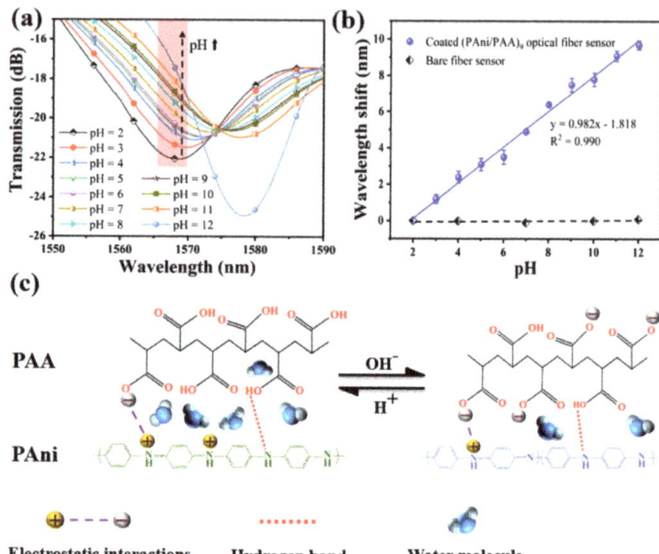

Figure 6. (**a**) Transmission spectra of the SNS sensor for different pH, ranging from 2 to 12 at 25 °C; (**b**) the fitting curve between resonant wavelength shift and pH; (**c**) schematic diagram shows the pH response of PAni/PAA sensitive film.

3.4. Response Time

As one of the most important features of pH sensors, response time should be investigated to evaluate its performance for practical applications. The response time of a sensor is measured by a spectrometer. In addition, the dip wavelength temporal shifts were recorded by continuously scanning the transmission spectrum. During the experiment for testing the response time, the time resolution of the spectrometer for sampling was set as 1 s, and the data points in Figure 7 were recorded with an interval of 1 s. As shown in Figure 7, the pH response time from pH = 2 to pH = 4 was measured for a sensing probe coated with PAni/PAA–sensitive film. The sensor required only 6 s to achieve a stable spectral change from pH = 2 to pH = 4, and, correspondingly, the recovery time was 4 s faster than the response time. Under alkaline conditions, the response time was recorded to be about 10 s when the solution pH was adjusted from pH = 10 to pH = 12.

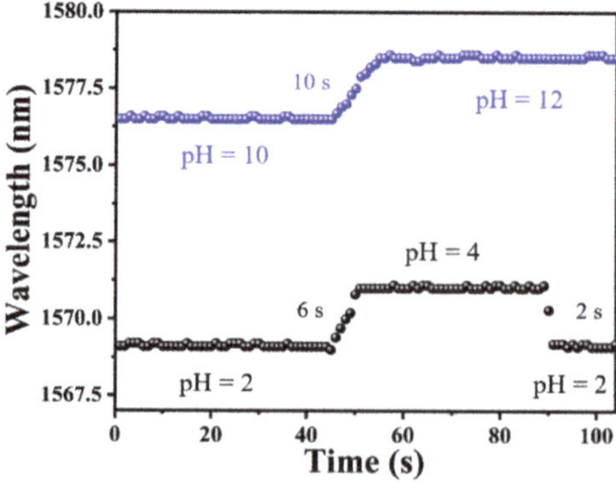

Figure 7. The response time of PAni/PAA–coated SNS sensor.

3.5. Stability and Repeatability of the pH Sensor

To test the stability of the sensor, it was immersed in solutions with a different pH for 1 h. Figure 8a shows the fluctuation of the sensor spectrum at a pH of 2, 4, 7, 10, and 12, at 10 min intervals for 1 h. The fluctuation of the resonance wavelength is 0.1 nm in 30 min, considering the sensitivity of the sensor is about 0.982 nm/pH, which results in a measurement error of ±0.102 pH. When the pH is 12, the maximum fluctuation of the resonance wavelength in 1 h is 0.3 nm. Thus, it can be considered that the sensor's stability curve remains flat within 1 h. Under other pH conditions, the fluctuation of the resonance wavelength is only 0.2 nm. This shows that the sensor has good stability.

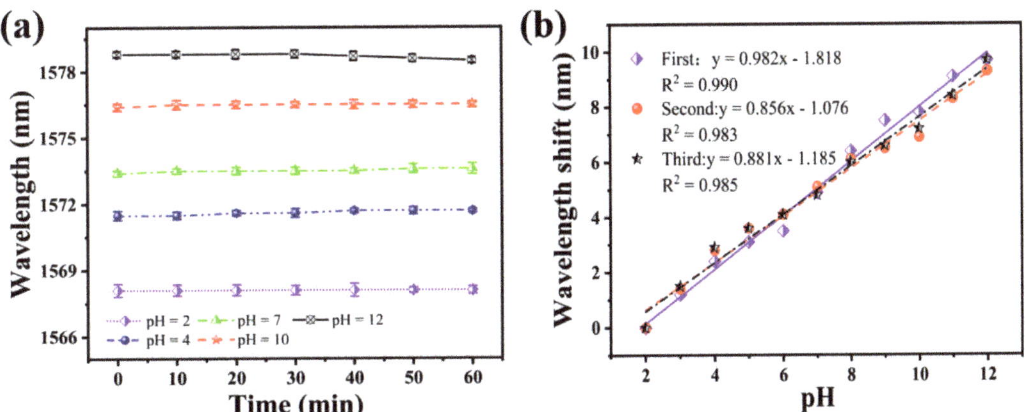

Figure 8. (**a**) Stability of the SNS sensor at different pH within 1 h; (**b**) shift of the resonance wavelength three pH tests on the same sensor, in the order of increasing pH.

Subsequently, to examine the repeatability of this sensor, we performed three pH tests in ascending order, using the same sensor. The results of the three repeated tests are shown in Figure 8b. The fitting curves of the three test results have a good coincidence. The average sensitivities were 0.982, 0.856, and 0.881 nm/pH, respectively. The similarity of the response characteristics and measurements' detection sensitivity confirms the same sensor's repeatability.

3.6. Effect of the Temperature

Many polymer–based pH sensors are subject to temperature crosstalk, considering that the ambient temperature may affect the pH of the solution and the physicochemical properties of PAni/PAA. Therefore, we investigated the response of temperature on the pH sensor. The experimental results are plotted in Figure 9. It can be noticed that the transmission spectrum shows a slight red shift with the increase in temperature. The main reason for this phenomenon is the gradual increase in the pH of the solution with the increase in the temperature. The increase in pH will cause the ionic cross–linking between PAni and PAA to increase, causing the shrinkage of the sensitive film. This causes an increase in the RI of the pH–sensitive film, which eventually causes a red shift in the interference spectrum. By calculation, the temperature sensitivity of the proposed pH sensor is only 0.01 nm/°C. At this point, the temperature cross–sensitivity of the sensor is approximately 0.0102 pH/°C. The temperature sensitivity of the proposed sensor is 1–2 orders of magnitude lower than the pH sensitivity, and its temperature crosstalk is lower than that of the previously reported optical fiber pH sensors. It is still applicable to some measurements without significant temperature variations.

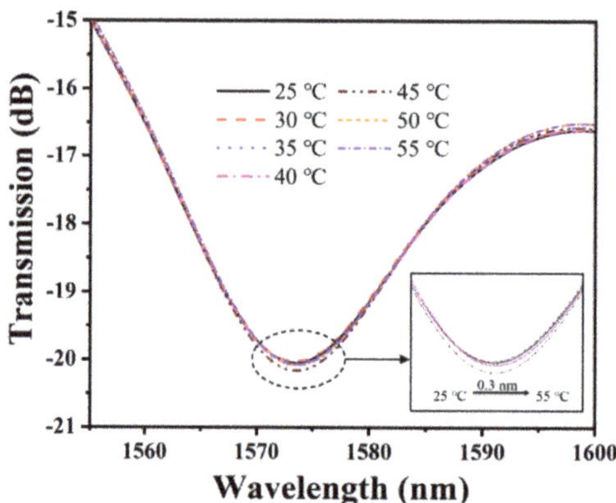

Figure 9. Temperature response of the pH sensor.

4. Discussion

Here, the proposed sensors have been compared with other optical fiber pH sensors reported in previous studies, as shown in Table 1. The SNS pH sensor has a wide pH test range (pH~2–12), which meets the needs of most test environments, and has an excellent linear response to pH variation. The minimum response time of this sensor is only 2 s, which is better than those obtained by most pH sensors. Although the pH sensitivity of the PAA/CS based sensor [17] is larger, it also has higher temperature crosstalk and a lesser pH range. The sensor has excellent stability over a pH range of 3.18–11.84, and the measurement accuracy of about 0.16 pH is affected. Although the hydrogel-based LPFG sensor [16] has a shorter response time, its pH sensitivity is low, and its temperature cross-sensitivity is high. Obviously, it does not accurately detect the pH of the liquid environment. The GO/PVA-based sensor [18] cascaded the MZI and FBG to form a temperature self-calibrating pH sensor with excellent stability and good repeatability. Although this reduces temperature crosstalk during testing, it cannot ensure detection for universal pH conditions. The cascade complicates the sensor fabrication, which also limits its practical application. The PAni-based tilted fiber Bragg grating (TFBG) sensor [25] is adapted to wide-range pH measurements with a rapid response. However, the sensor has very low sensitivity, long stabilization times, and a large hysteresis. The pH-measurement range of the PAni-based tapered fiber sensor [26] is very narrow, and the sensitivity is low. The PVA/PAA-based photonic crystal fiber interferometer (PCFI) sensor [39] has a good repeatability but a narrow pH-measurement range. In other words, the sensor proposed in this work achieves a linear response, a fast response, and low temperature crosstalk in universal pH conditions, which obtains the trade-off performance when used for pH measurement.

Table 1. Comparison of different optical fiber pH sensors.

Technique	pH Range	pH Sensitivity (nm/pH)	Minimum Response Time (s)	Temperature Sensitivity (nm/°C)	Stability	Repeatability
LPFG and hydrogel [16]	2–12	0.66	Less than 2	0.8	/	/
SPR and PAA/CS [17]	3.18–11.84	32.31	/	0.463	excellent	excellent
MZI cascading FBG and GO/PVA [18]	4–9.85	0.69	6	0.15 (MZI) 0.009 (FBG)	excellent	good
TFBG and PAni [25]	2–12	0.082	8	/	bad	/
Tapered fiber and PAni [26]	4–7 7–12	−0.54 0.28	/	/	/	/
PCFI and PVA/PAA [39]	2.5–6.5	0.9	12	−0.021 (pH = 3.52) −0.041 (pH = 4.68) 0.012 (pH = 5.82)	/	good
SNS and PAni/PAA (present study)	2–12	0.985	2	0.01	good	good

5. Conclusions

In conclusion, this paper proposes a PAni/PAA-coated SNS optical fiber pH sensor, which was experimentally demonstrated with excellent trade-off performance when used for pH detection. To improve the sensors' performance trade-off, PAni was synthesized by oxidative polymerization and in situ deposited a uniform PAni/PAA film on the surface of NCF by the self-assembly method, which successfully avoided their exfoliation and ensured the sensors' stability and reliability. The experimental results show that the self-assembled membrane-based pH sensor has a good linear response in a universal pH range from 2 to 12, with R^2 of 0.991 and large pH sensitivity of 0.985 nm/pH. The sensor's response time and recovery time are less than 10 s. In addition, the proposed sensor has also been proven to have good repeatability and stability. Compared with other optical fiber sensors, the SNS sensor based on multimode interference avoids complicated fabrication processes, complex optical fiber structures, and delicate sensitive materials. Moreover, the simple fabrication and excellent performance make it attractive for pH measurement in various application fields.

Author Contributions: Writing—original draft, investigation, writing—review & editing, G.L.; supervision, writing—review and editing, L.W.; writing—review and editing, formal analysis, B.X.; investigation, validation, C.Z.; methodology, supervision, K.N.; investigation, methodology, J.H.; validation, conceptualization, D.L.; funding acquisition, investigation, L.L.; investigation, supervision, F.Z.; investigation, validation, funding acquisition, writing - review & editing, N.W. All authors have read and agreed to the published version of the manuscript.

Funding: This work is supported by the Open Projects Foundation (No. SKLD 2001) of the State Key Laboratory of Optical Fiber and Cable Manufacture Technology, China (YOFC) and the State Key Laboratory of Advanced Optical Communication Systems Networks, China (2020GZKF007).

Institutional Review Board Statement: Not applicable.

Informed Consent Statement: Not applicable.

Data Availability Statement: Not applicable.

Conflicts of Interest: The authors declare no conflict of interest.

References

1. Pathak, A.K.; Chaudhary, D.K.; Singh, V.K. Broad range and highly sensitive optical pH sensor based on Hierarchical ZnO microflowers over tapered silica fiber. *Sens. Actuators A Phys.* **2018**, *280*, 399–405. [CrossRef]
2. Orouji, A.; Abbasi-Moayed, S.; Ghasemi, F.; Hormozi-Nezhad, M.R. A wide-range pH indicator based on colorimetric patterns of gold@silver nanorods. *Sens. Actuators B Chem.* **2022**, *358*, 131479. [CrossRef]
3. Wang, Y.; Zhang, H.; Cui, Y.; Duan, S.; Lin, W.; Liu, B. A complementary-DNA-enhanced fiber-optic sensor based on microfiber-assisted Mach-Zehnder interferometry for biocompatible pH sensing. *Sens. Actuators B Chem.* **2021**, *332*, 129516. [CrossRef]
4. Cai, Y.; Wang, M.; Liu, M.; Zhang, J.; Zhao, Y. A Portable Optical Fiber Sensing Platform Based on Fluorescent Carbon Dots for Real-Time pH Detection. *Adv. Mater. Interfaces* **2021**, *9*, 2101633. [CrossRef]
5. Semwal, V.; Gupta, B.D. Highly sensitive surface plasmon resonance based fiber optic pH sensor utilizing rGO-Pani nanocomposite prepared by in situ method. *Sens. Actuators B Chem.* **2019**, *283*, 632–642. [CrossRef]
6. Noman, A.A.; Dash, J.N.; Cheng, X.; Leong, C.Y.; Tam, H.Y.; Yu, C. Hydrogel based Fabry-Perot cavity for a pH sensor. *Opt. Express* **2020**, *28*, 39640–39648. [CrossRef] [PubMed]
7. Yang, D.R.; Lee, S.; Sung, S.W.; Lee, J. Improving Dynamics of Glass pH Electrodes. *IEEE Sens. J.* **2009**, *9*, 1793–1796. [CrossRef]
8. Khalid, A.; Peng, L.; Arman, A.; Warren-Smith, S.C.; Schartner, E.P.; Sylvia, G.M.; Hutchinson, M.R.; Ebendorff-Heidepriem, H.; McLaughlin, R.A.; Gibson, B.C.; et al. Silk: A bio-derived coating for optical fiber sensing applications. *Sens. Actuators B Chem.* **2020**, *311*, 127864. [CrossRef]
9. Kurniawan, D.; Anjali, B.A.; Setiawan, O.; Ostrikov, K.K.; Chung, Y.G.; Chiang, W.H. Microplasma Band Structure Engineering in Graphene Quantum Dots for Sensitive and Wide-Range pH Sensing. *ACS Appl. Mater. Interfaces* **2022**, *14*, 1670–1683. [CrossRef]
10. Li, D.-Y.; Zhang, H.; Li, Z.; Zhou, L.-W.; Zhang, M.-D.; Pu, X.-Y.; Sun, Y.-Z.; Liu, H.; Zhang, Y.-X. High sensitivity pH sensing by using a ring resonator laser integrated into a microfluidic chip. *Opt. Express* **2022**, *30*, 4106–4116. [CrossRef]
11. Rezaei, N.; Yahaghi, A. A High Sensitivity Surface Plasmon Resonance D-Shaped Fiber Sensor Based on a Waveguide-Coupled Bimetallic Structure: Modeling and Optimization. *IEEE Sens. J.* **2014**, *14*, 3611–3615. [CrossRef]
12. Tang, Z.; Gomez, D.; He, C.; Korposh, S.; Morgan, S.P.; Correia, R.; Hayes-Gill, B.; Setchfield, K.; Liu, L. A U-Shape Fibre-Optic pH Sensor Based on Hydrogen Bonding of Ethyl Cellulose With a Sol-Gel Matrix. *J. Lightwave Technol.* **2021**, *39*, 1557–1564. [CrossRef]

13. Cheng, X.; Bonefacino, J.; Guan, B.O.; Tam, H.Y. All-polymer fiber-optic pH sensor. *Opt. Express* **2018**, *26*, 14610–14616. [CrossRef] [PubMed]
14. Khan, M.R.R.; Watekar, A.V.; Kang, S.-W. Fiber-Optic Biosensor to Detect pH and Glucose. *IEEE Sens. J.* **2018**, *18*, 1528–1538. [CrossRef]
15. Zhao, L.; Li, G.; Gan, J.; Yang, Z. Hydrogel Optical Fiber Based Ratiometric Fluorescence Sensor for Highly Sensitive Ph Detection. *J. Lightwave Technol.* **2021**, *39*, 6653–6659. [CrossRef]
16. Mishra, S.K.; Zou, B.; Chiang, K.S. Wide-Range pH Sensor Based on a Smart-Hydrogel-Coated Long-Period Fiber Grating. *IEEE J. Sel. Top. Quantum Electron.* **2017**, *23*, 284–288. [CrossRef]
17. Li, L.; Zhang, Y.-N. Fiber-Optic SPR pH Sensor Based on MMF–NCF–MMF Structure and Self-Assembled Nanofilm. *IEEE Trans. Instrum. Meas.* **2021**, *70*, 9502509. [CrossRef]
18. Yan, R.; Sang, G.; Yin, B.; Wu, S.; Wang, M.; Hou, B.; Gao, M.; Chen, R.; Yu, H. Temperature self-calibrated pH sensor based on GO/PVA-coated MZI cascading FBG. *Opt. Express* **2021**, *29*, 13530–13541. [CrossRef]
19. Lei, M.; Zhang, Y.-N.; Han, B.; Zhao, Q.; Zhang, A.; Fu, D. In-Line Mach–Zehnder Interferometer and FBG With Smart Hydrogel for Simultaneous pH and Temperature Detection. *IEEE Sens. J.* **2018**, *18*, 7499–7504. [CrossRef]
20. Gong, J.; Tanner, M.G.; Venkateswaran, S.; Stone, J.M.; Zhang, Y.; Bradley, M. A hydrogel-based optical fibre fluorescent pH sensor for observing lung tumor tissue acidity. *Anal. Chim. Acta* **2020**, *1134*, 136–143. [CrossRef]
21. Lavine, B.K.; Kaval, N.; Oxenford, L.; Kim, M.; Dahal, K.S.; Perera, N.; Seitz, R.; Moulton, J.T.; Bunce, R.A. Synthesis and Characterization of N-Isopropylacrylamide Microspheres as pH Sensors. *Sensors* **2021**, *21*, 6493. [CrossRef] [PubMed]
22. Podrazký, O.; Mrázek, J.; Proboštová, J.; Vytykáčová, S.; Kašík, I.; Pitrová, S.; Jasim, A. Ex-Vivo Measurement of the pH in Aqueous Humor Samples by a Tapered Fiber-Optic Sensor. *Sensors* **2021**, *21*, 5075. [CrossRef] [PubMed]
23. Antohe, I.; Jinga, L.I.; Antohe, V.A.; Socol, G. Sensitive pH Monitoring Using a Polyaniline-Functionalized Fiber Optic-Surface Plasmon Resonance Detector. *Sensors* **2021**, *21*, 4218. [CrossRef]
24. Ehrlich, K.; Choudhary, T.R.; Ucuncu, M.; Megia-Fernandez, A.; Harrington, K.; Wood, H.A.C.; Yu, F.; Choudhury, D.; Dhaliwal, K.; Bradley, M.; et al. Time-Resolved Spectroscopy of Fluorescence Quenching in Optical Fibre-Based pH Sensors. *Sensors* **2020**, *20*, 6115. [CrossRef] [PubMed]
25. Lopez Aldaba, A.; González-Vila, Á.; Debliquy, M.; Lopez-Amo, M.; Caucheteur, C.; Lahem, D. Polyaniline-coated tilted fiber Bragg gratings for pH sensing. *Sens. Actuators B Chem.* **2018**, *254*, 1087–1093. [CrossRef]
26. Sun, D.; Xu, S.; Fu, Y.; Ma, J. Fast response microfiber-optic pH sensor based on a polyaniline sensing layer. *Appl. Opt.* **2020**, *59*, 11261–11265. [CrossRef]
27. Ge, C.; Armstrong, N.R.; Saavedra, S.S. pH-sensing properties of poly (aniline) ultrathin films self-assembled on indium–tin oxide. *Anal. Chem.* **2007**, *79*, 1401–1410. [CrossRef]
28. Wong, W.C.; Chan, C.C.; Hu, P.; Chan, J.R.; Low, Y.T.; Dong, X.; Leong, K.C. Miniature pH optical fiber sensor based on waist-enlarged bitaper and mode excitation. *Sens. Actuators B Chem.* **2014**, *191*, 579–585. [CrossRef]
29. Abdzaid, T.A.; Taher, H.J. No-core fiber interferometry pH sensor based on a polyvinyl alcohol/polyacrylic acid and silica/polyvinyl alcohol/polyacrylic acid hydrogel coating. *Appl. Opt.* **2021**, *60*, 1587–1594. [CrossRef]
30. Boomi, P.; Prabu, H.G.; Manisankar, P.; Ravikumar, S. Study on antibacterial activity of chemically synthesized PANI-Ag-Au nanocomposite. *Appl. Surf. Sci.* **2014**, *300*, 66–72. [CrossRef]
31. Tai, Q.; Chen, B.; Guo, F.; Xu, S.; Hu, H.; Sebo, B.; Zhao, X.-Z. In situ prepared transparent polyaniline electrode and its application in bifacial dye-sensitized solar cells. *ACS Nano* **2011**, *5*, 3795–3799. [CrossRef] [PubMed]
32. Urrutia, A.; Goicoechea, J.; Ricchiuti, A.L.; Barrera, D.; Sales, S.; Arregui, F.J. Simultaneous measurement of humidity and temperature based on a partially coated optical fiber long period grating. *Sens. Actuators B Chem.* **2016**, *227*, 135–141. [CrossRef]
33. Tang, Z.; Wu, J.; Liu, Q.; Zheng, M.; Tang, Q.; Lan, Z.; Lin, J. Preparation of poly(acrylic acid)/gelatin/polyaniline gel-electrolyte and its application in quasi-solid-state dye-sensitized solar cells. *J. Power Sources* **2012**, *203*, 282–287. [CrossRef]
34. Xia, Y.; Zhu, H. Polyaniline nanofiber-reinforced conducting hydrogel with unique pH-sensitivity. *Soft Matter* **2011**, *7*, 9388–9393. [CrossRef]
35. Zhao, Y.-B.; Liu, H.-P.; Li, C.-Y.; Chen, Y.; Li, S.-Q.; Zeng, R.-C.; Wang, Z.-L. Corrosion resistance and adhesion strength of a spin-assisted layer-by-layer assembled coating on AZ31 magnesium alloy. *Appl. Surf. Sci.* **2018**, *434*, 787–795. [CrossRef]
36. Hino, T.; Namiki, T.; Kuramoto, N. Synthesis and characterization of novel conducting composites of polyaniline prepared in the presence of sodium dodecylsulfonate and several water soluble polymers. *Synth. Met.* **2006**, *156*, 1327–1332. [CrossRef]
37. Tang, Z.; Liu, Q.; Tang, Q.; Wu, J.; Wang, J.; Chen, S.; Cheng, C.; Yu, H.; Lan, Z.; Lin, J.; et al. Preparation of PAA-g-CTAB/PANI polymer based gel-electrolyte and the application in quasi-solid-state dye-sensitized solar cells. *Electrochim. Acta* **2011**, *58*, 52–57. [CrossRef]
38. Zhang, Y.; Zhou, A.; Qin, B.; Deng, H.; Liu, Z.; Yang, J.; Yuan, L. Refractive index sensing characteristics of single-mode fiber-based modal interferometers. *J. Lightwave Technol.* **2014**, *32*, 1734–1740. [CrossRef]
39. Hu, P.; Dong, X.; Wong, W.C.; Chen, L.H.; Ni, K.; Chan, C.C. Photonic crystal fiber interferometric pH sensor based on polyvinyl alcohol/polyacrylic acid hydrogel coating. *Appl. Opt.* **2015**, *54*, 2647–2652. [CrossRef]